Tropane Alkaloids

T. Pullaiah • Parvatam Giridhar
Zhe-Sheng Chen
Editors

Tropane Alkaloids

Sources, Chemistry, Pharmacology and Biotechnology

Springer

Editors
T. Pullaiah
Department of Botany
Sri Krishnadevaraya University
Anantapur, Andhra Pradesh, India

Parvatam Giridhar
Plant Cell Biotechnology Dept
Central Food Technological Res Inst
Mysuru, Karnataka, India

Zhe-Sheng Chen
College of Pharmacy and Health Sciences
St. John's University
New York, NY, USA

ISBN 978-981-96-5804-6 ISBN 978-981-96-5805-3 (eBook)
https://doi.org/10.1007/978-981-96-5805-3

This Springer imprint is published by the registered company Springer Nature Singapore Pte Ltd.
The registered company address is: 152 Beach Road, #21-01/04 Gateway East, Singapore 189721, Singapore

Contents

Chapter 1
Tropane Alkaloids: A Biologically Important Group of Nitrogen-Containing Heterocyclic Phyto-secondary Metabolites

Lepakshi Md. Bhakshu, K. Venkata Ratnam, and T. Pullaiah

Abstract Nature provides many medicinal plants that act as biological factories of diverse groups of metabolites, including alkaloids. One of the well-known and earliest reported classes of natural secondary metabolites belongs to alkaloids, nitrogen-containing heterocyclic compounds, with diverse biological and pharmacologically active, especially neurological disorders. The major source for the tropane alkaloids belongs to Solanaceae, a botanical family with many subgroups. Tropane alkaloids have been thoroughly investigated, utilized in numerous pharmacological procedures, and employed in the creation of beneficial medicinal substances. As parasympatholytics that competitively inhibit acetylcholine, tropane alkaloids are advantageous. These alkaloids are based on the bicyclic ring of the tropane moiety, and the majority of tropane alkaloids are ester derivatives that are substituted on the atom C-3 of the tropane ring. The present revision focuses on the chemistry of tropane alkaloids and their possible biological resources that might be helpful to scientists to study further and develop nontoxic and safe medicaments.

Keywords Tropane alkaloids · Solanaceae · Scopolamine · Hyoscyamine · Recreational drugs · Anticholinergic medicaments

1.1 Introduction

Nature provides many medicinal plants that act as biological factories of diverse groups of metabolites, including alkaloids. One of the well-known and earliest reported classes of natural secondary metabolites belongs to alkaloids,

L. Md. Bhakshu
Dr. YSR Government Degree College, Vedurukuppam, Andhra Pradesh, India

K. V. Ratnam
Department of Botany, Rayalaseema University, Kurnool, Andhra Pradesh, India

T. Pullaiah (✉)
Department of Botany, Sri Krishnadevaraya University, Anantapur, Andhra Pradesh, India

T. Pullaiah et al. (eds.), *Tropane Alkaloids*,
https://doi.org/10.1007/978-981-96-5805-3_1

nitrogen-containing heterocyclic compounds, with diverse biological and pharmacologically active, especially neurological disorders. Alkaloids are a well-known and historically documented class of naturally occurring secondary metabolites. These heterocyclic molecules, including nitrogen, have various biological and pharmacological activities, particularly in treating neurological illnesses (Srinivasan and Smolke 2019). The primary source of alkaloids is the Solanaceae family of plants, which has numerous subgroups. Tropane alkaloids (TAs) are one of these subgroups and have been thoroughly explored, utilized in numerous pharmacological processes, and employed in creating valuable medicinal entities. As parasympatholytics that competitively oppose acetylcholine, tropane alkaloids are helpful. These alkaloids are based on the bicyclic ring of the tropane moiety, with most tropane alkaloids being substituted on atom C-3 of the tropane ring as ester derivatives (Alsamarrai 2019). The nightshade family (Solanaceae) genera *Duboisia*, *Atropa*, *Datura*, *Hyoscyamus*, and other genera produce them mostly as anticholinergics. The World Health Organization lists several TAs, including atropine, hyoscyamine, and scopolamine, as critical medications for treating organophosphate and nerve agent poisoning, cardiac arrhythmia, gastrointestinal spasms, and Parkinson's disease symptoms (Grynkiewicz and Gadzikowska 2008; WHO 2015).

1.2 Source of Tropane Alkaloids

All compounds with a tropane ring structure are classified as tropane alkaloids (TAs), a particular class of alkaloids. TAs can be divided into three types and are either esters of 3α-tropanole (tropine) or, to a lesser extent, 3β-tropanole (pseudotropine): There are approximately 200 different TAs that have been described, including TAs from Solanaceae plants like scopolamine and hyoscyamine, coca alkaloids like cocaine from *Erythroxylum coca*, and the recently identified calystegines group, which are polyhydroxylated nortropane alkaloids (NTAs) that are primarily found in Convolvulaceae, Solanaceae, Moraceae, Erythroxylaceae, and Brassicaceae, among others (Kohnen-Johannsen and Kayser 2019).

The secondary metabolites of plants in the Solanaceae family are grouped together by the tropane skeleton and further separated into two classes: derivatives of tropine and ecgonine. Researchers continue to draw inspiration from the first category, which is represented by the well-known alkaloids atropine and scopolamine, in their efforts to create more selective muscarinic receptor antagonists. These drugs are considered model anticholinergic drugs. The second class includes cocaine, one of the most often abused narcotics. The synthesis of much-needed cocaine antagonists has not been very successful despite extensive investigation. Consequently, novel strategies for treating cocaine rely on utilizing immunology and biotechnology. More than 20 potential medicinal compounds with tropane as the central skeleton in their structure are produced by the modern pharmaceutical industry and are used as bronchodilators, mydriatics, antiemetics, and antispasmodics.

Esters of tropane have been identified in the following plant families: Proteaceae, Rhizophoraceae, Euphorbiaceae, Convolvulaceae, and Erythroxylaceae, in addition to the Solanaceae family. Some of the more than 100 genera and 3000 species in the Solanaceae family—such as *Anisodus*, *Atropa*, *Brugmansia*, *Datura*, *Hyoscyamus*, *Przewalskia*, *Scopolia*, *Physochlaina*, *Mandragora*, *Anthotroche*, *Cyphantera*, and *Duboisia*—are known to be rich in alkaloids containing the tropane moiety (Griffin and Lin 2000). The term "tropane alkaloids" (TA, Fig. 1.1) refers to a unique family of harmful alkaloids that can contaminate food and feed and naturally emerge from a wide range of plant groups, including Brassicaceae, Erythroxylaceae, and Solanaceae. (−)-Hyoscyamine and (−)-scopolamine are the two naturally occurring alkaloids that are most common in TA-producing plants. Conversely, when (−)-hyoscyamine is purified, atropine is produced, which is a racemic mixture of (+)- and (−)-hyoscyamine. Globally, temperate and tropical regions are home to a large distribution of TA-producing weeds, which can occasionally contaminate agricultural products. However, the usage of these bio-active compounds is highly restricted with specific recommendations by physicians during severe conditions of pain. Certain tropane alkaloids such as cocaine and scopolamine are neurotoxic and showed psychoactive effects, related usage, and cultural associations. Research has been initiated toward in vitro biosynthesis and bio-engineering of production of nontoxic tropane alkaloids as effective medicines.

In addition to chemical synthesis, natural compounds extracted from domesticated plants—primarily scopolamine and atropine from Australian *Duboisia*—serve as the two main sources of raw materials for this commercial activity (Grynkiewicz

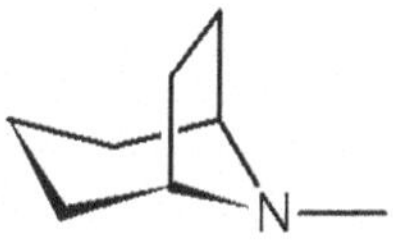

Tropane Skeleton

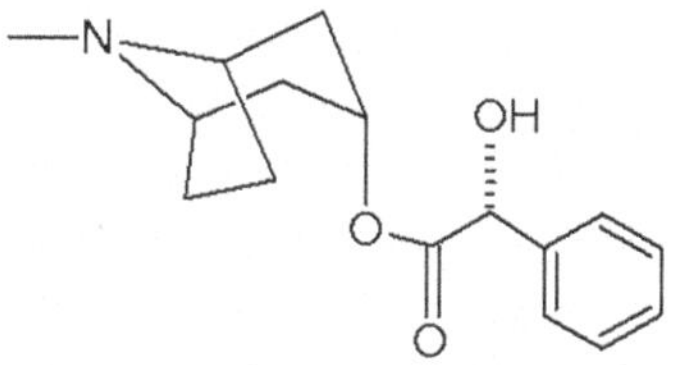

L - Hyocymine

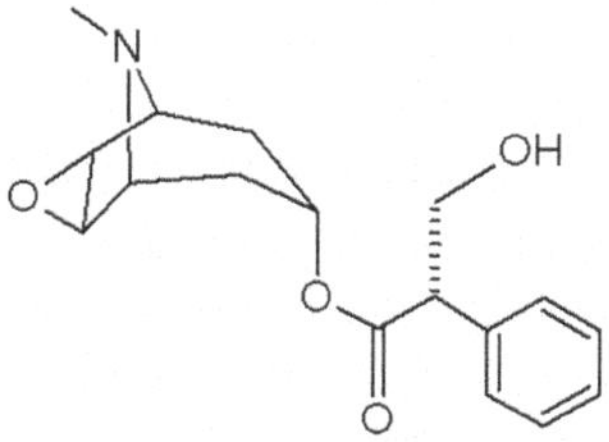

L - Scopalamine

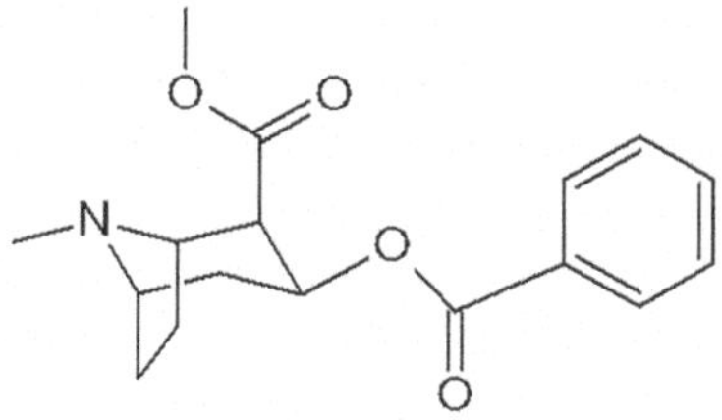

L - Cocaine

Fig. 1.1 Few popular examples of tropane alkaloids

and Gadzikowska 2008). Many plants with a moderate climate are listed in the ethnopharmacological tradition, whose extracts have been used for centuries as poisons and magic potions for various purposes (such as pain relievers, hallucinogens, etc.). It is now known that these extracts contain significant amounts of tropane alkaloids, such as up to 2% in ripe *Datura stramonium* seeds. Among these, there are species such as the poisonous nightshade (*Atropa belladonna*), mandrake (*Atropa mandragora*), henbane (*Hyoscyamus niger*, *Hyoscyamus albus*), scopola (*Scopolia carniolica*), and jimsonweed, which is also known as thorn-apple (*Datura stramonium*), stated as poisons (Griffin and Lin 2000).

1.3 Historical Aspects

Individual alkaloid compounds began to be isolated from 1830 onward, including atropine from *Atropa belladonna* L., hyoscyamine from *Hyoscyamus niger* L., and scopolamine. Meissner, who first used the term "alkaloid," was the first to recognize that the active principles of these toxic plants are alkaline and can, therefore, be isolated by extractive techniques in 1819. Tons of atropine and scopolamine derived from genetically modified cultivars are used in modern medicine, and new chemical and biotechnological manufacturing techniques are improved by constantly rising demand. Concurrently, cocaine derived from two species of *Erythroxylon*, which has limited medical applications due to its highly addictive features (Weil 1981; Rocio et al. 2004), was used illegally to manufacture and traffic "recreational drugs," which had negative socioeconomic and health effects worldwide (Grynkiewicz and Gadzikowska 2008).

Tropane alkaloids (TAs) are another type of hazardous alkaloids that have been shown to contaminate food and feed products. Plants belonging to the Brassicaceae, Solanaceae, and Erythroxylaceae families are among those that spontaneously produce these secondary metabolites. (−)-Hyoscyamine and (−)-scopolamine are the two naturally occurring alkaloids that are most common in plants that produce TA. However, when (−)-hyoscyamine is purified, atropine is produced, which is a racemic mixture of (+)- and (−)-hyoscyamine. Weeds that produce TA are found throughout the world in both temperate and tropical regions, and they can occasionally contaminate agricultural products. According to a 2013 report by the EFSA CONTAM panel, TAs are bad for human health, and weeds that contain TAs are widespread in both temperate and tropical regions, which may unintentionally contaminate agricultural products, tea, and herbal combinations (https://www.eurofins.in/food-testing/blog/tropane-alkaloids-in-tea/). The primary toxicological effects of atropine and scopolamine are due to excessive antimuscarinic activity in the central and autonomic nervous systems. The EFSA Panel has set a group acute reference dosage of 0.016 μg/kg body weight for the whole of the relevant TAs because of these impacts on human health (EFSA 2013; León et al. 2022; Martinello et al. 2022). Similar to PAs, TAs have the ability to contaminate teas, herbal blends, animal and agricultural products, and feeds. Consequently, studies on TAs have looked at their prevalence in plants, cow's milk, seeds, bread, leafy vegetables, cereal-based baby food, honey and its derivatives, teas, and herbal infusion matrices (León et al. 2022; Martinello et al. 2022; Romera-Torres et al. 2020; Marin-Saez et al. 2019; Du et al. 2019; Klein et al. 2022; Kowalczyk and Kwiatek 2022; Versilovskis et al. 2020; Gonzalez-Gomez et al. 2022; Gumus 2023).

1.4 Industrial Applications

Tropane alkaloids (TAs), a unique class of secondary metabolites produced by medicinal plants in the Solanaceae family, have potent anticholinergic effects. Scopolamine and hyoscyamine are two tropane alkaloids and anticholinergic drugs that are used medicinally. These drugs are used in the treatment of functional gastrointestinal disorders, pain, motion sickness, Parkinson's syndrome, and asthma. Hyoscyamine is an essential component in the industrial manufacturing of ipratropium bromide and scopolamine for tiotropium bromide. Tiotropium bromide and ipratropium are required drugs for the treatment of chronic obstructive pulmonary disease. As a result, the two alkaloids are in high demand. On the other hand, scopolamine is very profitable to produce and grow due to its wide range of medical applications.

Consequently, there is an increasing global need for these chemicals, driving up commercial demand. Additionally, the World Health Organization (WHO) lists scopolamine as one of the most significant pharmaceuticals. Scopolamine and hyoscyamine come from massive plantations of *Duboisia* plants in Queensland, Australia. The synthesis of scopolamine at high levels is becoming more difficult for the food and pharmaceutical sectors because of new biotic and abiotic factors brought about by climate change (Gonzalez-Gomez et al. 2022; Kohnen-Johannsen and Kayser 2019).

1.5 Chemical Nature and its Skeleton

The bicyclic saturation structure (N-methyl-8-azabicyclo [3.2.1] octane-1) with a nitrogen bridge connecting a cycloheptane ring is called tropane. Tropane alkaloids belong to a family of about 300 alkaloids that are easily separated based on stereochemical characteristics and the amount of carbons in the tropane skeleton (Zhang et al. 2023). Despite having a similar biogenetic reaction sequence that produces the tropane skeleton and only branches at the tropinone reduction step, tropine and ecgonine derivatives are separated in scientific literature for another crucial reason. The sole well-known member of the ecgonine class is cocaine, a neurostimulant that is known to cause severe addiction (Grynkiewicz and Gadzikowska 2008). Nearly

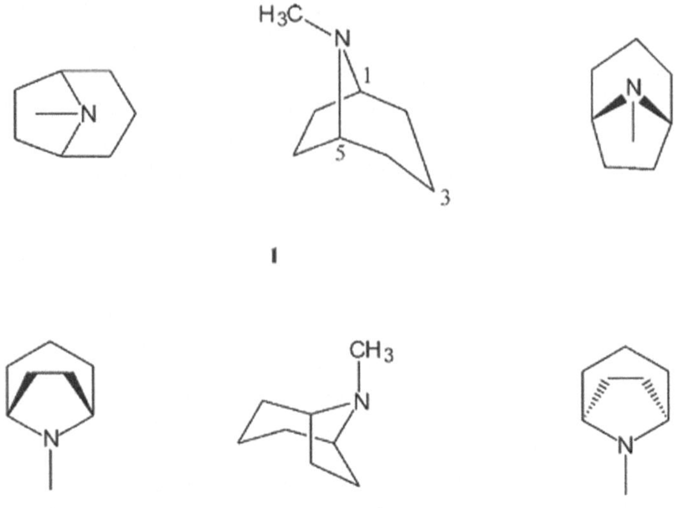

Fig. 1.2 Conventional representations of tropane structure

all TAs are structurally esters of different organic acids, such as tiglic, isovaleric, cinnamic, benzoic, and tropic acids conjugated to hydroxylated tropane derivatives (Huang et al. 2021, Fig. 1.2).

1.6 Model for the Synthesis of Hyoscyamine

The main alkaloid found in the plants is hyoscyamine, which is the ester of tropine with levorotatory {l(−)} or (S)-tropic acid. It is thought to be produced biogenetically by skeletal rearrangement of phenyllactic acid, which is created from L-phenylalanine. The 1,2-carboxy group shift, which has been widely accepted by recent treatises on alkaloid biosynthesis, is a component of the metabolic pathway from tropine phenyl lactic ester, littorine, to tropic acid ester, hyoscyamine (Scheme 3) which has recently disputed (Patterson and O'Hagan 2002; Fig. 1.3).

6-Hydroxytropine, also called anisodamine, is produced by the hyoscyamine enzyme. 6-Hydroxylase, also known as 6-(R)hydroxylase or H6H, is a precursor of scopine, another alkamine that is a basic component of hyoscine and has an extra epoxide ring. Tropine, nortropine, or hydroxylated tropines are alkamines found in minor tropane alkaloids. A range of acids, including benzoic, cinnamic, tiglic, truxillic, isovaleric, methylbutyric, and others, can esterify them at position C-3, with the noteworthy exception of anisodamine (3-tropoyl-6-hydroxytropine), which is not handled as medicines (Poupko et al. 2006).

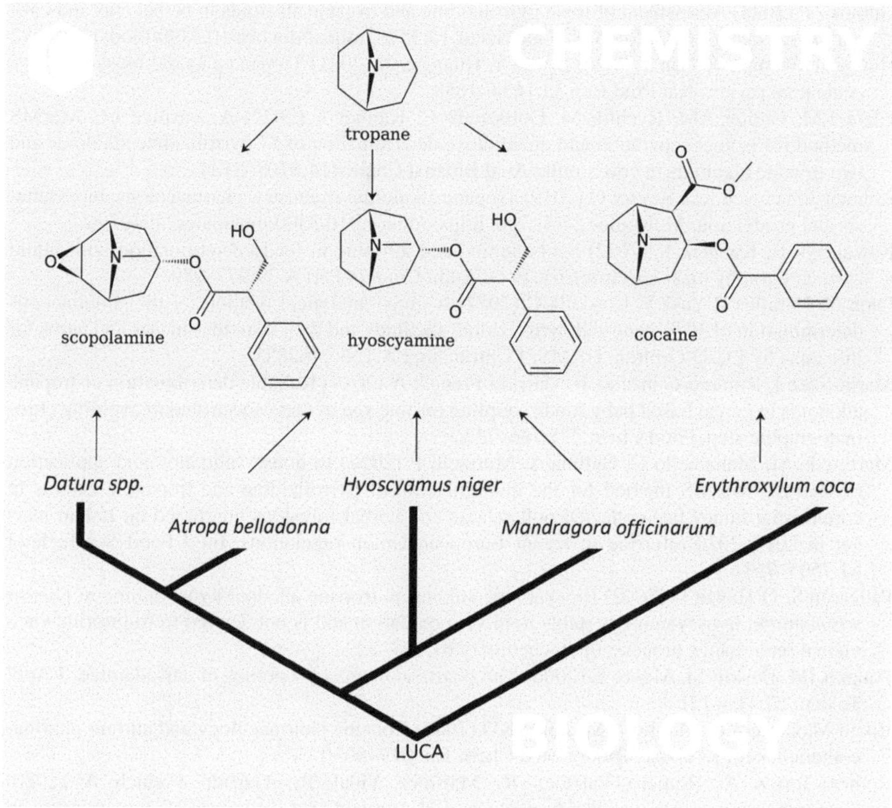

Fig. 1.3 Chemical structure and phylogeny of tropane alkaloids demonstrated the path of synthesis and their occurrence in five plant species. (Source: By Mplanine—Own work, CC BY-SA 4.0, https://commons.wikimedia.org/w/index.php?curid=115787548)

References

Alsamarrai ASH (2019) Synthesis of tropane derivatives. In: Kurek J (ed) Alkaloids—their importance in nature and human life. IntechOpen. https://doi.org/10.5772/intechopen.83382

Du N, Zhou W, Jin H, Liu Y, Zhou H, Liang X (2019) Characterization of tropane and cinnamamide alkaloids from *Scopolia tangutica* by high-performance liquid chromatography with quadrupole time-of-flight tandem mass spectrometry. J Sep Sci 42:1163–1173

EFSA Panel on Contaminants in the Food Chain (CONTAM) (2013) Scientific opinion on tropane alkaloids in food and feed. EFSA J 11:3386

Gonzalez-Gomez L, Morante-Zarcero S, Pereira JAM, Câmara JS, Sierra I (2022) Improved analytical approach for determination of tropane alkaloids in leafy vegetables based on μ-QuEChERS combined with HPLC-MS/MS. Toxins 14:650

Griffin WJ, Lin GD (2000) Chemotaxonomy and geographical distribution of tropane alkaloids. Phytochemistry 53:623–637

Grynkiewicz G, Gadzikowska M (2008) Tropane alkaloids as medicinally useful natural products and their synthetic derivatives as new drugs. Pharmacol Rep 60:439–446

Gumus ZP (2023) Assessment of toxic pyrrolizidine and tropane alkaloids in herbal teas and culinary herbs using LC-Q-ToF/MS. Food Secur 12:3572. https://doi.org/10.3390/foods12193572

Huang JP, Wang Y-J, Tian T, Wang L, Yan Y, Huang S-X (2021) Tropane alkaloid biosynthesis: a centennial review. Nat Prod Rep 38:1634–1658

Klein LM, Gabler AM, Rychlik M, Gottschalk C, Kaltner F (2022) A sensitive LC–MS/MS method for isomer separation and quantitative determination of 51 pyrrolizidine alkaloids and two tropane alkaloids in cow's milk. Anal Bioanal Chem 414:8107–8124

Kohnen-Johannsen KL, Kayser O (2019) Tropane alkaloids: chemistry, pharmacology, biosynthesis and production. Molecules 24(4):796. https://doi.org/10.3390/molecules24040796

Kowalczyk E, Kwiatek K (2022) Scopolamine and atropine in feeds–determination with liquid chromatography mass spectrometry. Food Addit Contam Part A 39:977–989

León N, Miralles P, Yusà V, Coscollà C (2022) A green analytical method for the simultaneous determination of 30 tropane and pyrrolizidine alkaloids and their N-oxides in teas and herbs for infusions by LC-Q-Orbitrap HRMS. J Chromatogr A 1666:462835

Marin-Saez J, Romero-González R, Garrido Frenich A (2019) Reliable determination of tropane alkaloids in cereal based baby foods coupling on-line spe to mass spectrometry avoiding chromatographic step. Food Chem 275:746–753

Martinello M, Manzinello C, Gallina A, Mutinelli F (2022) In-house validation and application of UHPLC-MS/MS method for the quantification of pyrrolizidine and tropane alkaloids in commercial honey bee-collected pollen, teas and herbal infusions purchased on Italian market in 2019–2020 referring to recent European Union regulations. Int J Food Sci Technol 57:7505–7516

Patterson S, O'Hagan D (2002) Biosynthetic studies on tropane alkaloid hyoscyamine in *Datura stramonium*; hyoscyamine is stable to in vivo oxidation and is not derived from littorine via a vicinal interchange process. Phytochemistry 61:323–329

Poupko JM, Baskin SI, Moore E (2006) The pharmacological properties of anisodamine. J Appl Toxicol 27:116–121

Rocio M, Carrera A, Meijler MM, Janda KD (2004) Cocaine pharmacology and current pharmacotherapies for its abuse. Bioorg Med Chem 12:5019–5030

Romera-Torres A, Romero-González R, Martínez Vidal JL, Garrido Frenich A (2020) Comprehensive tropane alkaloids analysis and retrospective screening of contaminants in honey samples using liquid chromatography-high resolution mass spectrometry (Orbitrap). Food Res Int 133:109–130

Srinivasan P, Smolke CD (2019) Engineering a microbial biosynthesis platform for de novo production of tropane alkaloids. Nat Commun 10:3634. https://doi.org/10.1038/s41467-019-11588-w

Versilovskis A, Mulder PPJ, Pereboom-de Fauw DPKH, de Stoppelaar J, de Nijs M (2020) Simultaneous quantification of ergot and tropane alkaloids in bread in The Netherlands by LC-MS/MS. Food Addit Contam Part B Surveill 13:215–223

Weil AT (1981) The therapeutic value of coca in contemporary medicine. J Ethnopharmacol 3:367–376

WHO (World Health Organization) (2015) WHO model list of essential medicines—19th list. WHO, Geneva

Zhang F, Qiu F, Zeng J et al (2023) Revealing evolution of tropane alkaloid biosynthesis by analyzing two genomes in the Solanaceae family. Nat Commun 14:1446. https://doi.org/10.1038/s41467-023-37133-4. https://www.eurofins.in/food-testing/blog/tropane-alkaloids-in-tea/

Chapter 2
Plant Sources of Tropane Alkaloids

T. Pullaiah

Abstract Tropane alkaloids are known to exist in plants such as those belonging to Solanaceae, Convolvulaceae, Erythroxylaceae, Brassicaceae, Euphorbiaceae, Rhizophoraceae, Moraceae, and Proteaceae. This chapter provides information on the plant sources of these alkaloids and the distribution of these plants. The valid and accepted names have been obtained from Plants of the World Online. Additionally, the text includes information on the tropane alkaloids found in different species.

Keywords Tropane alkaloids · *Atropa* · *Datura* · *Duboisia* · *Hyoscyamus* · *Erythroxylum* · *Schizanthus* · Hyoscyamine · Scopolamine

2.1 Introduction

Tropane alkaloids are a fascinating group of compounds defined by their unique structure, which includes both a pyrrolidine ring and a piperidine ring, creating the foundational skeleton of tropane. These compounds are particularly noteworthy for their anticholinergic properties, which can offer various applications in medicine and pharmacology. Understanding their structure and function can lead to further insights into their potential uses and benefits (Lockery et al. 2021). The distribution of tropane alkaloids in plants such as Solanaceae, Convolvulaceae, Erythroxylaceae, Brassicaceae, Euphorbiaceae, Rhizophoraceae, Moraceae, and Proteaceae has been reported (Griffin and Lin 2000). The chemical synthesis of tropane alkaloids remains a challenge, as there are no commercially viable methods available. However, these valuable compounds can be effectively obtained through extraction from medicinal plants, highlighting an opportunity for further research and development in this area (Hu et al. 2019; Kohnen-Johannsen and Kayser 2019). Savita et al. (2021) gave an account of the chemotaxonomic significance of tropane alkaloids.

T. Pullaiah (✉)
Department of Botany, Sri Krishnadevaraya University, Anantapur, Andhra Pradesh, India

T. Pullaiah et al. (eds.), *Tropane Alkaloids*,
https://doi.org/10.1007/978-981-96-5805-3_2

They listed the occurrence of tropane alkaloids in different plant families. In this chapter, plant sources of the alkaloids and the distribution of these plants have been given. Plants of the World Online has been followed for valid and accepted names. Under each family, plant species are arranged in alphabetical sequence.

2.2 Solanaceae

Solanaceae is the main source of tropane alkaloids.

2.2.1 *Alkekengi*

***Alkekengi officinarum* Moench** (Synonym: *Physalis alkekengi* L.)

This species' native range is SE Europe to China. It is a perennial, rhizomatous herb. Basey and Woolley (1973) reported tropane alkaloids in *Alkekengi officinarum* (synonym: *Physalis alkekengi* L.).

2.2.2 *Anisodus*

There are four species of *Anisodus*, all distributed in Himalayas and China.

***Anisodus acutangulus* C.Y. Wu & C. Chen**
A. acutangulus, known in Chinese as san fen san, is a perennial herb native to China. *A. acutangulus* is rich in tropane alkaloids (Geng et al. 2013). To clarify its alkaloid constituents, Geng et al. (2013) investigated *A. acutangulus* and discovered one dimeric tropane alkaloid, namely bis hyoscyamine.

***Anisodus stramoniifolius* (Wall.) G. Don** (Synonyms: *Scopolia lurida* (Link) Dunal; *Scopolia anomala* (Link & Otto) Airy Shaw; *Hyoscyamus luridus* E.H.L. Krause).

It is distributed in Himalayas and China. Air-dried leaves and stems are reported to contain 0.32% of alkaloids, comprising hyoscyamine, himaline, atropine, scopolamine, cuscohygrine, hellaradine, tropine, and scopine.

In most of the tropane alkaloid literature, this species was treated under *Scopolia lurida*.

***Anisodus tanguticus* (Maxim.) Pascher**
It grows in Nepal and Central China. *A. tanguticus* is a tropane alkaloids-producing herb (Chen et al. 2022a, b). Its roots are rich in anisodine, anisodamine, and atropine.

There are four species in *Anisodus* of which *Anisodus carniolicoides* (C.Y. Wu & C.C. Chen) D'Arcy & Zhi Y. Zhang has not been studied for tropane alkaloids.

2.2.3 *Anthocercis*

Evans and Ramsey (1983) reported tropane alkaloids in aerial parts of *Anthocercis anisantha*, *A. genistoides*, *A. gracilis*, *A. ilicifolia*, *A. intricata*, and *A. viscosa*.

***Anthocercis anisantha* Endl.**
This species' native range is Australia (Western to South). It is a shrub and grows primarily in the desert or dry shrubland.

***Anthocercis fasciculata* F. Muell.**
It is a shrub that grows primarily in the subtropical SW. Western Australia. *Anthocercis fasciculata* contains hyoscyamine (Cannon et al. 1969).

***Anthocercis genistoides* Miers**
It is a shrub that grows primarily in the desert or dry shrubland of Southwest Australia. Aponorhyoscine is indicated as a component of the roots of *Anthocercis genistoides* (Evans and Ramsey 1981).

***Anthocercis gracilis* Benth.**
This species is native to Southwest of Western Australia and primarily grows as a subshrub or shrub in the subtropical biome.

***Anthocercis ilicifolia* Hook.**
It is a shrub and grows in subtropical Western Australia. *Anthocercis ilicifolia* ssp. *ilicifolia* contain considerable amounts of both nicotine and nornicotine, together with lower levels of the tropane alkaloid hyoscyamine (<0.2% DW) (Ryan et al. 2015).

***Anthocercis intricata* F. Muell.**
It is a shrub native to subtropical Western Australia.

***Anthocercis littorea* Labill.**
It is a shrub and grows in the subtropical W. and S. Western Australia. *A. littorea* contains littorine and meteloidine (Cannon et al. 1969) and tropane derivatives mono- and ditigloyl esters of teloidine (Evans and Treagust 1973).

***Anthocercis viscosa* R.Br.**
It is a shrub and native to subtropical SW. Australia. *Anthocercis viscosa* contains hyoscyamine (Cannon et al. 1969).

2.2.4 *Anthotroche*

There are three accepted species of *Anthotroche*, and tropane alkaloids have been reported in all the three species. Aponoratropine has been detected in the aerial parts of *A. myoporoides*, *A. pannosa*, and *A. walcottii* (Evans and Ramsey 1981, 1983).

Anthotroche myoporoides **C.A. Gardner**
It is a shrub and native to subtropical Western Australia.

Anthotroche pannosa **Endl.**
It is a shrub and grows in the desert or dry shrubland of W. Western Australia. Bremner and Cannon (1968) isolated hyoscyamine from *Anthotroche pannosa*.

Anthotroche walcottii **F. Muell.**
It is a shrub and grows in the subtropical W. Western Australia.

2.2.5 *Atropa*

Atropa is a genus with seven species distributed in temperate forests. *A. belladonna* contains hyoscyamine, scopolamine, and apoatropine as the principal alkaloids (Evans 1979). Tropane alkaloids of *A. belladonna* and *A. acuminata* were investigated by Ashtiana and Sefidkonb (2011). Banihashemi et al. (2020) investigated the production of scopolamine in hairy roots via *Agrobacterium rhizogenes* in *Atropa komarovii*. Fatemeh and Fatemeh (2011) determined the tropane alkaloid content (atropine and scopolamine) in different parts of the wild and cultivated *Atropa belladonna* and wild *Atropa acuminata* in Iran using high-performance liquid chromatography (HPLC) method.

Atropa acuminata **Royle ex Lindl.**
This perennial herb grows in the temperate areas and has a native range from E. Iran to W. Himalaya.

Atropa baetica **Willk.**
This perennial herb grows in CE. and S. Spain and Morocco. Zárate et al. (1997) reported tropane alkaloids in *Atropa baetica*.

Atropa belladonna **L.**
It is a perennial herb and grows in Europe to N. Iran and NW. Africa. Baralle and Gros (1969) and Van Haga (1954) reported the presence of cuscohygrine as a constituent alkaloid of *A. belladonna*.

Atropa komarovii **Blin. & Shalyt**
It is a perennial herb and grows in NE. Iran to S. Turkmenistan.

The following three species of *Atropa* have not been studied for the presence of tropane alkaloids: *Atropa indobelladonna* Karthik. & V.S.Kumar, *Atropa* × *martiana* Font Quer, and *Atropa pallidiflora* Schönb.-Tem.

Brugmansia

Brugmansia is a genus of seven species and two hybrids. Tropane alkaloids have been reported in most of the species. The aerial parts contain scopolamine and hyoscyamine, while the roots contain teloidine (Griffin and Lin 2000).

***Brugmansia arborea* (L.) Sweet** (Synonym: *Datura arborea* L.)

This species' native range is Ecuador to N. Chile. It is a shrub or tree that grows in the montane tropics. Cultivated in hilly regions as ornamental (Fig. 2.1).

***Brugmansia aurea* Lagerh**

It is a tree and its native range is the wet tropical Colombia to Ecuador.

***Brugmansia insignis* (Barb.Rodr.) Lockwood ex R.E.Schult.**

It is a shrub or tree that grows in the seasonally dry tropical West South America to Northwest Brazil.

***Brugmansia sanguinea* (Ruiz. & Pav.) D. Don**

This species' native range is from West South America to North Chile. It is a shrub or tree that grows in the montane tropics.

***Brugmansia suaveolens* (Humb. & Bonpl. ex Willd.) Sweet**

The native range of this species is Brazil (S. Bahia to N. Rio Grande do Sul). It is a shrub or tree and grows in the seasonally dry tropics (Fig. 2.2).

Fig. 2.1 *Brugmansia arborea* (L.) Sweet (Syn.: *Datura arborea* L.)

Fig. 2.2 *Brugmansia suaveolens* (Humb. & Bonpl. ex Willd.) Sweet

Brugmansia versicolor **Lagerh.**

This species' native range is W. Ecuador. It is a tree that grows in the wet tropics.

Brugmansia vulcanicola **(A.S.Barclay) R.E. Schult.**

This species' native range is from SW. Colombia to S. Central Ecuador. It is a shrub or tree that grows primarily in the montane tropics.

Brugmansia × candida **Pers.**

It is a hybrid between *B. aurea* × *B. versicolor*. This hybrid's native range is from South Colombia to Ecuador. It is a tree that grows in the seasonally dry tropics.

Brugmansia × rubella **(Saff.) Moldenke**

It is a hybrid between *B. arborea* × *B. sanguinea*. This hybrid's native range is Ecuador. It is a tree and grows in the seasonally dry tropics.

Crenidium

Crenidium spinescens primarily contains hyoscyamine as its main alkaloid but also has anabasine along with ursolic acid (El-Imam and Evans 1984). The genus *Crenidium* has only one species.

***Crenidium spinescens* Haegi**

This species' native range is S. Central Western Australia. It is a shrub that grows in the desert or dry shrubland.

Cyphanthera

The alkaloids of *Grammosolen odgersii* (synonym: *Cyphanthera odgersii*) and *Cyphanthera tasmanica* are consistent with those of other species of the genus. *Cyphanthera tasmanica* contains significant amounts of nicotine and nornicotine, along with lower levels of the tropane alkaloid hyoscyamine (<0.2% DW) (Ryan et al. 2015). Evans and Ramsey (1983) reported tropane alkaloids *in Cyphanthera albicans*, *C. anthocercidea*, *C. microphylla*, *C. racemosa*, and *C. scabrella*.

There are eight accepted species in the genus *Cyphanthera*.

***Cyphanthera albicans* (A. Cunn.) Miers** (Synonym: *Anthocercis albicans* A. Cunn.)

Cyphanthera albicans, or *Anthocercis albicans*, is a shrub found in southeastern Queensland to northeastern Victoria. It contains butyryl esters of tropine and 6β-hydroxytropine, its predominant alkaloids (El-Imam and Evans 1984). In addition to known tropane alkaloids, 3α-*iso*-butyryloxytropan-6β-ol and 3α-*n*-butyryloxytropane have been characterized as respective components of the aerial parts of *Cyphanthera albicans* (synonym: *Anthocercis albicans*) (Evans and Ramsey 1981).

***Cyphanthera anthocercidea* (F. Muell.) Haegi** (synonym: *Anthocercis anthocercidea* (F.Muell.) Druce).

Cyphanthera anthocercidea is a shrub distributed in the subtropics of Victoria (Australia). It contains hyoscyamine, scopolamine, and nicotine (which is more abundant).

Cyphanthera microphylla **Miers**

This species is a shrub and its native range is SW Australia.

***Cyphanthera myosotidea* (F. Muell.) Haegi**

This subshrub, which grows in the subtropics, is native to SE South Australia to NW Victoria.

***Cyphanthera racemosa* (F. Muell.) Haegi**

It is a shrub and its native range is W. and SW. Western Australia.

***Cyphanthera scabrella* (Benth.) Miers**

This species' native range is New South Wales (Blue Mountains). It is a scrambling shrub that grows in the subtropics.

***Cyphanthera tasmanica* Miers**

This species' native range is E. Tasmania. It is a shrub or tree that grows in the subtropics.

Cyphomandra

The primary compounds derived from the roots of *Cyphomandra betacea* include N,N′ O—bis-(4-dimethylaminobutyl) hexamide, tropinone, cuscohygrine, hyoscyamine, tigloidine, tropine, and pseudotropine. The accepted scientific name for this species is *Solanum betaceum* Cav. It is also known by the synonym *Cyphomandra betacea* (Cav.) Sendtn. This species is a tree found in the tropical regions of South America.

Datura

Datura is a genus of about 14 species of poisonous flowering plants (POWO 2024). The taxonomy of *Datura* is infamously unclear. Several other described species were later reclassified. For example, the name *D. meteloides* is ambiguous; it is usually classified as *D. wrightii*, but sporadically as *D. innoxia* (Haegi 1976; Hammer et al. 1983). *D. lanosa* was reclassified as either *D. wrightii* or *D. innoxia*, whereas *D. velutinosa* and *D. guyaquilensis* are generally accepted to be *D. innoxia* (Hammer et al. 1983).

Cinelli and Jones (2021) gave an excellent review of the alkaloids of the genus *Datura*. *Datura* species contain hyoscyamine, scopolamine, and littorine, while the roots yield valtropine. Tropane alkaloids have been reported in *Datura ceratocaula*, *Datura discolor*, *Datura ferox*, *Datura innoxia*, *Datura metel*, *Datura stramonium*, and *Datura wrightii*. Other species of *Datura* have not been studied for the presence of tropane alkaloids; these include *Datura arenicola*, *Datura kymatocarpa*, *Datura lanosa*, *Datura leichhardtii*, *D. pruinosa*, *D. quercifolia*, and *D. reburra*. Gerlach (1948) reported that *Datura innoxia* can be a potential commercial source of scopolamine.

***Datura ceratocaula* Ortega**

It is an annual herb or helophyte and grows in the seasonally dry tropical Mexico.

***Datura discolor* Bernh.**

It is an annual herb that grows in the subtropical W. Central USA to Honduras.

***Datura ferox* L.**

It is an annual herb, native to Mexico (Central, Northeast, Southwest) and Texas (USA), and introduced elsewhere.

***Datura innoxia* Mill.**

It is an annual or perennial herb—native to Arizona to Texas and Mexico—and introduced in North and South Africa (Fig. 2.3).

***Datura metel* L.** (Syn.: *Datura alba* F. Muell.; *Datura fastuosa* L.)

The native range of this species is from Texas to Colombia. It is an annual herb widespread in Old World Tropics and introduced in New World Tropics (Figs. 2.4 and 2.5).

Fig. 2.3 *Datura innoxia* Mill. (Source: A. Lalithamba with permission)

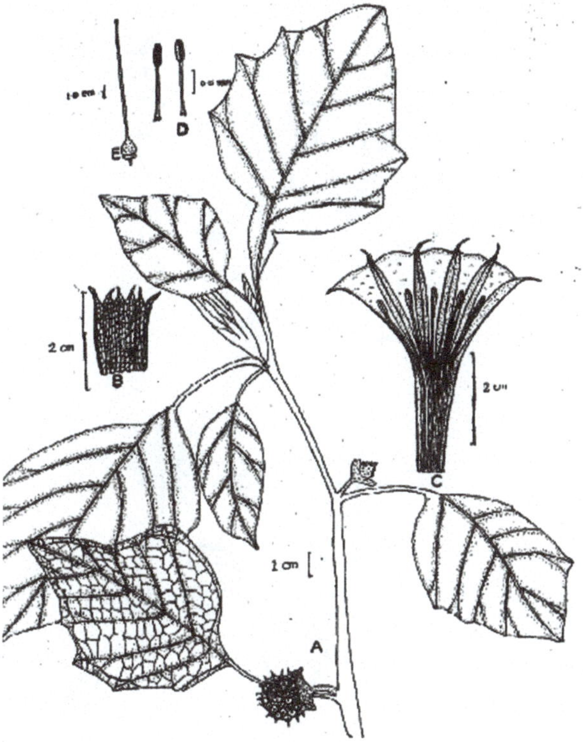

Fig. 2.4 *Datura metel* L.

Datura quercifolia H.B.K.

It is an annual, erect, small herb, up to 50 cm tall; native to Southwestern USA and Mexico; and grown as ornamental in many other countries (Fig. 2.6).

Fig. 2.5 *Datura metel* L. (Source: K. Raja Kullayiswamy with permission)

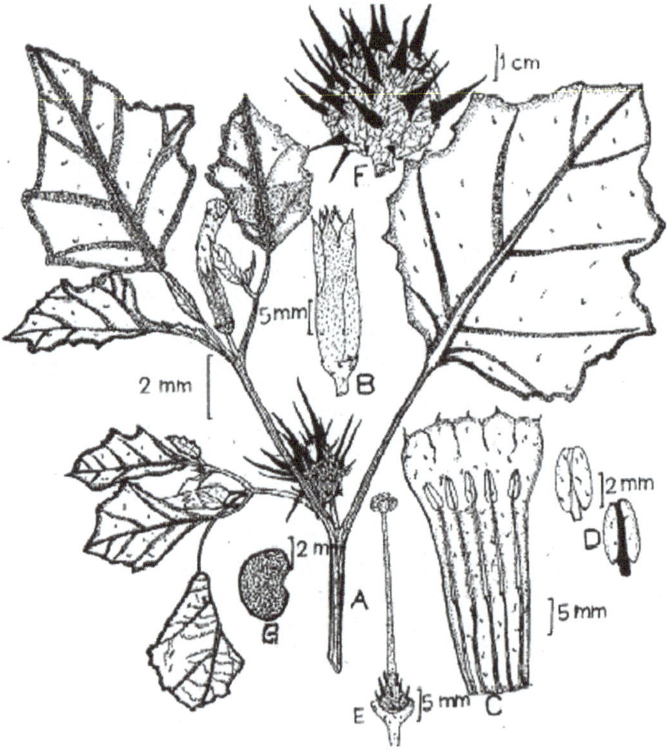

Fig. 2.6 *Datura quercifolia* H.B.K.

Datura stramonium L.

It is an annual erect herb and grows in subtropical Texas to Central America and the Caribbean and common in many warm temperate countries (Fig. 2.7).

Fig. 2.7 *Datura stramonium* L. (Source: K. Raja Kullayiswamy with permission)

Fig. 2.8. *Duboisia* (ABC Southern QLD: Jon Daly) https://www.abc.net.au/news/2021-11-13/sleepy-town-secretive-drug-centre-big-pharma-criminal-cartel/100572628

Datura wrightii Regel

This species' native range is SW. and S. Central USA to N. Mexico. It is an annual herb or subshrub that grows in subtropical regions.

Duboisia (Fig. 2.8)

There are four species of *Duboisia* and all of them contain tropane alkaloids. All the species of *Duboisia* are endemic to Australia. *D. myoporoides* contains scopolamine; *D. leichhardtii* contains hyoscyamine, scopolamine, and calystegine; *D.*

arenitensis contains hyoscyamine and scopolamine (Evans and Ramsey 1981). Luanratana and Griffin (1982) reported tropane alkaloids of *Duboisia hopwoodii*.

***Duboisia arenitensis* Craven, Lepschi, & Haegi**

It is a shrub and grows in the seasonally dry tropical N. northern territory of Australia.

***Duboisia hopwoodii* (F. Muell.) F. Muell.**

It is a shrub and grows in the desert or dry shrubland in Australia. A naturally occurring hybrid, *Duboisia hopwoodii* × *Grammosolen dixonii* contained a total alkaloid yield of 0.29% in which main alkaloids were nornicotine and scopolamine (Evans 1986).

***Duboisia leichhardtii* (F.Muell.) F.Muell.**

It is a shrub or tree and grows primarily in the desert or dry shrubland of S. Queensland in Australia.

***Duboisia myoporoides* R.Br.**

This species' native range is E. Australia and New Caledonia. It is a shrub or tree that grows in the subtropics.

Grammosolen

The alkaloids of *Grammosolen odgersii* (synonym: *Cyphanthera odgersii*) are consistent with those of other species of the genus *Cyphanthera*. Evans and Ramsey (1983) reported tropane alkaloids in aerial parts of *Grammosolen odgersii* (synonym: *Cyphanthera odgersii*) and *Grammosolen dixonii*.

***Grammosolen dixonii* (F. Muell. & Tate) Haegi**

This species' native range is CS. and SE. South Australia. It is a shrub that grows in the subtropics.

***Grammosolen odgersii* (F. Muell.) Haegi** (Synonym: *Cyphanthera odgersii*).

It is a shrub and grows in the desert or dry shrubland of SW. Australia.

Hyoscyamus

Hyoscyamus is rich in tropane alkaloids. The alkaloid composition of *H. boveanus*, *H. desertorum*, *H. muticus*, and *H. albus* was measured by El-Shazly et al. (1997). Altogether 39 alkaloids of the tropane and pyrrolidine types were identified. Many alkaloids were found in different species, e.g., 22 alkaloids in *H. boveanus*, 13 in *H. desertorum*, 24 in *H. muticus*, and 4 in *H. albus*. 3ß-Hyoscyamine, dehydrohyoscyamine, phyllalbine, 3α-phenylacetyl tropane, and phenylacetylscopine as well as phygrine and 2,3-diacetonyl-N-methyl-pyrrolidine are seemingly new for the genus *Hyoscyamus*. The aerial parts of *H. albus* contain scopolamine, atropine, and hyoscyamine (Ashraf et al. 2019).

***Hyoscyamus albus* L.**

It is an annual or biennial herb and grows in the subtropical Macaronesia and Mediterranean to N. Iraq and Arabian Peninsula.

***Hyoscyamus aureus* L.**

It is a biennial or perennial herb and grows in the subtropical E. Mediterranean to NW. Iraq.

***Hyoscyamus boveanus* (Dunal) Asch. & Schweinf.**

It is a semisucculent subshrub and grows in the desert or dry shrubland Egypt to Sinai.

***Hyoscyamus desertorum* (Asch. ex Boiss.) Täckh.**

It is an annual herb and grows in the desert or dry shrubland of Syria to Egypt and NE. Saudi Arabia.

***Hyoscyamus muticus* L.**

It is a perennial herb or subshrub and grows in the desert or dry shrubland from Sahara to India.

***Hyoscyamus niger* L.**

It is an annual or biennial herb and grows in the temperate regions from Mediterranean to Iran and cultivated in many other countries.

***Hyoscyamus pusillus* L.**

It is an annual herb and grows in the temperate S. European Russia to Mongolia and Arabian Peninsula.

***Hyoscyamus reticulatus* L.**

This species' native range is E. Mediterranean to Central Asia and Iran. It is an annual or biennial herb that grows in the subtropical biome.

***Hyoscyamus senecionis* Willd.**

It is a perennial herb and grows in the temperate N. Iraq to Afghanistan.

There are 31 species of *Hyoscyamus*. Of these the following species have not yet been studied for the presence of tropane alkaloids: *Hyoscyamus afghanicus* Pojark, *Hyoscyamus arachnoideus* Pojark, *Hyoscyamus bornmuelleri* Khat., *Hyoscyamus coelesyriacus* Bornm., *Hyoscyamus flaccidus* C. Wright, *Hyoscyamus gallagheri* A.G.Mill. & Biagi, *Hyoscyamus grandiflorus* Franch., *Hyoscyamus insanus* Stocks, *Hyoscyamus kotschyanus* Pojark, *Hyoscyamus kurdicus* Bornm., *Hyoscyamus leptocalyx* Stapf, *Hyoscyamus leucanthera* Bornm. & Gauba, *Hyoscyamus longipedunculatus* C.C.Towns., *Hyoscyamus malekianus* Parsa, *Hyoscyamus multicaulis* Rech.f. & Edelb., *Hyoscyamus nutans* Schönb.-Tem., *Hyoscyamus orthocarpus* Schönb.-Tem., *Hyoscyamus pojarkovae* Schönb.-Tem., *Hyoscyamus rosularis* Schönb.-Tem., *Hyoscyamus squarrosus* Griff., *Hyoscyamus tenuicaulis* Schönb.-Tem., and *Hyoscyamus tibesticus* Maire.

***Latua pubiflora* (Griseb.) Baill.**

There is only one species in this genus, i.e., *Latua pubiflora* (Griseb.) Baill. which is endemic to Chile. Tropane alkaloids from *Latua pubiflora* were investigated by Muñoz and Casale (2003).

Mandragora

There are four species of *Mandragora*, all of them contain tropane alkaloids. *Mandragora* spp. contain hyoscyamine, scopolamine, cuscohygrine, apoatropine, 3α-tigloyloxytropane, and 3,6-ditigloyloxytropane (Jackson and Berry 1973). Schlesinger et al. (2019) reported that *M. autumnalis* accumulates hyoscyamine, anisodamine, and scopolamine; *M. officinarum* and *M. turcomanica* lack anisodamine and scopolamine but show up to tenfold more hyoscyamine levels as compared with *M. autumnalis*. The roots of *M. caulescens* contain higher concentrations of two tropane alkaloids, viz., hyoscine and anisodamine (Zhang et al. 1994).

***Mandragora autumnalis* Bertol.**

It is a perennial herb and grows in the temperate Mediterranean to Western Iran.

***Mandragora caulescens* C.B. Clarke**

It is a perennial herb and native to Nepal to China (W. Sichuan, NW. Yunnan) and Northern Myanmar.

***Mandragora officinarum* L.** (Synonym: *Mandragora vernalis* Bertol).

It is a perennial herb with tuberous roots and indigenous to Mediterranean region.

***Mandragora turcomanica* Mitzgir.**

It is a perennial herb and native to NNE Iran to Southern Turkmenistan.

Nicandra

Nicandra species does not contain esters of tropic acid. However, tropine was isolated from the roots of *Nicandra physalodes* (Parr 1992). There are three species of *Nicandra*, among which only *Nicandra physalodes* showed tropine.

***Nicandra physalodes* (L.) Gaertn**

It is an annual, erect herb and grows in the seasonally dry tropical Peru to NW Argentina.

Physalis

There are 94 species of the genus *Physalis*. The roots of *Physalis* plants are known to contain alkaloids, especially tropane and pyrrolidine alkaloids. Tropane alkaloids have been reported in only in *Physalis peruviana* (Sahal and Ray 1980). *Physalis alkekengi*, in its roots, contains tigloidine, 3α-tigloyloxytropane, cuscohygrine, and phygrine (hygrine dimer) (Basey and Woolley 1973). Valid name of this species is *Alkekengi officinarum* Moench.

Fig. 2.9 *Physalis peruviana* L.

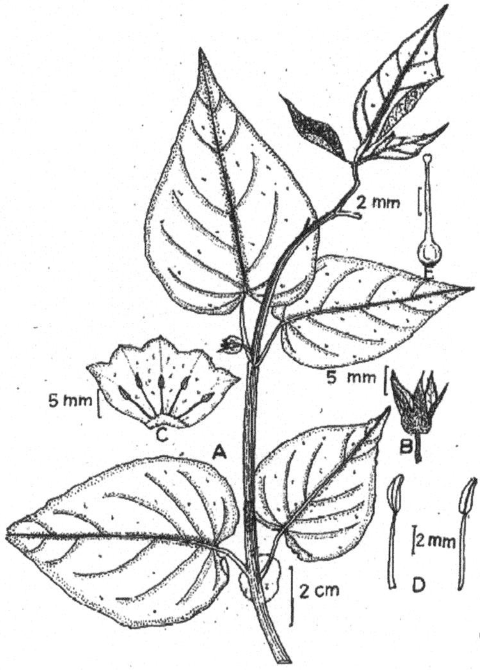

Physalis peruviana **L.**

It is a perennial herb and grows in the seasonally dry tropics of Bolivia to W. Brazil, widely naturalized elsewhere in the tropics (Figs. 2.9 and 2.10).

Physochlaina

The genus *Physochlaina* G. Don comprises six species: *Physochlaina orientalis*, *P. physaloides*, *P. infundibularis*, *P. alaica*, *P. praealta*, and *P. officinalis*. These plants are perennial herbs. The first five species contain tropane alkaloids, mainly apo atropine, apo hyoscine, hyoscyamine, scopolamine, 6-hydroxyatropine, and physochlaine (Gorinova et al. 1999).

Physochlaina alicia **Korotkova**

It is perennial and native to Central Asia. Tropane alkaloids reported in this species include atropine, apotropine, anisodamine, α-belladonnine, β-belladonnine, hyoscyamine, 6-hydroxyatropine, 3,6-dihydroxyatropine, 3 α-tigloyloxy-6-acetoxytrpane, scopolamine, apohyoscine, and tropine (Zhao et al. 2013; Zang et al. 2021; Tantray et al. 2024).

Physochlaina infundibularis **Kuang.**

It is a herb, native to China. *Physochlaina Radix* (called Huashanshen (华山参) in Chinese) is the root of *Physochlaina infundibularis* Kuang. Tropane alkaloids have been reported in this plant (Li et al. 2024). Tropane alkaloids reported in this species include atropine, anisodamine, hyoscyamine, scopoline, scopolamine, and apohyoscine (Zhao et al. 2013; Zang et al. 2021; Tantray et al. 2024).

Fig. 2.10 *Physalis peruviana* L.

Physochlaina orientalis **(M.Bieb.) G. Don**

It occurs across the Caucasus and Middle Asia. Tropane alkaloids represented about 90% of the total alkaloid mixture. The maximum content of alkaloids in the underground parts during flowering was found to be 0.718%. Tropane alkaloids reported in this species include anisodamine, hyoscyamine, hygrine, 3 α-tigloyloxy-6-acetoxytropane, 3 α-tigloyloxy-6-isobutyryloxytropane, apohyoscine, tropinone, and tropine (Zhao et al. 2013; Zang et al. 2021; Tantray et al. 2024).

Physochlaina physaloides **(L.) G. Don**

It is distributed in Kazakhstan, Mongolia, Russia, and China's Hebei, Heilongjiang, Inner Mongolia, and Xinjiang. In a phytochemical report seven alkaloids (cuscohygrine, anisodamine, L-hyoscyamine, atropine, scopolamine, α-belladonnine, and β-belladonnine) have been reported in *Physochlaina physaloides* (Xie et al. 2011; Wu et al. 2016; Xi et al. 2019). The study of the whole plants of *Physochlaina physaloides* by Zang et al. (2021) has led to the isolation of 8 alkaloids, 14 flavonoids, 2 coumarins, and 1 iridoid. Tropane alkaloids reported in this species include atropine, apotropine, anisodamine, α-belladonnine, β-belladonnine, cuscohygrine, hyoscyamine, scopolamine, and scopolamine-N-oxide (Zhao et al. 2013; Zang et al. 2021; Tantray et al. 2024).

Physochlaina praealta **(Decne) Miers**

It is a herb native to China, Tibet, Nepal, West Himalaya. Tropane alkaloids reported in this species include atropine, apoatropine, anisodamine, hyoscyamine, and scopolamine, (Zhao et al. 2013; Zang et al. 2021; Tantray et al. 2024).

Przewalskia
Przewalskia tangutica **Maxim.**

Przewalskia tangutica, a plant of Tibetan traditional medicine, contains tropane alkaloids, tropine, cuscohygrine, (−)-6beta-hydroxyhyoscyamine, hyoscyamine, apoatropine, scopolamine, and daturamine (Peizen and Liyi 1982). The roots of this plant contain significant amounts of hyoscyamine (1.67–3.82%) and total alkaloids (2.06–4.01%) (Da-Wa et al. 2016). It is native to Tibet and Central China. Shao et al. (2013) isolated hyoscyamine, anisodamine, hyoscine, and tropine from *Przewalskia tangutica.*

Salpichroa
There are 19 species of *Salpichroa*, among which tropane alkaloids have been reported only in *Salpichroa origanifolia*. Roots of *S. origanifolia* contain small quantities of cuscohygrine, pseudotropine, tropine, and possibly hyoscyamine (Evans et al. 1972).

***Salpichroa origanifolia* (Lam.) Baill.**

It is a scrambling subshrub and grows in the subtropical regions in Peru to S. Brazil and S. South America.

Schizanthus
S. pinnatus contains tropane derivatives such as schizanthines A and B which are 6β-senecioyloxytropan-3β-ol esters of dibasic mesaconic acid. *S. littoralis* contains hydroxytropane esters, hygrolines, and the tropane diester of itaconic acid. *S. grahamii* contains schizanthines C, D, and E and trimeric tropane alkaloids (San-Martin et al. 1987). Reina et al. (2010) isolated 6β-hydroxy-3α-senecioyloxytropane, 3α-hydroxy-6β-tigloyloxytropane, 3α-hydroxy-6β-senecioyloxytropane, and 3α-hydroxy-6β-angeloyloxytropane from *Schizanthus grahamii* and *Schizanthus pinnatus*. Humam et al. (2011) reported tropane alkaloids in *Schizanthus litoralis* var. *humilis* (synonym: *Schizanthus tricolor*). Tropane alkaloids have been reported in *Schizanthus hookeri* by Griffin and Lin (2000) and Cretton et al. (2017). Gambaro et al. (1982) reported the presence of 3a-hydroxytropane, 3a-hydroxy-6b-angeloyloxytropane, and 3a-senecioyloxy-6b-hydroxytropane in *Schizanthus alpestris*. Christen et al. (2019) gave an excellent review of the tropane alkaloids of different species of *Schizanthus* and the biological activities of these alkaloids. Two tropane alkaloids, schizanthines A and B, were isolated and identified from a methanol extract of the leaves and stems of *S. pinnatus* (Ripperger 1979). Schizanthine A is known as 3a-ethylmesaconyloxy-6b-senecioyloxytropane, while schizanthine B features a dicarboxylic mesaconic acid linking the two tropanol monomers. Additionally, seven other 3a,6b-dihydroxytropane derivatives—schizanthines F, G, H, I, K, L, and M—were later isolated (De la Fuente et al. 1988). Other compounds identified include 3a-hydroxytropane, 3a-hydroxy-6b-angeloyloxytropane, 3a-hydroxy-6b-tigloyloxytropane, and 3a-senecioyloxytropane. So far, schizanthines A, B, G, K, L, and M have only been identified in this particular species. From the leaves of *S. porrigens*, four tropane alkaloids have been isolated and characterized (Muñoz and Cortes 1998). Two of these are monoesterified tropane derivatives: 3a-hydroxy-6b-angeloyloxytropane and 3a-senecioyloxytropan-6b-hydroxytropane. The other two compounds are ditropane alkaloids named schizanthine Y and Z; schizanthine Y contains an itaconic acid residue instead of the mesaconic acid residue found between the two tropane skeletons. In total, 20 alkaloids have been detected or isolated from the roots and

aerial parts of *S. grahamii*. Among those identified are 3α-hydroxytropane, hygrolines A and B, 3α-senecioyloxytropane, 3α-hydroxy-6β-angeloyloxytropane, and 3α-senecioyloxytropan-6β-hydroxytropane (Gambaro et al. 1982; San-Martin et al. 1987).

***Schizanthus alpestris* Poepp.**

It is an herb, 20–40 cm high that grows in the Atacama and Coquimbo regions of Chile, between 900 and 2900 m of altitude.

***Schizanthus grahamii* Gillies**

This annual or biennial herb grows in the Andes between 700 and 2500 m. Its native range is central Chile to western central Argentina.

***Schizanthus hookeri* Gillies ex Graham**

This plant flourishes on the western slopes of the Andes, located in relatively dry environments at elevations ranging from 1500 to 3000 m. It is commonly found in the regions of Coquimbo and La Araucanía in central Chile, as well as in the provinces of Mendoza and Neuquén in Argentina. As an annual or biennial herb, it contributes to the biodiversity of these areas and plays a significant role in the local ecosystems. Angeloyloxytropan-3α-ol,3α-senecioyloxytropan-6 ~ -ol and 6α-tigloyloxytropan-3α-01, tropine, and two diastereomeric hygrolines were isolated from *Schizanthus hookeri* (Gambaro et al. 1983). A new hygroline and tropane alkaloid, 4-hydroxybenzenepropanoylhygroline, has been isolated from the aerial parts of *S. hookeri*, a plant indigenous from Chile.

***Schizanthus integrifolius* Phil.**

It is an annual herb and grows in the Atacama and Coquimbo regions of Chile between 600 and 2700 m.

***Schizanthus litoralis* Phil.**

It is an annual herb and grows in the littoral zone of the Coquimbo and Valparaiso regions of Chile.

***Schizanthus litoralis* var. *humilis* (Lindl.) V. Morales & Muñoz-Schick** (Synonym: *Schizanthus tricolor* Grau & Gronbach).

This variety's native range is Chile (Valparaíso). It is an annual herb and grows principally in the subtropical biome. A new hydroline and tropane 3α,4β-dihydroxy-6β-angeloyoxytropane have been isolated from the aerial parts of *S. tricolor*.

***Schizanthus pinnatus* Ruiz & Pav.**

It is an annual or biennial herb of about 60 cm high. It is indigenous to Chile, in the hills of coastal areas of the Andes between 50 and 1400 m.

Schizanthus porrigens **Graham ex Hook.**

It is an annual herb of up to 80 cm high and grows in the hills of coastal areas of Chile.

There are 15 accepted species of *Schizanthus*. The following eight species have not been studied for the presence of tropane alkaloids: *Schizanthus candidus* Lindl., *S. carlomunozii* V. Morales & Muñoz-Schick, *S. coccineus* (Phil.) J.M.Watson, *S. lacteus* Phil., *S. laetus* Phil., *S. nutantiflorus* J. Chinga & Lavandero, *S. parvulus* Sudzuki, and *S. splendens* Sudzuki.

Scopolia

There are three species of *Scopolia*. Tropane alkaloids have been reported in two species—*Scopolia carniolica* and *Scopolia japonica,* while *Scopolia caucasica* has not been studied for tropane alkaloid content. Fatur et al. (2021) investigated the alkaloids present in *Scopolia carniolica* var. *hladnikiana*.

Scopolia carniolica **Jacq.**

It is a perennial, rhizomatous herb and grows E. Central & SE. Europe to W. Caucasus.

Scopolia caucasica **Kolesn. ex Kreyer**

It is an herb and indigenous to Caucasus.

Scopolia japonica **Maxim.** (Synonym.: *Scopolia parviflora* (Dunn) Nakai).

It is indigenous to Korea and Central and S. Japan. It is a perennial herb.

Solandra

The genus *Solandra* contains atropine, littorine, hyoscyamine, and their derivatives as principal alkaloids. Five *Solandra* species (*Solandra grandiflora*, *S. guttata*, *S. hartwegii*, *S. hirsuta*, and *S. macrantha*) have been investigated by Evans et al. (1972) for their alkaloid content. Alkaloids isolated or detected are atropine, hyoscyamine, noratropine (principal alkaloids), littorine, hyoscine, norhyoscine in one instance, tigloidine, 3α-tigloyloxytropane, 3α-acetoxytropane, valtropine, tropine, ψ-tropine, and cuscohygrine.

Solandra grandiflora **Sw.** (Synonym: *Solandra hirsuta* Dunal).

It is an epiphyte and grows in the wet tropical Central Mexico to Venezuela, Caribbean, and E. and S. Brazil.

Solandra guttata **D. Don**

It is a liana and grows in the seasonally dry tropical Mexico.

Solandra longiflora **Tussac** (Synonym: *Solandra macrantha* Dunal).

It is indigenous to Cuba to Hispaniola, Venezuela, Ecuador to Peru. It is a scrambling tree.

Fig. 2.11 *Withania somnifera* (L.) Dunal. (Source: A. Lalithamba with permission)

Solandra maxima **(Moc. & Sessé ex Dunal) P.S. Green** (Synonym: *Solandra hartwegii* N.E. Br. ex C.F. Ball).

It is indigenous to Mexico to Colombia. It is a scrambling shrub.

The following species of *Solandra* have not been studied for tropane alkaloid presence: *S. boliviana* Britton, *S. brachycalyx* Kuntze, *S. brevicalyx* Standl., *S. guerrensis* Martinez, *S. nizandensis* Matuda, and *S. paraensis* Huber ex Ducke.

Solanum

Solanum species contain calystegine A3 along with calystegine B2 in the leaves (Evans 1979).

Symonanthus

There are two species of *Symonanthus*. Tropane alkaloids have been investigated only in *Symonanthus aromaticus* but not in *Symonanthus bancroftii* (F.Muell.) Haegi.

Symonanthus aromaticus **(C.A. Gardner) Haegi**

It is indigenous to SW. Western Australia. It is a shrub and grows in the subtropical biome. *S. aromaticus* contains mono- and ditigloyl esters. Scopolamine and its derivative aposcopolamine are found as the main alkaloids in both aerial parts and roots (Evans and Ramsey 1983).

Withania

There are 19 accepted species of *Withania*, among which tropane alkaloids have been reported in *Withania somnifera*. *W. somnifera* contains cuscohygrine, 3 α-tigloyloxytropane, tropine, and pseudotropine (Leary et al. 1963).

Withania somnifera **(L.) Dunal**

It is a subshrub or shrub (Fig. 2.11) and distributed in S. Europe to Central China, Africa to Myanmar, and cultivated in many other tropical countries.

The minor genus *Acristus* contain hyoscyamine and tigloidine (Evans 1979). This genus is not mentioned in the web.

2.3 Erythroxylaceae

2.3.1 Erythroxylum

There are 270 species of *Erythroxylum* in the tropics, with two notable ones: *Erythroxylum coca* and *Erythroxylum novogranatense*. These species have been cultivated for thousands of years by South American societies and are now sources of cocaine, which is often diverted into illegal trade. Restrepo et al. (2019) gave a detailed account on different aspects of the genus *Erythroxylum*.

Tropane alkaloids have been identified in 35 species. Among these, *Erythroxylum mamacoca* contains 3β-benzoyloxytropane, along with nortropacocaine (3β-benzoyloxynortropane) (El-Imam et al. 1985). *Erythroxylum coca*, *Erythroxylum novogranatense*, and its variety contain dihydrocuscohygrine and cuscohygrine. In *Erythroxylum cataractarum*, cuscohygrine is the major alkaloid present. A new alkaloid, tropacocaine, has been discovered in *Erythroxylum ulei*, as well as in *Erythroxylum mamacoca* and *Erythroxylum argentinum*, but it is absent in the sections *Macrocalyx* and *Rhabdophyllum*. Additionally, cocaine and its cis- and trans-isomers of cinnamoyl cocaine have been reported in *Erythroxylum coca* and *Erythroxylum novogranatense* var. *truxillense* (Rivier 1981; Zuanazzi et al. 2001).

Erythroxylum argentinum O. E. Schulz, *E. cataractarum* Spruce ex Peyr, *E. cumanense* Kunth, *E. glaucum* O. E. Schulz, *E. mamacoca* Mart., and *E. shatona* Macbride contain total alkaloids in the range 0.06–0.20% (El-Imam et al. 1985).

Lv et al. (2022) gave a fascinating review of the phytochemistry of the genus *Erythroxylum*. Oliveira et al. (2010) comprehensively summarized structures of 186 TAs found in 35 species of *Erythroxylum*. As an update, Lv et al. (2022) found 11 more new TAs reported in studies since then. Among these newly identified TAs, two members were isolated from *E. pungens* (Sena-Filho et al. 2010) and *E. caatinga* (de Oliveira et al. 2011), respectively; 6β,7β-dibenzoyloxytropan-3α-ol was obtained from *E. subsessile* (Soares Cruz et al. (2016); 7β-acetoxy-6βbenzoyloxy-3α-hydroxytropane was isolated from the twigs of *E. macrocalyx* (da Silva et al. 2021); six members named as erythrobezerrines A–F were isolated from the stem bark of *E. bezerrae* (Brito et al. 2020); and 7β-acetoxy-3β,6β-dibenzoyloxytropane was isolated from the leaves of *E. rimosum* (Ribeiro et al. 2013). Aynilian et al. (1974) reported cocaine in *Erythroxylum campestre*, *E. deciduum*, *E. novogranatense*, *E. panamense*, *E. pelleterianum*, and *E. pulchrum*.

***Erythroxylum alternifolium* A. Rich.**
It is indigenous to Cuba. It is a shrub or tree and grows in the wet tropical regions. Tropane alkaloid alaternifoline was reported by Payo-Hill et al. (2000).

***Erythroxylum argentinum* O.E. Schulz**
It is a shrub or tree and grows in the subtropical Bolivia to Brazil and N. Argentina. Zuanazzi et al. (2001) reported 3b,6b-Bis(tigloyloxy)nortropane, methylecgonidine, and tropacocaine in *E. argentinum*.

***Erythroxylum australe* F. Muell.**

It is shrub and indigenous to Queensland to NE. New South Wales (Australia). *E. australe* root-bark contained the new alkaloids 3α-(4-hydroxyphenylacetoxy)tropane and 3-cinnamoyloxytropane-6,7-diol, methylecgonidine as principal base, and other benzoyl- and cinnamoyl-esters (El-Imam et al. 1988).

***Erythroxylum bezerrae* Plowman**

It is a shrub indigenous to Brazil (Piauí). Six previously undescribed tropane alkaloids, designated as erythrobezerrines A–F, were isolated from the stem bark of *E. bezerrae* (Brito et al. 2020).

***Erythroxylum caatingae* Plowman**

It is a shrub and grows in dry tropics of NE Brazil. Three tropane alkaloids were isolated from *E. caatingae,* i.e., 6β-benzoyloxy-3α-[(4-hydroxy-3,5-dimethoxybenzoyl)oxy]tropane, a new tropane alkaloid, along with the known alkaloids 3α,6β-dibenzoyloxytropane and 6β-benzoyloxy-3α-[(3,4,5-trimethoxybenzoyl)oxy]tropane (catuabine B) (De Oliveira et al. 2011).

***Erythroxylum campestre* A. St.-Hil.**

It is a shrub and grows from Bolivia to Brazil and Paraguay.

***Erythroxylum cataractarum* Spruce ex Peyr.**

It is a shrub or tree and grows in S. Tropical America.

***Erythroxylum citrifolium* A. St. Hil.**

It is a shrub or tree and grows in Central and S. Tropical America.

***Erythroxylum coca* Lam.**

It is a shrub and grows in W. South America. Johnson (1996) reported 3-[(Z)-cinnamoyloxy]cocaine, cocaine, methylecgonine, 3-[(E)-cinnamoyloxy]cocaine, tropacocaine, and tropinone. Alkaloids, nicotinoyl-, 2′-pyrroloyl-, and 2′- and 3′-furanoylecgonine methyl ester, were detected by Moore and Casale (1997). Two additional alkaloids 3α- and 3*β*-3′,4′,5′-trimethoxy-*cis*-cinnamoyloxytropane were reported by Casale and Moore (1996a). Casale and Moore (1996b) also reported butyroyl-, 2-methylbutyroyl-, isovaleroyl-, senecioyl-, tigloyl-, *trans*-4-hexenoyl-, hexanoyl-, *trans*-3-hexenoyl-, *trans*-2-hexenoyl-, *trans*-3-heptenoyl-, and *trans*-2,4,-hexadienoyl-ecgonine methyl esters. Methylecgonidine, tropine, 3α-acetoxytropane, ecgonine methyl ester, cuscohygrine, N-norbenzoyltropine, benzoyltropine, hexanoylecgonine methyl ester, cocaine, *cis*-cinnamoylcocaine, and *trans*-cinnamoylcocaine were reported by Casale et al. (2005) from the seeds of *E. coca* var. *coca.* Sauvain et al. (1997) reported cocaine from the leaves of *E. coca* var. *coca.*

***Erythroxylum cuneatum* (Miq.) Kurz**

It is a tree and its native range is Indo-China to New Guinea. From the leaves of *E. cuneatum*, a new alkaloid (±)-3α,6β-dibenzoyloxytropane was isolated; four other benzoyl and tigloyl esters were obtained, together with nicotine. Three of these bases were also detected in the stem-bark (El-Imam et al. 1988).

***Erythroxylum deciduum* A. St.-Hil.**
It is a shrub or tree and grows in Peru to Brazil and NE. Argentina. Cocaine, methylecgonidine, and tropacocaine have been reported from *E. deciduum* by Zuanazzi et al. (2001).

***Erythroxylum dekindtii* (Engl.) O.E. Schulz**
It is a shrub and indigenous to Angola (Huíla). 1α*H*, 5α*H*-Tropan-3α-yl isovalerate is the principal alkaloid of the root-bark of *Erythroxylum dekindtii*; other new bases characterized are 1α*H*, 5α*H*-tropan-3α-yl phenylacetate and 1α*H*, 5α*H*-tropan-3α-yl 2-furoate. The root-bark also contains methylecgonidine, valeroidine, poroidine, isoporoidine, and tropine (Al-Yahya et al. 1979).

***Erythroxylum ecarinatum* Hochr.**
It is a tree and grows in the region Central Malesia to N. Queensland. Tropacocaine was the principal alkaloid of *E. ecarinatum* leaves (El-Imam et al. 1988).

***Erythroxylum ellipticum* R.Br. ex Benth.**
It is a shrub or tree and its native range is N. Australia. 3a-(3,4,5-trimethoxycinnamoyloxy)tropane has been reported by Al-Said et al. (1986a) in *E. ellipticum.*

***Erythroxylum emarginatum* Thonn.**
It is a shrub or tree and indigenous to tropical and S. Africa. Anhydroecgonine methyl ester and anhydroecgonine methyl ester N-oxide have been reported by Nishiyama et al. (2007) from *E. emarginatum.*

***Erythroxylum glaucum* O. E. Schulz**
It is a shrub and grows in tropical Ecuador and Peru.

Erythroxylum havanense **Jacq.** (Synonym: *Erythroxylum cumanense* Kunth).

It is a shrub or tree and grows in S. Florida, Caribbean, and Mexico to Guyana. Anhydroecgonine has been reported by Dominicis and Férnandez (1991).

***Erythroxylum hypericifolium* Lam.**
It is a shrub or tree and indigenous to Mascarenes. The root-bark of *E. hypericifolium* contains the new bases tropan-3α-yl 3-hydroxyphenylacetate, (+)-tropane-3α,6β-diol 3-phenylacetate, tropane-3α,6β-diol 3-phenylacetate 6-acetate, tropane-3α,6β,7β-triol 3-phenylacetate, and nortropan-3-yl phenylacetate; tropan-3α-yl phenylacetate was also isolated (Al-Said et al. 1986a, b). Al-Said et al. (1989b) reported tropane alkaloids in the leaves of *E. hypericifolium.* Thirteen bases were characterized from the stem-bark of *E. hypericifolium*; hygrine is the principal component (Al-Said et al. 1989a). Alkaloids reported for the first time are 3α-phenylacetoxynortropan-6β-ol, 6β-acetoxy-3α-benzoyloxytropane and 3-acetoxy-6-phenylacetoxytropane.

***Erythroxylum macrocalyx* Mart.**
It is a shrub or tree and indigenous to E. Brazil (Bahia, Mijnas Gerais). A new tropane 7β-acetoxy-6β-benzoyloxy-3α-hydroxytropane was isolated by da Silva et al.

(2021) from the twigs of *E. macrocalyx* along with the known substances: 6β,7β-dibenzoyloxy-3α-hydroxytropane, 6β,7β-dihydroxy-3α-(phenylacetoxy)tropane, 3α-benzoyloxy-6β,7β-dihydroxytropane, 6β-benzoyloxy-3α-(4-hydroxy-3,5-dimetoxybenzoyloxy)tropane, ombuin-3-rutinoside-5-glucoside, lupeol, taraxerol, and lupenone.

***Erythroxylum macrocarpum* O.E. Schulz**
It is a shrub or tree and indigenous to Mauritius. 3α-(benzoyloxy)nortropan-6b-ol, 3α-(benzoyloxy)nortropane, 3α-(cinnamoyloxy)nortropane, pseudotropine, and tropacocaine have been reported from *E. macrocarpum* by Al-Said et al. (1986a, b).

***Erythroxylum macrophyllum* Cav.** (Synonym: *Erythroxylum lucidum* Kunth).

It is a shrub or tree and grows in tropical S. Mexico to S. Tropical America. 3α-(acetoxy)tropane, littorine, tropine, and tropinone were isolated by Brachet et al. (1997) from the stem bark of *E. macrophyllum*.

***Erythroxylum mamacoca* Mart.**
It is a shrub and indigenous to Peru (Huánuco).

***Erythroxylum microphyllum* A. St.-Hil.**
It is a subshrub or shrub and grows in dry tropical SE. and S. Brazil to NE. Argentina biome. Methylecgonidine and tropacocaine were reported by Zuanazzi et al. (2001) in *E. microphyllum*.

***Erythroxylum monogynum* R. Br.**
It is a shrub (Fig. 2.12) and grows in dry tropics of India to Myanmar, Sri Lanka. 3-(phenylacetoxy)nortropane, 3-(2-methylbutyryloxy)tropane-6,7-diol, 3-(2-methylbutyryloxy) tropane, 3-oxotropan-6β-ylbenzoate, 3α-(3,4,5-trimethoxybenzoyloxy)tropane-6b,7b-diol, 3α-(3,4,5-trimethoxybenzoyloxy)tropane, 3α-(3,4,5-trimethoxybenzoyloxy) tropan-6β-ol, 6β-(benzoyloxy)-3α-(3,4,5-trimethoxycinnamoyl oxy)tropane, 3α-[(E)-3,4,5-trimethoxycinnamoyloxy]tropane, 3α-(4-methylvaleroyloxy)tropane, 3α-[(Z)-cinnamoyloxy]tropane, 3α-(isobutyryloxy)nortropane, 3α-(phenylacetoxy)tropane, 3α-(phenylacetoxy)tropan-6β-ol, 3b-(phenylacetoxy)tropane, 6β-(benzoyloxy) tropan-3α-ol, 6βb-hydroxytropan-3-one, butropine, convolamine, isoporoidine, pseudotropine, tropacocaine, tropane-3α,6β-diol, 3α-(3,4,5-trimethylcinnamoyloxy)tropan-6b-ol benzoate, tropane-3α,6β-diol6-benzoate, tropane-3α,6β,7β-triol, tropane-3α,6β,7β-triol-(3,4,5-trimethoxybenzoate), tropan-3α-yl 3,4,5-trimethoxybenzoate, tropan-3α-yl3,4,5-trimethoxycinnamate, tropine, tropinone, and valeroidine have been reported from *E. monogynum* by Christen et al. (1995) and Agar and Evans (1976).

***Erythroxylum moonii* Hochr.**
It is a shrub and grows in tropical S. India and Sri Lanka. Four new tropane alkaloids were isolated by Khattak et al. (2002) from the leaves of *E. moonii* and identified as 3α-isobutyryloxy-7β-hydroxynortropane, 3α-hydroxy-7β-phenylacetoxynortropane, 3α-cis-cinnamoyloxytropane, and 3α-hydroxy-6β-(3′-hydroxy-2′-methyl-3′-phenylpropionyloxy)-7β-hydroxytropane. Other alkaloids isolated for the first time

Fig. 2.12 *Erythroxylum monogynum* R. Br. (Source: A. Lalithamba with permission)

from *E. moonii* were 3α-benzoyloxytropane, 3α-phenylacetoxytropane, 3α-trans-cinnamoyloxytropane, and 3α-phenylacetoxy-6β,7β-dihydroxynortropane. Two new dimeric tropane alkaloids, mooniines A and B, were identified from the leaves of *E. moonii* by Atta-Ur-Rahman et al. (1998). Khattak et al. (2003) reported nortropane-3α,6β,7β-triol 3-benzoate7-(2-hydroxy 3-phenylpropanoate), nortropane-3α,7β-diol3-(2-methylpropanoate)7-[(Z)-3,4,5-trimethoxycinnamate), nortropane-3α,7β-diol7-benzoate 3-(2-methylpropanoate), nortropane-3a,7b-diol7-[(E)-cinnamate]3-propanoate, tropane-3α,6β,7β-triol 3-benzoate, tropane-3a,7b-diol3-benzoate, tropane-3α,7β-diol 3-phenylacetate, tropane-3α,7β-diol7-benzoat, and tropane-3α-yl 3-(3,4,5-trimethoxybenzoate) from *E. moonii*.

Erythroxylum novogranatense **(D. Morris) Hieron.**
It is an evergreen shrub and distributed from Colombia to NW. Venezuela and Peru. Tamprateep et al. (1963) and Balick et al. (1982) reported the presence of cocaine in *E. novogranatense*. 1-Hydroxytropacocaine was isolated by Moore et al. (1994) from leaves of greenhouse-cultivated *E. novogranatense* var. *novogranatense*. *E. novogranatense* var. *novogranatense*, a species cultivated for cocaine production, contained 0.2% total calystegines in dry leaves (Brock et al. 2005).

Erythroxylum ovalifolium **Peyr.**
E. ovalifolium is a woody shrub extensively distributed along Brazil's coastal plains. Along with other species in the genus, *E. ovalifolium* can produce tropane alkaloids (Restrepo et al. 2019).

Erythroxylum panamense **Turcz.**
It is a shrub or tree and grows in wet tropical Mexico (Veracruz), Costa Rica to N. Venezuela.

***Erythroxylum pelleterianum* A. St. Hil.**
It is a shrub and grows in tropical Bolivia to Brazil and Paraguay. Methylecgonidine and tropacocaine have been reported by Zuanazzi et al. (2001) in *E. pelleterianum.*

***Erythroxylum pervillei* Baill.**
It is a shrub and endemic to Madagascar. Silva et al. (2001) isolated pervilleine A, B, C, D, E, F, tropane-3α,6β,7β-triol 3-phenylacetate, and tropan-3a-yl3,4,5-trimethoxybenzoate. From the stem-bark of *Erythroxylum pervillei*, Chin et al. (2006) isolated four tropane aromatic ester alkaloids, namely, pervilleines G and H, and *cis*-pervilleines B and F.

***Erythroxylum pulchrum* A. St. Hil.**
It is a shrub or tree and grows in E. Brazil.

***Erythroxylum pungens* O.E. Schulz**
It is a shrub or tree and grows in NE. Brazil. Sena-Filho et al. (2010) isolated pungencine: a new tropane alkaloid from the roots of *E. pungens.*

***Erythroxylum rimosum* O.E. Schulz**
It is a shrub and native to Brazil. A new tropane alkaloid, 7β-acetoxy-3β,6β-dibenzoyloxytropane, was isolated from *E. rimosum* leaves (Ribeiro et al. 2013).

***Erythroxylum rotundifolium* Lunan**
It is a shrub or tree and grows in wet tropical Mexico to Central America and the Caribbean. Payo-Hill et al. (2000) reported eythrorotundine in this species. Chávez et al. (2002) reported 6β-(Benzoyloxy)-3α-[(E)-3,4,5-trimethoxy cinnamoyloxy]tropane, 6β-(benzoyloxy)-3α-[(E)-3,4,5-trimethoxy cinnamoyloxy]tropan-7β-ol, 6β-(Benzoyloxy)-3α-[(Z)-3,4,5-trimethoxy cinnamoyloxy]tropane, 6β-(Benzoyloxy)-3a-hydroxytropane, 7β-(Acetoxy)-6β-(benzoyloxy)-3a-[(E)-3,4,5-trimethoxycinnamoyloxy]tropane, and 7β-Hydroxy-6β-(3,4,5-trimethoxybenzoyloxy) 3αs-[(E)-3,4,5-trimethoxycinnamoyloxy]tropane in *E. rotundifolium.*

***Erythroxylum rufum* Cav.**
It is a shrub or tree and grows in the wet tropical Caribbean to Brazil. Cocaine was reported in this species (Oliveira et al. 2010).

***Erythroxylum shatona* J.F. Macbr.**
It is a tree and native to Northern Peru.

***Erythroxylum sideroxyloides* Lam.**
It is a shrub and grows in the wet tropical Mauritius, Réunion. 3a-(Benzoyloxy) nortropan-6β-ol, 3α-(benzoyloxy)nortropane, 3α-(benzoyloxy)tropane, and 3α-(benzoyloxy)tropan-6β-ol have been reported in this species by Al-Said et al. (1986a, b).

***Erythroxylum subsessile* (Mart.) O.E. Schulz**
It is a shrub and grows in dry tropical SE. Brazil (Rio de Janeiro). Soares Cruz et al. (2016) isolated a novel alkaloid, 6β,7β-dibenzoyloxytropan-3α-ol, together with the

two known 3α-benzoyloxynortropan-6β-ol and 3α,6β-dibenzoyloxytropane alkaloids from the leaves of *E. subsessile*.

Erythroxylum ulei **O.E. Schulz**
It is a shrub or tree that grows in the wet tropics from West South America to Brazil.

Erythroxylum vaccinifolium **Mart.**
It is a shrub and native to Central Brazil. (3aH)-Catuabine H; 7α-hydroxycatuabineH; 7β-hydroxycatuabineD; 7β-hydroxycatuabineE; 7β-hydroxycatuabineF; 7β-acetylcatuabineE; 7β-hydroxycatuabineH; 7β-hydroxycatuabinel; catuabine A, B, C, D, E, F, G, H, and I; catuabine E N-oxide; and vaccinine A and B have been reported in this species by Zanolari et al. (2003a, b, 2005) and Graf and Lude (1977, 1978).

Erythroxylum zambesiacum **N. Robson**
It is a shrub or tree and grows in dry tropical S. Tanzania to N. Botswana. Christen et al. (1993) and El-Iman et al. (1987) reported 3-(2-methylbutyryloxy)tropane-6,7-diol, 3-(2-methylbutyryloxy)tropane, 3α-[(E)-3,4,5-trimethoxycinnamoyloxy]tropane, 3a-(cinnamoyloxy)tropan-6b-ol, 3-(phenylacetoxy)nortropane, 3α-(3,4,5-trimethoxybenzoyloxy)nortropan-6β-ol, 3α-(3,4,5-trimethoxybenzoyloxy)nortropane, 3α-(3,4,5-trimethoxybenzoyloxy) tropane-6β,7β-diol, 3α-(3,4,5-trimethoxybenzoyloxy)tropan-6β-ol, 3α-(3,4,5-trimethoxybenzoyloxy)tropane, 7β-acetoxy-6β-(benzoyloxy)-3α-(3,4,5-trimethoxy cinnamoyloxy)tropane, 6β-(benzoyloxy)-3α-(3,4,5-trimethoxycinnamoyl oxy)tropane, 6β-(benzoyloxy)-3α-[(E)-3,4,5-trimethoxycinnamoyl oxy]tropan-7β-ol, 3α-[(E)-3,4,5-trimethoxycinnamoyloxy)tropane, 3α-(phenylacetoxy)tropan-6β-ol, 3α-(phenylacetoxy)tropane, 6β-(isovaleroyloxy)tropan-3β-ol, 6β-(benzoyloxy) tropan-3-one, 6β-(benzoyloxy)tropane-3α,7β-diol, 6β-(benzoyloxy)tropan-3α-ol, 6β-(tigloyloxy)tropan-3α-ol, 7β-(acetoxy)-6β-(benzoyloxy)-3α-(3,4,5-trimethoxy cinnamoyloxy)tropane, convolamine, pseudotropine, tropane-3α,6β-diol, 6β-hydroxytropan-3-one, tropine, tropinone, and valeroidine in *E. zambesiacum*.

Erythroxylum zeylanicum **O.E. Schulz**
It is a shrub or tree and grows in Sri Lanka. Bringmann et al. (2000) reported tropane alkaloids 3α-(3,4,5-trimethoxybenzoyloxy)tropane and erythrozeylanine A, B(Z), B(E), C(Z) and C(E) from *E. zeylanicum*.

2.4 Proteaceae

Agastachys odorata **R.Br.**
It is a shrub and endemic to Tasmania. *Agastachys odorata* contains 6β-acetoxy-3α-tigloyloxytropane α-(p-hydroxybenzoyloxy) trop-6-ene (Bick et al. 1979d).

Bellendena montana **R.Br.**
A monotypic plant endemic in Tasmania. Bellendine has been isolated by Bick et al. (1971) from the flowers of *B. montana*. Fourteen alkaloids have been isolated from

B. montana, the structures of 10 of these, bellendine, isobellendine, darlingine, 5,11-dihydroisobellendine, 2,3-dihydrobellendine, 2,3-epidihydrobellendine, 2,3-dihydrodarlingine, 3α-acetoxy-6β-isobutoxytropane, 6β-acetoxy-3α-isobutoxytropane, and 3α-acetoxy-6β-hydroxytropane, have been established (Bick et al. 1979a).

***Darlingia darlingiana* (F. Muell.) L. A. S. Johnson**

It is a Queensland (Australia) tree. Darlingine as the major alkaloid was isolated from *D. darlingiana*. The structures of 11 alkaloids isolated from *D. darlingiana* have been determined. They comprise a tropene, two pyranotropanes, and eight pyrrolidine alkaloids (Bick et al. 1979b).

***Darlingia ferruginea* J. F. Bailey**

It is a tree and grows in N. and NE. Queensland. *Darlingia ferruginea* contains the new tropane alkaloids ferrugine and 3α-benzoyloxy-2α-hydroxybenzyltropane and darlingine and ferruginine (Bick et al. 1979c).

***Eucarpha* (Syn.: *Knightia*)**

Alkaloids of a 2-benzyltropine type were obtained from genus *Knightia*, *K. deplanchei* Vieill ex Brongn. & Gris, and *K. strobilina* R. Br. (Lounasmaa et al. 1975). D-strobiline is the main alkaloid of *K. strobilina* R. Br., together with 3α-cinnamoyloxytropan-6β-ol, 3α-acetoxy-2α-acetoxybenzyltropane (acetylknightionl), 3α-acetoxy-2α-benzyltropan-6β-ol(knightoline), 6β-benzoyloxytropan-3α-ol, 2α-hydroxybenzyl-3α-acetoxytropane (knightinol), dihydrostrobiline, strobamine B, chalcostrobamine, strobolamine, knightalbinol, and knightolamine.

***Eucarpha deplanchei* (Vieill. ex Brongn, ex Gris) P.H. Weston & Mabb.** (Synonym: *Knightia deplanchei* Vieill. ex Brongn, ex Gris)

It is a shrub or tree, native to New Caledonia.

***Eucarpha strobilina* (Labill.) P.H. Weston & Mabb**. (Synonym: *Knightia strobilina* (Labill.) R.Br. ex Meisn)

Eucarpha strobilina contains strobiline as the major alkaloid, but it also contains 3α-cinnamoyloxytropan-6β-ol and 3α-acetoxy-2α-acetoxybenzyltropane (acetylknightionl), 3α-acetoxy-2α-benzyltropan-6β-ol (knightoline), 6 β-benzoyloxytropan-3α-ol, 2α-hydroxybenzyl-3α-acetoxytropane (knightinol), and dihydrostrobiline (Lounasmaa et al. 1980).

***Triunia montana* (C.T.White) Foreman**

It is an Australian rain forest tree. Four tropane alkaloids, including a new compound triuniamine A, a known compound 2,3-dihydrodarlingine, with previously unassigned relative configuration and two known compounds darlingine, 10-hydroxydarlingine, were obtained from the stems of *T. montana* (Yang et al. 2017).

2.5 Elaeocarpaceae

***Peripentadenia mearsii* (C.T.White) L.S.Sm.**
It is a tree endemic to Queensland (Australia). *P. mearsii* contains the tropane alkaloids tropacocaine (3β-benzoyloxytropane), 2α-benzoyloxy-3β-hydroxytropane, and 3 α-acetoxy-6β-hydroxytropane (Johns et al. 1971).

2.6 Euphorbiaceae

***Croton megalobotrys* Müll.Arg.** (Synonym: *Croton gubouga* S. Moore)

It is a tree and grows in the desert or dry shrubland SW. Tanzania to S. Africa. Pyrrolidine and tropane alkaloids have been reported from *Croton megalobotrys* (Syn.: *C. gubouga*) (Savita et al. 2021).

***Margaritaria discoidea* (Baill.) G.L. Webster** (Synonym: *Phyllanthus discoideus* Müll.Arg.)

It is a shrub or tree and grows in dry tropical and S. Africa. Similar tropane alkaloids were also isolated from *Margaritaria discoidea* (synonym: *Phyllanthus discoideus*) (Seigler 1977).

2.7 Rhizophoraceae

***Bruguiera exaristata* Ding Hou**
It is a tree and grows in Lesser Sunda Islands (Timor), New Guinea, N. Western Australia to E. Queensland. *Bruguiera exaristata* contains tropine esters of acetic, benzoic, n-butyric, isobutyric, propionic, and isovaleric acids (Loder and Russel 1969).

***Bruguiera sexangula* (Lour.) Poir.**
It is a tree and distributed in India to Hainan and Bismarck Archipelago, N. Northern Territories to N. and NE. Queensland. *Bruguiera sexangula* (a tropical mangrove) contains brugine (tropine 1,2-dithiolane-3-carboxylate) (Loder and Russel 1966).

***Crossostylis biflora* J.R. Forst. & G. Forst.**
It is a shrub or tree and indigenous to Samoa to Marquesas. Hygrine and four known tropane alkaloids tropanone 3α-tropanol, brugine, and tropan 3α-yl benzoate have been isolated from *C. biflora* (Medina et al. 1983).

***Crossostylis multiflora* Brongn. & Gris**
It is a tree, endemic to New Caledonia. Hygrine and four known tropane alkaloids tropanone, 3α-tropanol, brugine, and tropan 3α-yl benzoate have been isolated from *C. multiflora* (Medina et al. 1983).

***Crossostylis sebertii* Pancher ex Brongn. & Gris**
It is a tree, endemic to New Caledonia. Hygrine and four known tropane alkaloids tropanone, 3α-tropanol, brugine, and tropan 3α-yl benzoate have been isolated from *C. sebertii* (Medina et al. 1983).

***Pellacalyx axillaris* Korth**
It is a tree and distributed from W. Malesia to the Philippines (Mindanao). Extraction of the fresh leaves and terminal branches of *Pellacalyx axillaris* yielded 8-azabicyclo[3.2.1]oct-endo-3-yl (E)-3-phenylprop-2-enoate (nortropinyl cinnamate) (Arbain et al. 1991).

***Pellacalyx saccardianus* Korth.**
It is a tree and distributed from W. Malesia to the Philippines (Mindanao). Seven new tropane alkaloids, including five monomeric, one dimeric, and one trimeric, 3α-nortropane ester, along with two known monomeric nortropane alkaloids, were isolated from the leaves and bark of *P. saccardianus* (Chan et al. 2021).

2.8 Convolvulaceae

Tropane alkaloids found in this family include convolvine, convolidine, confoline, and convolamine N-oxide (Sharova et al. 1980; Aripova 1985). One hundred twenty-nine species of Convolvulaceae belonging to 29 genera (12 tribes) revealed the occurrence of 1 to 6 polyhydroxy alkaloids of the nortropane type (calystegines).

***Calystegia sepium* (L.) R. Br.**
The native range of this species is temperate and subtropics. It is a climbing perennial. Roots and aerial parts of *C. sepium* contain calystegines B1, B2, and A3 (Scholl et al. 2001). Calystegines are products of the tropane alkaloid pathway. Calystegines are nortropane alkaloids bearing between three and five hydroxyl groups in various positions.

***Convolvulus arvensis* L.**
It is an annual trailing or twining herb, pantropical in distribution. *Convolvulus arvensis* was found to contain the tropane alkaloids tropine, pseudotropine, and tropinone and the pyrrolidine alkaloids cuscohygrine and hygrine (Todd et al. 1995).

***Convolvulus krauseanus* Regel & Schmalh.**
It is a subshrub and native of Kirgizstan to N. Tadzhikistan. Aripova (1985) isolated N-oxide of convolamine from *Convolvulus krauseanus*.

***Convolvulus dorycnium* subsp.** *subhirsutus* (Synonym: *Convolvulus subhirsutus* Regel & Schmalh.)

It is a shrub and its distribution is NE. Iran to Central Asia and Afghanistan. Convolvine, convolamine, convolidine, phyllalbine, and the new alkaloid phyllalbine *N*-oxide and the aminoalcohol nortropine were isolated from *Convolvulus subhirsutus* (Sharova et al. 1980; Gapparova et al. 2007).

***Erycibe elliptilimba* Mer. & Chun**
It is a liana and distributed in SE. China to Indo-China. It contains erycibelline (2β,7β-dihydroxynortropane) (Chen et al. 1986; Lu et al. 1986; Wang et al. 1990).

***Erycibe hainanensis* Merr.**
It is a climbing shrub and its distribution is from SE. China to Vietnam. It contains erycibelline (2β,7β-dihydroxynortropane) (Chen et al. 1986; Lu et al. 1986; Wang et al. 1990).

***Erycibe obtusifolia* Benth.**
It is a liana and its distribution is from SE. China to Vietnam. *Erycibe obtusifolia* contains baotongteng A, (2β-hydroxy-6β-acetoxynortropane) (Yao et al. 1981) and baotongteng B (Seigler 1977).

***Evolvulus sericeus* Sw.**
It is a perennial and its distribution is tropical and subtropical America. It is used as a medicine. *E. sericeus* contains convoline, convolamine, and convolidine (Conselo and Alejandra 1972).

Merremia
The occurrence and distribution of tropane and biogenetically related pyrrolidine alkaloids in 18 *Merremia* species have been studied by Jenett-Siems et al. (2005). The valid names of these former *Merremia* species are given below.

***Distimake cissoides* (Lam.) A.R.Simões & Staples** (Synonym: *Merremia cissoides* (Lam.) Hallier f.)

It is a climbing annual or perennial, and its distribution is Mexico to tropical America. Merresectine A, B, C, and E; merresectine E β-D glucoside; kurameric acid; and merredissine are isolated from the root (Olatunji et al. 2021).

***Distimake dissectus* (Jacq.) A.R.Simões & Staples** (Synonym: *Merremia dissecta* (Jacq.) Hallier f.)

It is a climbing herb and is distributed in tropical and subtropical America. Datumetine isolated from root (Olatunji et al. 2021).

***Distimake guerichii* (A.Meeuse) A.R.Simões & Staples** (Synonym: *Merremia guerichii* A. Meeuse)

It is a tuberous geophyte and endemic to Namibian desert. Merresectine D and merresetine β-D glucoside are isolated from the root (Olatunji et al. 2021).

Distimake quinatus **(R.Br.) A.R.Simões & Staples** (Synonym: *Merremia quinata* (R.Br.) Ooststr.)

It is a scrambling subshrub or climbing shrub, and its distribution is tropical and subtropical Asia to N. Australia. Consabatine was isolated from the root (Olatunji et al. 2021).

Distimake quinquefolius **(L.) A.R.Simões & Staples** (Synonym: *Merremia quinquefolia* (L.) Hallier f.)

It is a climbing annual or perennial and its distribution is from Mexico to tropical America. Merresectine A, B, and C; merresectine E β-D glucoside; kurameric acid; and merredissine have been isolated from the root (Olatunji et al. 2021).

2.9 Brassicaceae

The investigation of the ten *Cochlearia* species revealed the presence of cochlearine, tropine, and pseudotropine in all samples. Notably, calystegines—nortropane alkaloids derived from pseudotropine—were also consistently identified across these species, with accumulation reaching up to 0.5% of the dry mass in their leaves. Furthermore, a comprehensive analysis of Brassicaceae species from all major lineages of the family demonstrated that of the 43 species studied, 18 successfully accumulated various structures of calystegines. This research highlights the potential for further exploration into the ecological roles and applications of these compounds within the family (Brock et al. 2006).

Cochlearia groenlandica **L.** (Synonym: *Cochlearia arctica* Schltdl. ex DC.)

It is a biennial or perennial herb, and grows in Subarctic to N. Japan and N. California. Cochlearine, tropine, and pseudotropine were found in *C. groenlandica* and ten other species of *Cochlearia* (Platonova and Kusovkov 1963; De Simone and De Feo Margarucci 2008).

Cochlearia officinalis **L.**
It is a biennial or perennial herb and grows in N. and W. Europe. Cochlearine, the 3-hydroxybenzoate ester of tropine, was found in *Cochlearia officinalis* (Kim et al. 2021).

2.10 Moraceae

Morus alba **L.**
It is native of China, now cultivated in many tropical and subtropical regions. *Morus alba* contains tropane alkaloids. Root bark of *M. alba* contains penthydroxylated

nortropane calystegine C1, while fruits contain calystegines A8 and A9 (Savita et al. 2021).

2.11 Endophytic Fungi and Bacteria: Microbial Sources of Tropane Alkaloids

Naik et al. (2017) for the first time reported tropane alkaloid producing endophytes, namely, *Colletotrichum boninense*, *Phomopsis* sp., *Fusarium solani*, *Colletotrichum incarnatum*, *Colletotrichum siamense*, and *Colletotrichum gloeosporioides*, that are found in *Datura metel* and possess the enzymes PMT, TR-I, and H6H. These fungi produce higher amount of scopolamine (4.1 mg L 1) and hyoscyamine (3.9 g L 1). It may be possible to identify the biosynthetic pathways of TAs in these hosts, which could either be optimized in the fungi themselves and enhanced or transferred into heterologous microbial hosts (Naik et al. 2017).

Wang et al. (2023a) found that endophytic fungi associated with geographic regions enhance the accumulation of tropane alkaloids in *Anisodus tanguticus*. They studied 55 wild populations on the Tibetan Plateau, analyzing tropane alkaloid content and the diversity of root-associated bacteria and fungi using HPLC and sequencing methods. The results showed that tropane alkaloid content varied geographically. Anisodine levels positively correlated with latitude, while anisodamine and atropine levels negatively correlated with it. Variation partition analysis revealed that root endophytes accounted for 14% of the variation in alkaloid production, making them the most significant factor. Soil properties explained 5%, and climate variables accounted for 1%. Endophytic fungi contributed 11%, while bacteria contributed 5%. Additionally, random forest analysis and Mantel tests indicated that regionally enriched endophytic fungi have a more significant effect on tropane alkaloid accumulation than the overall community of endophytic fungi.

Wang et al. (2023b) found that core root bacteria enhance the growth and anisodine accumulation in *Anisodus tanguticus*. Their study showed that, despite being sampled from 1500 km away, the roots selectively recruit specific bacteria. They identified 102 core operational taxonomic units (OTUs) of root bacteria, which comprised only 2.99% of the total community but accounted for 73% of its abundance. Analyses revealed a strong link between the composition of these core microbiomes and the plant's anisodine content, aboveground biomass, and nitrogen levels. Key genera such as *Rhizobacter*, *Variovorax*, *Polaromonas*, and *Mycobacterium* were found to be significantly enriched in the roots. Furthermore, nitrogen-cycling and pathogenic bacteria were associated with the plant's growth metrics. Overall, Wang et al. (2023b) concluded that the selective recruitment of core bacteria plays a crucial role in the anisodine content, biomass, and nitrogen levels of *Anisodus tanguticus*.

2.12 Conclusion

The tropane alkaloids are special category of alkaloids which are having diversified beneficial as well as toxic properties. Few plants were able to contaminate the foods when they grow as weeds nearby agricultural fields. Hence knowledge on the bioresources, details of such metabolites, and their pharmacological properties to the specified use is very essential. More over the upcoming chapters clearly explains the diversity of tropane alkaloids and their chemical structures. Their importance in the chemotaxonomy and their relationship with the other species in the botanical families are also being discussed.

References

Agar JTH, Evans WS (1976) Alkaloids of the genus *Erythroxylum*. Part 1. *E. monogynum* Roxb roots. J Chem Soc Perkin Trans 1 14:1550–1558

Al-Said MS, Evans WC, Grout RJ (1986a) Alkaloids of the genus *Erythroxylum*. Part 5. *E. hypericifolium* Lam. Root-bark. J Chem Soc Perkin Trans 1 6:957

Al-Said MS, Evans WC, Grout RJ (1986b) Alkaloids of *Erythroxylum macrocarpum* and *E. sideroxyloides*. Phytochemistry 25:851–853

Al-Said MS, Evans WC, Grout RJ (1989a) Alkaloids of *Erythroxylum hypericifolium* stem bark. Phytochemistry 28:671–673

Al-Said MS, Evans WC, Grout RJ (1989b) Alkaloids of *Erythroxylum hypericifolium* leaves. Phytochemistry 28:3211–3215

Al-Yahya MAI, Evans WC, Grout RJ (1979) Alkaloids of the genus *Erythroxylum*. Part 2. *E. dekindtii* (Engl.) O. E. Schultz. J Chem Soc Perkin Trans 8:2130

Arbain D, Wiryani RD, Sargent MV (1991) A new tropane alkaloid from *Pellacalyx axillaris*. Aust J Chem 44:1013–1015

Aripova SF (1985) N-Oxide of convolamine from *Convolvulus krauseanus*. Khimiya Prirodnykh Soedinenii 2:275

Ashraf M, Ahmad M, Ahmad H, Ahmad S (2019) Phytochemical investigation of *Hyoscyamus albus*. Pak J Pharm Sci 32(2):661–667

Ashtiana F, Sefidkonb F (2011) Tropane alkaloids of *Atropa belladonna* L. and *Atropa acuminata* Royle ex Miers plants. J Med Plants Res 5(29):6515–6522

Atta-Ur-Rahman, Khattak KF, Nighat F, Shabbir M, Hemalal KD, Tillekeratne LM (1998) Dimeric tropane alkaloids from *Erythroxylum moonii*. Phytochemistry 48:377–383

Aynilian GH, Duke JA, Gentner WA, Farnsworth NR (1974) Cocaine content of *Erythroxylum* species. J Pharm Sci 63:1938

Balick MJ, Rivier L, Plowman T (1982) The effects of field preservation on alkaloid content of fresh coca leaves (*Erythroxylum* spp.). J Ethnopharmacol 6:287–291

Banihashemi O, Khavari-Nejad R-A, Yassa N, Najafi F (2020) Raise up of scopolamine in hairy roots via *Agrobacterium rhizogenes* ATCC15834 as compared with untransformed roots in *Atropa komarovii*. Iran J Pharm Res 19(1):46–56

Baralle FE, Gros EG (1969) Biosynthesis of cuscohygrine in *Atropa belladonna* from sodium acetate-2-14C*. Phytochemistry 8(5):853–855

Basey K, Woolley JG (1973) Alkaloids of *Physalis alkekengi*. Phytochemistry 12:2557–2559

Bick IRC, Bremer JB, Gillard JW (1971) Methyl(p-hydroxybenzoyl)acetate and an alkaloid, bellendine, from *Bellendena montana*. Phytochemistry 10(2):475–477

Bick IRC, Gillard JW, Lew H-M (1979a) Alkaloids of *Bellendena montana*. Aust J Chem 32(8):1827–1840

Bick IRC, Gillard JW, Lew H-M (1979b) Alkaloids of *Darlingia darlingiana*. Aust J Chem 32:2537

Bick IRC, Gillard JW, Lew H-M (1979c) Alkaloids of *Darlingia ferruginea*. Austral J Chem 32(11):2537–2543

Bick IRC, Gillard JW, Huck-Meng L, Preston W (1979d) Alkaloids of *Agastachys odorata*. Aust J Chem 32:2071

Brachet A, Muñoz O, Gupta M, Veuthey J-L, Christen P (1997) Alkaloids of *Erythroxylum lucidum* stem-bark. Phytochemistry 46:1439–1442

Bremner JB, Cannon JR (1968) Isolation of (−)-Hyoscyamine from *Anthotroche pannosa* Endl. Austral J Chem 21(5):1369–1370

Bringmann G, Günther C, Mühlbacher J, Gunathilake MDLP, Wickramasinghe A (2000) Tropane alkaloids from *Erythroxylum zeylanicum* O.E. Schulz (Erythroxylaceae). Phytochemistry 53(3):409–416

Brito LSDO, Pinto FDCL, de Filho MOM, Rocha DD, Mendoza MFM, Ayala AP, Bezerra BP, Loiola MIB, Canuto KM, Silveira ER, Pessoa ODL (2020) Tropane alkaloids from the stem bark of *Erythroxylum bezerrae*. Phytochemistry 178:112458

Brock A, Bieri S, Christen P, Dräger B (2005) Calystegines in wild and cultivated *Erythroxylum* species. Phytochemistry 66:1231

Brock A, Herzfeld T, Paschke R, Koch M, Dräger B (2006) Brassicaceae contain nortropane alkaloids. Phytochemistry 67(18):2050–2057

Cannon JR, Joshi KR, Meehan GV, Williams J (1969) The tropane alkaloids from three western Australian *Anthocercis* species. Aust J Chem 22:221

Casale JF, Moore JM (1996a) Lesser alkaloids of cocaine-bearing plants II. 3-Oxo-substituted tropane esters: detection and mass spectral characterization of minor alkaloids found in South American *Erythroxylum coca* var. *coca*. J Chromatogr A 749:173–180

Casale JF, Moore JM (1996b) Lesser alkaloids of cocaine-bearing plants III. 2-carbomethoxy-3-oxo substituted tropane esters: detection and gas chromatographic-mass spectrometric characterization of new minor alkaloids found in South American *Erythroxylum coca* var. *coca*. J Chromatogr A 756:185–192

Casale JF, Toske SG, Colley VL (2005) Alkaloid content of the seeds from *Erythroxylum coca* var. *coca*. J Forensic Sci 50:1402

Chan Z-Y, Krishnan P, Modaresi M, Hii L-W, Mai C-W, Lim W-M, Leong C-O, Low Y-Y, Wong S-K, Yong K-T, Leong AZ-X, Lee MK, Ting K-N, Lam K-H (2021) Monomeric, dimeric, and trimeric tropane alkaloids from *Pellacalyx saccardianus*. J Nat Prod 84(8):2272–2281

Chávez D, Cui B, Chai H-B, Garcia R, Meja M, Farnsworth NR, Cordell GA, Pezzuto JM, Kinghorn AD (2002) Reversal of multidrug resistance by tropane alkaloids from the stems of *Erythroxylum rotundifolium*. J Nat Prod 65:606–610

Chen Z, Xu P, Yao T (1986) YaoxueXuebao, 21, 513 (Ch.), via CA, 106, 153016

Chen C, Li JJ, Xiong F, Wang B, Xiao YM, Zhou GY (2022a) Multivariate statistical analysis of tropane alkaloids in *Anisodus tanguticus* (Maxim.) Pascher from different regions to trace geographical origins. Acta Chromatogr 34:422–429. https://doi.org/10.1556/1326.2021.00952

Chen C, Wang B, Li JJ, Xiong F, Zhou GY (2022b) Multivariate statistical analysis of metabolites in *Anisodus tanguticus* (Maxim.) Pascher to determine geographical origins and network pharmacology. Front. Plant Sci 13:927336. https://doi.org/10.3389/fpls.2022.927336

Chin Y-W, Jones WP, Waybright TJ, McCloud TG, Rasoanaivo P, Cragg GM, Cassady JM, Kinghorn AD (2006) Tropane aromatic ester alkaloids from a large-scale recollection of *Erythroxylum pervillei* stem bark obtained in Madagascar. J Nat Prod 69:414–417

Christen P, Roberts MF, Phillipson JD, Evans WC (1993) Alkaloids of *Erythroxylum zambesiacum* stem-bark. Phytochemistry 34:1147–1151

Christen P, Roberts MF, Phillipson JD, Evans WC (1995) Alkaloids of *Erythroxylum monogynum* root-bark. Phytochemistry 38:1053–1056

Christen P, Cretton S, Humum M, Bieri S, Muñoz O, Joseph-Nathan P (2019) Chemistry and biological activity of alkaloids from the genus *Schizanthus*. Phytochem Rev 19:615–641

Cinelli MA, Jones AD (2021) Alkaloids of the genus *Datura*: review of a rich resource for natural product discovery. Molecules 26:2629. https://doi.org/10.3390/molecules26092629

Conselo FL, Alejandra SG (1972) Rev. Colomb. Cienc. Quim,-Farm., 2, 27 (Span.), via C.A., 79, 15798

Cretton S, Muñoz O, Tapia J, Marcourt L, Maes L, Christen P (2017) Two new hygroline and tropane alkaloids isolated from *Schizanthus hookeri* and *S. tricolor* (Solanaceae). Nat Prod Commun 12(3):355–358

da Silva LJC, Alves LA, Silva VR, Santos LS, Bezerra DP, Soares MBP, Doriguetto AC, Barbosa LCA, do Nascimento JC, GEL M, Queiroz RF, de Paula VF (2021) A new tropane alkaloid and other metabolites from *Erythroxylum macrocalyx* (Erythroxylaceae) and their antiproliferative activities. Phytochem Lett 41:168–174. https://doi.org/10.1016/j.phytol.2020.11.019

Da-Wa ZM, Min Z, Guo DL, Fang DM, Yan Z (2016) Chemical constituents from *Przewalskia tangutica*. J Chin Med Mater 39(9):2013–2015

De la Fuente G, Reina M, Muñoz O, San Martin A, Girault JP (1988) Tropane alkaloids from *Schizanthus pinnatus*. Heterocycles 27:1887–1897. https://doi.org/10.3987/COM-88-4562

de Oliveira SL, Tavares JF, Castello Branco MVS, Lucena HFS, Barbosa Filho JM, Agra MDF, do Nascimento SC, Aguiar JDS, da Silva TG, de Simone CA, de Araujo-Junior JX, da Silva MS (2011) Tropane alkaloids from *Erythroxylum caatingae* Plowman. Chem Biodivers 8:155–165

De Simone L, De Feo Margarucci V (2008) Tropane alkaloids: an overview. Pharmacol. Online 1:70–89

Dominicis ME, Férnandez H (1991) Rev Cubana Farm 25:137

El-Imam YMA, Evans WC (1984) Tropane alkaloids of species of *Anthocercis, Cyphanthera* and *Crenidium*. Planta Med 50(1):86–87. https://doi.org/10.1055/s-2007-969628

El-Imam YMA, Evans WC, Plowman T (1985) Alkaloids of some South American *Erythroxylum* species. Phytochemistry 24:2285–2289

El-Imam YMA, Evans WC, Grout RJ (1988) Alkaloids of *Erythroxylum cuneatum*, *E. ecarinatum* and *E. australe*. Phytochemistry 27:2181–2184

El-Iman YMA, Evans WC, Grout RJ, Ramsey KPA (1987) Alkaloids of *Erythroxylum zambesiacum* root-bark. Phytochemistry 26:2385–2389

El-Shazly A, Tei A, Witte L, El-Domiaty M, Wink M (1997) Tropane Alkaloids of *Hyoscyamus boveanus, H. desertorum, H. muticus* and *H. albus* from Egypt. Z Naturforsch 52c:729–739

Evans WC, Treagust PG (1973) Alkaloids of *Datura pruinosa*. Phytochemistry 12(8):2077–2078

Evans WC (1979) Tropane alkaloids of the Solanaceae. In: Hawkes JG, Lester RN, Skelding AD (eds) The biology and taxonomy of the Solanaceae, Linnean Soc Symp Series No. 7. Academic, New York, p 241

Evans WC (1986) Hybridisation and secondary metabolism. In: D'Arcy WG (ed) Solanaceae: biology and systematics. Columbia University Press, New York, p 179

Evans WC, Ramsey KPA (1981) Tropane alkaloids from *Anthocercis* and *Anthotroche*. Phytochemistry 20(3):497–499

Evans WC, Ramsey KPA (1983) Alkaloids of the Solanaceae tribe Anthocercideae. Phytochemistry 22(10):2219–2225

Evans WC, Ghani A, Woolley VA (1972) Alkaloids of *Solandra* species. Phytochemistry 11(1):470–472

Fatemeh A, Fatemeh S (2011) Tropane alkaloids of *Atropa belladonna* L. and *Atropa acuminata* Royle ex Miers plants. J Med Plants Res 5(29):6515–6522. https://doi.org/10.5897/JMPR11.482

Fatur K, Ravnikar M, Kreft S (2021) *Scopolia carniolica* var. *hladnikiana*: Alkaloidal analysis and potential taxonomical implications. Plan Theory 10:1643. https://doi.org/10.3390/plants10081643

Gambaro V, Labbe C, Castillo M (1982) Alkaloids from *Schizanthus* (Solanaceae). Bol Soc Chil Quim 27:296–298

Gambaro V, Labbe C, Castillo M (1983) Angeloyl-, tigloyl-, and senecioyloxytropane alkaloids from *Schizanthus hookeri*. Phytochemistry 22:1838–1839. https://doi.org/10.1016/s0031-9422(00)80289-7

Gapparova AM, Razzakov NA, Aripova SF (2007) Alkaloids of *Convolvulus subhirsutus* from Uzbekistan. Chem Nat Compd 43:291–292

Geng C-A, Ma Y-B, Chen J-J (2013) Bishyoscyamine, one unusual dimeric tropane alkaloid from *Anisodus acutangulus*. Chin Chem Lett 24(3):236–238

Gerlach GH (1948) *Datura innoxia* – A potential commercial source of scopolamine. Econ Bot 2:436–454

Gorinova NI, Atanassov AI, Velcheva MP (1999) XXI. *Physochlaina* species; In vitro culture and the production of physoclaine and other tropane alkaloids. In: Bajaja YPS (ed) Biotechnology in agriculture and forestry. Springer, Berlin/Heidelberg, pp 350–363

Graf E, Lude W (1977) Alkaloids from *Erythroxylum vacciniifolium* Martius, I: isolation of catuabine A, B, and C. Arch Pharm 310:1005–1010

Graf E, Lude W (1978) Alkaloids from *Erythroxylum vacciniifolium* Martius, II: The structures of catuabine A, B, and C. Arch Pharm 311(2):139–152. https://doi.org/10.1002/ardp.19783110211

Griffin WJ, Lin GD (2000) Chemotaxonomy and geographical distribution of tropane alkaloids. Phytochemistry 53:623–637

Haegi L (1976) Taxonomic account of *Datura* L. (Solanaceae) in Australia with a note on *Brugmansia* Pers. Aust J Bot 24:415–435

Hammer K, Romeike A, Tittel C (1983) Vorarbeiten zur monographichen Darstellung von Wildpflanzensortimenten: *Datura* L., sections *Dutra* Bernh., *Ceratocaulis* Bernh. et *Datura*. Kulturpflanze 31:13–75

Hu LS, Xu ZP, Wang MJ, Fan R, Yuan DJ, Wu BD et al (2019) The chromosome-scale reference genome of black pepper provides insight into piperine biosynthesis. Nat Commun 10:4702. https://doi.org/10.1038/s41467-019-12607-6

Humam M, Kehrli T, Jeannerat D, Munoz O, Hosettmann K, Christen P (2011) Schizanthines N, O, and P, tropane alkaloids from the aerial parts of *Schizanthus tricolor*. J Nat Prod 74:50–53

Jackson BP, Berry MI (1973) Hydroxytropane tiglates in the roots of *Mandragora* species. Phytochemistry 12:1165–1166

Jenett-Siems K, Weigl R, Böhm A, Mann P, Tofern-Reblin B, Ott SC, Ghomian A, Kaloga M, Siems K, Witte L, Hilker M, Müller F, Eich E (2005) Chemotaxonomy of the pantropical genus *Merremia* (Convolvulaceae) based on the distribution of tropane alkaloids. Phytochemistry 66(12):1448–1464

Johns SR, Lamberton JA, Sioumis AA (1971) New tropane alkaloids, (+)-(3R, 6R)-3α-acetoxy-6β hydroxytropane and (+)-2α-benzoyloxy-3β-hydroxynortropane, from *Peripentadenia mearsii* (Euphorbiaceae). Aust J Chem 24(11):2399–2403

Johnson EL (1996) Alkaloid content in *Erythroxylum coca* tissue during reproductive development. Phytochemistry 42:35–38

Khattak KF, Atta-ur-Rahman, Choudhary MI, Hemalal KD, Tillekeratne LM (2002) New tropane alkaloids from *Erythroxylum moonii*. J Nat Prod 25:929–931

Khattak KF, Atta-ur-Rahman, Choudhary MI (2003) Nortropane alkaloids from the leaves of *Erythroxylum moonii*. Heterocycles 60:917

Kim N, Chavez B, Stewart C Jr, D'Auria JC (2021) Structure and function of enzymes involved in the biosynthesis of tropane alkaloids. In: Srivastava V et al (eds) Tropane alkaloids. Springer, Singapore, pp 21–50

Kohnen-Johannsen KL, Kayser O (2019) Tropane alkaloids: chemistry, pharmacology, biosynthesis and production. Molecules 24:796. https://doi.org/10.3390/molecules24040796

Leary JD, Khanna KL, Schwarting AE (1963) Occurrence of cuscohygrine and 3a-tigloyloxytropane in *Withania somnifera*. Lloydia 26:44–48

Li J, Liu SJ, Huang ZL, Yu J (2024) *Physochlaina Radix*, a review of its phytochemistry, pharmacology, toxicity and medica processing. Phytochem Rev. https://doi.org/10.1007/s11101-024-09975-3

Lockery JE, Broder JC, Ryan J, Stewart AC, Woods RL, Chong TTJ et al (2021) A cohort study of anticholinergic medication burden and incident dementia and stroke in older adults. J Gen Intern Med 36:1629–1637. https://doi.org/10.1007/s11606-020-06550-2

Loder JW, Russel GB (1966) Tropine 1,2-dithiolane-3-carboxylate, a new alkaloid from *Bruguiera sexangula*. Tetrahedron Lett 51:6327–6329

Loder JW, Russel GB (1969) The alkaloids of *Bruguiera sexangula* and *Bruguiera exaristata* (Rhizophoraceae). Aust J Chem 22:1271

Lounasmaa M, Pusset J, Sevenet T (1980) The New Caledonian plants. Part 57. 2-Benzyl pseudotropanes and dihydropyranotropanes, alkaloids of *Knightia strobilina*. Phytochemistry 19(5):953–955

Lounasmaa M, Wovkulich PM, Wenkert E (1975) Carbon-13 nuclear magnetic resonance spectroscopy of naturally occurring substances. XXXVI. Structures of some *Knightia deplanchei* alkaloids. J Org Chem 40(25):694–697

Lu Y, Yao T, Chen Z (1986) Studies on the constituents of *Erycibe elliptilimba*. Yao xuexuebao 21(11):829–835

Luanratana O, Griffin WJ (1982) Alkaloids of *Duboisia hopwoodii*. Phytochemistry 21(2):449–451

Lv Y, Tian T, Wang Y-J, Huang J-P, Huang S-X (2022) Advances in chemistry and bioactivity of the genus *Erythroxylum*. Nat Prod Bioprospect 12:15. https://doi.org/10.1007/s13659-022-00338-z

Medina DHG, Pusset M, Pusset J, Husson HP (1983) Tropane alkaloids from *Crossostylis* sp. J Nat Prod 46(3):398–400

Moore JM, Casale JF (1997) Lesser Alkaloids of Cocaine-Bearing Plants. Part I: Nicotinoyl-, 2′-Pyrroloyl, and 2′- and 3′-Furanoylecgonine Methyl Ester—Isolation and Mass Spectral Characterization of Four New Alkaloids of South American *Erythroxylum coca* var. *coca*. J Forensic Sci 42:246

Moore JM, Hays PA, Cooper DA, Casale JF, Lydon J (1994) 1-Hydroxytropacocaine: An abundant alkaloid of *Erythroxylum novogranatense* var. *novogranatense* and var. *truxillense*. Phytochemistry 36:357–360

Muñoz O, Casale JF (2003) Tropane alkaloids from *Latua pubiflora*. Z Naturforsch C J Biosci 58(9–10):626–628. https://doi.org/10.1515/znc-2003-9-1003

Muñoz O, Cortes S (1998) Tropane alkaloids from *Schizanthus porrigens*. Pharm Biol 36:162–166. https://doi.org/10.1076/phbi.36.3.162.6341

Naik T, Vanitha SC, Rajvanshi PK, Chandrika M, Kamalraj S, Jayabaskaran C (2017) Novel microbial sources of tropane alkaloids: first report of production by endophytic fungi isolated from *Datura metel* L. Curr Microbiol 75:1–7

Nishiyama Y, Moriyasu M, Ichimaru M, Sonoda M, Iwasa K, Kato A, Juma FD, Mathenge SG, Chalo Mutiso PB (2007) Tropane alkaloids from *Erythroxylum emarginatum*. J Nat Med 61:56–58

Olatunji TL, Adetunji AE, Olisah C, Idris OA, Saliu OD, Siebert F (2021) Research progression of the genus *Merremia*: A comprehensive review on the nutritional value, ethnomedicinal uses, phytochemistry, pharmacology, and toxicity. Plants (Basel) 10(10):2070

Oliveira SL, da Silva MS, Tavares JF, Sena-Filho JG, Lucena HFS, Romero MAV, Barbosa-Filho JM (2010) Tropane alkaloids from *Erythroxylum* genus: distribution and compilation of 13C-NMR spectral data. Chem Biodivers 7:302–326

Parr AJ (1992) Alternative metabolic fates of hygrine in transformed root cultures of *Nicandra physalodes*. Plant Cell Rep 11:270–273

Payo-Hill AL, Dominguez RS, Suarez MO, Batista-Baez M, Castro HTV, Rastrelli L, Aquino R (2000) Tropane alkaloids from the leaves and stem bark of *Erythroxylon alternifolium* and *Erythroxylon rotundifolium*. Phytochemistry 54:927–932

Peizen X, Liyi H (1982) *Przewalskia tangutica* – a tropane alkaloid-containing plant. Planta Med 45(2):112–115

Platonova TF, Kusovkov AD (1963) Med. Ind. UdSSR 17, 19 in Romeike, A (1978). Bot Notiser 131:85

POWO (2024) Plants of the World Online. https://powo.science.kew.org/taxon/urn:lsid:ipni.org:names:30002853-2. Accessed on 24 Aug 2024

Reina M, Burgueño-Tapia E, Bucio MA, Joseph-Nathan P (2010) Absolute configuration of tropane alkaloids from *Schizanthus* species by vibrational circular dichroism. Phytochemistry 71(7):810–815

Restrepo DA, Saenz E, Jara-Muñoz OA, Calixto-Botía IF, Rodríguez-Suárez S, Zuleta P, Chavez BG, Sanchez JA, D'Auria JC (2019) *Erythroxylum* in focus: An interdisciplinary review of an overlooked genus. Molecules 24(20):3788. https://doi.org/10.3390/molecules24203788

Ribeiro EMDO, Lima LS, David JM, do Vale AE, Lopes LMX, David JP (2013) A new tropane alkaloid and other constituents of *Erythroxylum rimosum* (Erythroxylaceae). Phytochem Lett 6:232–235

Ripperger H (1979) Schizanthine A and B, two new tropane alkaloids from *Schizanthus pinnatus*. Phytochemistry 18:171–173

Rivier L (1981) Analysis of alkaloids in leaves of cultivated *Erythroxylum* and characterization of alkaline substances used during coca chewing. J Ethnopharmacol 3:313–335

Ryan SM, DeBoer KD, Hamill JD (2015) Alkaloid production and capacity for methyljasmonate induction by hairy roots of two species in tribe Anthocercidae, family Solanaceae. Funct Plant Biol 42(8):792–801. https://doi.org/10.1071/FP15045

Sahal M, Ray AB (1980) Secotropane alkaloids of *Physalis peruviana*. J Org Chem 45(16):3265–3268

San-Martin AS, Labbé C, Muñoz O, Castillo M, Reina M, de la Fuente B, González A (1987) Tropane alkaloids from *Schizanthus grahamii*. Phytochemistry 26:819–822

Sauvain M, Rerat C, Moretti C, Saravia E, Arrazola S, Gutierrez E, Lema A-M, Muñoz V (1997) A study of the chemical composition of *Erythroxylum coca* var. *coca* leaves collected in two ecological regions of Bolivia. J Ethnopharmacol 56:179–191

Savita, Srivastava A, Jain R, Nagpal AK, Pati PK (2021) Plant tropane alkaloids (TAs): chemotaxonomic significance, biogenesis, and biotechnological interventions. In: Srivastava V et al (eds) Tropane alkaloids, pp 3–19. https://doi.org/10.1007/978-981-33-4535-5_1

Schlesinger D, Rikanati RD, Volis S, Faigenboim A, Vendramin V, Cattanaro F, Hooper M, Oren E, Taylor M, Sitrit Y, Inbar M, Lewinsohn E (2019) Alkaloid chemodiversity in *Mandragora* spp. is associated with loss-of-functionality of *MoH6H*, a hyoscyamine 6β-hydroxylase gene. Plant Sci 283:301–310

Scholl Y, Höke D, Dräger B (2001) Calystegines in *Calystegia sepium* derive from the tropane alkaloid pathway. Phytochemistry 58(6):883–889

Seigler DS (1977) Plant systematics and alkaloids. In: Manske RHF (ed) The alkaloids, vol 16. Academic, New York, pp 1–82

Sena-Filho JG, da Silva MS, Tavares JF, Oliveira SL, Romero MAV, Xavier HS, Barbosa Filho JM, Braz Filho R (2010) Cytotoxic evaluation of pungencine: a new tropane alkaloid from the roots of *Erythroxylum pungens* O. E. Schulz. Helv Chim Acta 93:1742–1744

Shao JJ, Peng Y, He CN, Xu LJ, Xiao X, Li YS (2013) Isolation and identification the chemical constituents of *Przewalskia tangutica* Maxim. J Shenyang Pharm Univ 30(11):840–845

Sharova EG, Aripova SF, Yunusov SY (1980) Alkaloids of *Convolvulus subhirsutus*. Khimiya Prirodnykh Soedinenii 5:672–676

Silva GL, Cui B, Chávez D, You M, Chai H-B, Rasoanativo P, Lynn SM, Oneill MJ, Lewis JA, Besterman JM, Monks A, Farnsworth NR, Cordell GA, Pezzuto JM, Kinghorn AD (2001) Modulation of the multidrug-resistance phenotype by new tropane alkaloid aromatic esters from *Erythroxylum pervillei*. J Nat Prod 64:1514–1520

Soares Cruz RA, Almeida H, Fernandes CP, Joseph-Nathan P, Rocha L, Leitao GG (2016) A new tropane alkaloid from the leaves of *Erythroxylum subsessile* isolated by pH-zone-refining counter-current chromatography. J Sep Sci 39:1273–1277

Tamprateep P, Taylor EH, Ramstad E (1963) Biosynthesis of the alkaloids of *Erythroxylum novogranatense*. Lloydia 26:203–208

Tantray YR, Jan I, Mohiuddin I, Gorla P, Anand C, Arya JC, Bhat SA, Kumar V, Narasimhaji CV, Nagayya N, Acharya R (2024) A review on the traditional applications, phytochemistry, and pharmacology of the genus *Physochlaina* G. Don. Chem Biodivers 21(6):e202400394

Todd FD, Stermitz FR, Schultheis P, Knight AP, Traub-Dargatz J (1995) Tropane alkaloids and toxicity of *Convolvulus arvensis*. Phytochemistry 39(2):301–303

Van Haga RP (1954) Cuscohygrine, a normal constituent alkaloid of *Atropa belladonna*. Nature 174(4435):833–834

Wang P, Yao T, Chen Z (1990) Zhiwu Zhongyao Zazhi, 14: 739 (Ch.), via C.A., 113, 37699

Wang B, Chen C, Xiao Y, He Y, Gao Y, Kang Z, Wei X, Deng Y, Feng S, Zhou G (2023a) Geographically associated endophytic fungi contribute to the tropane alkaloids accumulation of *Anisodus tanguticus*. Front Plant Sci Sec Plant Symbiotic Interact 14. https://doi.org/10.3389/fpls.2023.1297546

Wang B, Chen C, Xiao Y, Chen K, Wang J, Wang L, Li J, Kang Z, Zhou G (2023b) A core root bacteria contribute to plant growth and anisodine accumulation of *Anisodus tanguticus*. BMC Plant Biol 23:655. https://doi.org/10.1186/s12870-023-04690-1

Wu YG, Bao ML, Ba GN (2016) Evolved investigation of Mongolian medicine common *Physochlain*a. J Med Pharm Chin Minor 4:58–60. https://doi.org/10.16041/j.cnki.cn15-1175.2016.04.037

Xi LTY, Sun XJ, Yan JF, Su ZX, Qin G, Liu J, Lin N, Li X, Si QB (2019) Advances in Mongolian medicine Vesicular Herb. J Chin Ethnic Med 25(11):34–37. https://doi.org/10.3969/j.issn.1006-6810.2019.11.021

Xie HX, Wang BZ, Sa RN (2011) Determination of total alkaloids in different part of *Physochlaina physaloides* Saxicola by spectrophotography. J Inner Mong Med Coll 33(1):46–49. https://doi.org/10.16343/j.cnki.issn.2095-512x.2011.01.002

Yang F, Zhao H, Carroll AR (2017) Tropane alkaloids from the Australian plant *Triunia montana* (Proteaceae). Tetrahedron Lett 58(8):736–739

Yao T, Chen Z, Yi D, Xu G (1981) YaoxueXuebao, 16, 582 (Ch.), via C.A., 96: 48972

Zang E-H, Chen Z-W, Zhang C-H, Li M-H (2021) Chemical constituents of *Physochlaina physaloides* (L.) G. Don (Solanaceae). Biochem Systemat Ecol 98:104332

Zanolari B, Wolfender L, Guilet D, Marston A, Queiroz EF, Paulo MQ, Hostettmann K (2003a) On-line identification of tropane alkaloids from *Erythroxylum vacciniifolium* by liquid chromatography–UV detection–multiple mass spectrometry and liquid chromatography–nuclear magnetic resonance spectrometry. J Chromatogr A 1020(1):75–89

Zanolari B, Guilet D, Marston A, Queiroz EF, Paulo Mde Q, Hostettmann K (2003b) Tropane alkaloids from the bark of *Erythroxylum vaccinifolium*. J Nat Prod 66(4):497–502. https://doi.org/10.1021/np020512m

Zanolari B, Guilet D, Marston A, Queiroz EF, Paulo Mde Q, Hostettmann K (2005) Methylpyrrole tropane alkaloids from the bark of *Erythroxylum vaccinifolium*. J Nat Prod 68(8):1153–1158. https://doi.org/10.1021/np040144h

Zárate R, Hermosin B, Cantos M, Troncoso A (1997) Tropane alkaloid distribution in *Atropa baetica* Plants. J Chem Ecol 23:2059–2066

Zhang Z, Lu A, William GD (1994) Solanaceae. In: Wu ZY, Raven PH (eds) Flora of China (Verbenaceae through Solanaceae), vol 17. Science Press/Missouri Botanical Garden Press, Beijing/St. Louis, pp 300–332

Zhao S-M, Chou G-X, Wang Z-T (2013) Chemical constituents from roots and rhizomes of *Physochlaina infundibularis*. Chin Tradit Herb Drug 67:938–941

Zuanazzi JAS, Tremea V, Limberger RP, Sobral M, Henriques AT (2001) Alkaloids of *Erythroxylum* (Erythroxylaceae) species from Southern Brazil. Biochem Syst Ecol 29:819–825

Chapter 3
Chemistry of Tropane Alkaloids: A Comprehensive Study of Biologically Important Group of Nitrogen-Containing Heterocyclic Phyto-secondary Metabolites

Lepakshi Md. Bhakshu and T. Pullaiah

Abstract Nature provides a plethora of medicinal plants that act as biological factories of diverse groups of metabolites, which include alkaloids. One of the well-known and earliest reported classes of natural secondary metabolites belongs to alkaloids, nitrogen-containing heterocyclic compounds with diverse biological and pharmacologically active, especially neurological disorders. The alkaloids major source is Solanaceae, a botanical family with many subgroups. Tropane alkaloids have been well studied and applied in many pharmacological processes and used in preparing applicable therapeutic entities. As parasympatholytics that competitively inhibit acetylcholine, tropane alkaloids are helpful. These alkaloids are based on the tropane moiety's bicyclic ring, and most are ester derivatives substituted on the atom C-3 of the tropane ring. The present revision focuses on the chemistry of tropane alkaloids and their possible plant resources, especially from botanical families such as Solanaceae, Erythroxylaceae, Convolvulaceae, etc., which will be helpful to the scientific world to take up further studies and develop nontoxic and safe medicaments.

Keywords Tropane alkaloids · Solanaceae · Erythroxylaceae · Cocaine · Calystegines · Atropine · Scopolamine · Hyoscyamine · Cuscohygrine · Catuabine · Anticholinergic

---

L. Md. Bhakshu (✉)
Dr. YSR Government Degree College, Vedurukuppam, Andhra Pradesh, India

T. Pullaiah
Department of Botany, Sri Krishnadevaraya University, Anantapur, Andhra Pradesh, India

T. Pullaiah et al. (eds.), *Tropane Alkaloids*,
https://doi.org/10.1007/978-981-96-5805-3_3

3.1 Introduction

Nature provides many medicinal plants that act as biological factories of diverse groups of metabolites, including alkaloids. One of the well-known and earliest reported classes of natural secondary metabolites belong to alkaloids, nitrogen-containing heterocyclic compounds with diverse biological and pharmacologically active, especially neurological disorders. According to Srinivasan and Smolke (2019), these heterocyclic compounds, which incorporate nitrogen, have a range of biological and pharmacological actions, especially when treating neurological conditions. The Solanaceae family, which includes many subgroups, is the main source of alkaloids. One of these subgroups, tropane alkaloids, has been extensively studied, is used in various pharmacological treatments and utilized in synthesizing useful therapeutic compounds. Tropane alkaloids (TAs) are helpful parasympatholytics that competitively counteract acetylcholine. Most tropane alkaloids are substituted on the atom C-3 of the tropane ring as ester derivatives, with the bicyclic ring of the tropane moiety serving as the base of these alkaloids (Alsamarrai 2019). They are generally produced as anticholinergics by the nightshade family (Solanaceae) genera *Duboisia*, *Atropa*, and others. Some TAs, such as atropine, hyoscyamine, and scopolamine, are listed by the World Health Organization as crucial medicines for treating Parkinson's disease (PD), cardiac arrhythmia, organophosphate as well as nerve poisoning agents, and gastrointestinal spasms (Grynkiewicz and Gadzikowska 2008; WHO 2015). The berries that *Atropa belladonna* produces are the most deadly component of the plant because they contain a lot of seeds, which causes intoxication due to scopolamine, atropine, and hyoscyamine among the various forms of TA's found in the leaves, roots, and berries (Qureshi and Sidhartha 2024).

3.2 Scope and Natural Resources

The secondary metabolites of plants in the Solanaceae family are grouped by the tropane skeleton and are further divided into two classes: derivatives of tropine and ecgonine. The first group, represented by the well-known alkaloids scopolamine and atropine, is still a source of inspiration for researchers looking to develop more selective muscarinic receptor antagonists. These medications are thought to be model anticholinergic medicines. One of the main substances that is abused, cocaine, is included in the second class. The synthesis of necessary cocaine antagonists has not been very effective despite extensive research. Consequently, new approaches to treating cocaine dependence are utilizing immunology and biotechnology. The contemporary pharmaceutical industry produces about 20 active pharmaceutical compounds containing a tropane moiety in their structure, which can be used as antispasmodics, mydriatics, bronchodilators, and antiemetics.

The two main sources of raw materials for this commercial activity, aside from chemical synthesis, are natural compounds derived from domesticated transgenic plants, particularly Australian *Duboisia*'s scopolamine and atropine (Grynkiewicz and Gadzikowska 2008). Since the extracts contain significant amounts of tropane alkaloids, such as up to 2% in ripe *Datura stramonium* seeds, many plants with a moderate climate are listed in the ethnopharmacological tradition. These plants have been used for centuries as poisons and magic potions for a variety of purposes (such as pain relievers, hallucinogens, etc.). Some species that are found among them are *Hyoscyamus niger*, *H. albus* (henbane), *Scopolia carniolica* (scopola), *Mandragora officinarum* (henbane-mandrake), *Atropa belladonna* (deadly-nightshade), and *Datura stramonium* (Jimson weed) were considered as poisons (Griffin and Lin 2000).

3.3 Historical Aspects

Meissner was the first to identify in 1819 that these deadly plants' active ingredients are alkaline and can be separated using extractive methods. Consequently, starting in 1830, specific alkaloid compounds were isolated, first scopolamine, then atropine from *A. belladona*, and finally hyoscyamine from *H. niger*. Many atropine and scopolamine derivatives from genetically modified cultivars are used in modern medicine, and the increasing demand for these drugs is driving the development of new chemical and biotechnological manufacturing processes.

According to Grynkiewicz and Gadzikowska (2008), cocaine, which is made from two species of *Erythroxylum* and has powerful addictive properties, is used in medicine only infrequently (Weil 1981; Rocio et al. 2004). It is illegally manufactured and trafficked as "recreational drugs," which can have harmful socioeconomic and health effects worldwide. Esters of tropane have been identified in the following plant families: Proteaceae, Rhizophoraceae, Euphorbiaceae, Convolvulaceae, and Erythroxylaceae, in addition to the Solanaceae family. Some of the more than 100 genera and 3000 species in the Solanaceae family—such as *Datura*, *Brugmansia*, *Hyoscyamus*, *Atropa*, *Scopolia*, *Anisodus*, *Przewalskia*, *Atropanthe*, *Physochlaina*, *Mandragora*, *Anthotroche*, *Cyphanthera*, and *Duboisia*—are known to be rich in alkaloids containing the tropane moiety (Griffin and Lin 2000). The word "tropane alkaloids" (TA) describes a special class of toxic alkaloids that naturally occur from a variety of plant groupings, including as Brassicaceae, Erythroxylaceae, and Solanaceae, and that can contaminate food and feed. The two naturally occurring alkaloids that are most prevalent in plants that produce TA are (−)-hyoscyamine and (−)-scopolamine. On the other hand, atropine, a racemic combination of (+)- and (−)-hyoscyamine, is created when (−)-hyoscyamine is purified. Weeds that produce TA are widely distributed across temperate and tropical parts of the world and can occasionally contaminate agricultural products.

In 2013, the EFSA CONTAM panel issued a study stating that TAs are toxic to humans and that weeds harboring TA are abundant in temperate and tropical regions. These weeds have the potential to contaminate tea, herbal mixtures, and agricultural products unintentionally. The central and autonomic nervous systems' increased antimuscarinic activity is the main cause of atropine and scopolamine's toxicological consequences. The EFSA Panel set a group acute reference dose of 0.016 μg/kg body weight for all pertinent TAs because of these health consequences (EFSA 2013; Leon et al. 2022; Martinello et al. 2022). TAs can contaminate agricultural and animal products, teas, herbal blends, and feeds. Scientific investigations on TAs confirm that they are found in teas and herbal infusion matrices, as well as in plants, infant food made from cereal, cow's milk, seeds, bread, green vegetables, and other bee products (Leon et al. 2022; Martinello et al. 2022; Romera-Torres et al. 2020; Marin-Saez et al. 2019; Du et al. 2019; Klein et al. 2022; Kaur and Pandey 2020; Versilovskis et al. 2020; Gonzalez-Gomez et al. 2022; Kowalczyk and Kwiatek 2022; Gumus 2023).

3.4 Industrial Applications

A particular group of the unique class of metabolites known as tropane alkaloids (TAs), generated by medicinal plants in the Solanaceae family, has strong anticholinergic effects. Hyoscyamine and scopoline are two tropane alkaloids used in medicine as anticholinergic drugs. Asthma, Parkinson's disease, pain, motion sickness, and functional gastrointestinal problems are all treated with these drugs. Hyoscyamine is an essential component in commercially synthesizing ipratropium bromide and scopolamine for tiotropium bromide. Chronic obstructive pulmonary disease requires the use of the drugs ipratropium and tiotropium bromide. This makes the two alkaloids very desirable. On the other hand, scopolamine is extremely profitable to grow and produce due to its numerous medicinal applications.

Consequently, the global need for these compounds is rising, leading to an increase in commercial demand. Additionally, scopolamine has been recognized by the World Health Organization (WHO) as one of the most significant drugs. The sources of scopolamine and hyoscyamine are the massive *Duboisia* crops in Australia. Because of the altered biotic and abiotic circumstances brought about by climate change, the food and pharmaceutical sectors are finding it more and more difficult to synthesize scopolamine in high quantities (Gonzalez-Gomez et al. 2022; Kohnen-Johannsen and Kayser 2019).

3.5 Chemical Nature and Its Skeleton

Tropane is a bicyclic amine that shares two carbon atoms and one nitrogen atom between its pyrrolidine and piperidine rings. All tropane alkaloids share this structural component (Lounasmaa and Tamminen 1993). The tropane is a bicyclic saturation structure (N-methyl-8-azabicyclo[3.2.1]octane-1), which consists of a cycloheptane ring connected by a nitrogen bridge. TAs belong to a class consisting

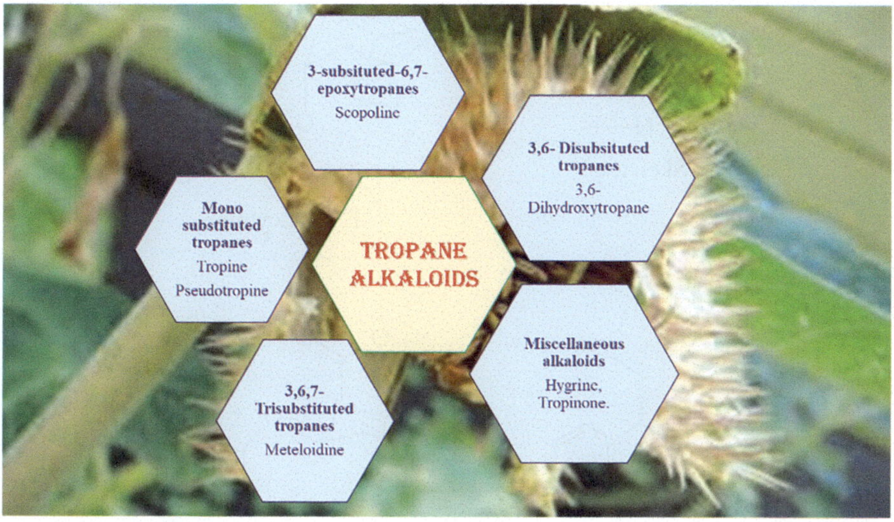

Fig. 3.1 Classification of tropane alkaloids. (Kaur and Pandey 2020)

of about 300 alkaloids that are easily separated based on stereochemical characteristics and the amount of carbons in the tropane skeleton (Zhang et al. 2023).

Another important factor that distinguishes the two classes in scientific literature is that, although the derivatives of ecgonine and tropine share a biogenetic reaction sequence that leads to the tropane skeleton and branches only at the tropinone reduction step, the ecgonine class is only strongly associated with its main member, cocaine, a neuro-stimulant that is known to cause severe addiction (Grynkiewicz and Gadzikowska 2008). For simplicity of comprehension, Kaur and Pandey (2020) have given a brief description of the categorization based on its structural complexity (Fig. 3.1). Esters of various organic acids, including tiglic, isovaleric, tropic, benzoic, and cinnamic acids conjugated to hydroxylated tropane derivatives (Fig. 3.2), provide the structural basis for almost all TAs (Huang et al. 2021).

3.6 Tropane Alkaloids of Solanaceous Plants

3.6.1 Tropane Alkaloids of Genus Anthocercis *and Other Little-Known Species*

Anthocercis littorea Labill., *Anthocercis viscosa* R. Br. and *A. fasciculata* F. Muell., and *Cyphanthera anthocercidea* Haegi (*Anthocercis frondosa*).

The two main alkaloids in *Anthocercis* species are scopolamine and hyoscyamine, often accompanied by their apo- and nor-derivatives. In addition to certain mono- and ditigloyl esters of toluidine (Evans and Treagust 1973), the species

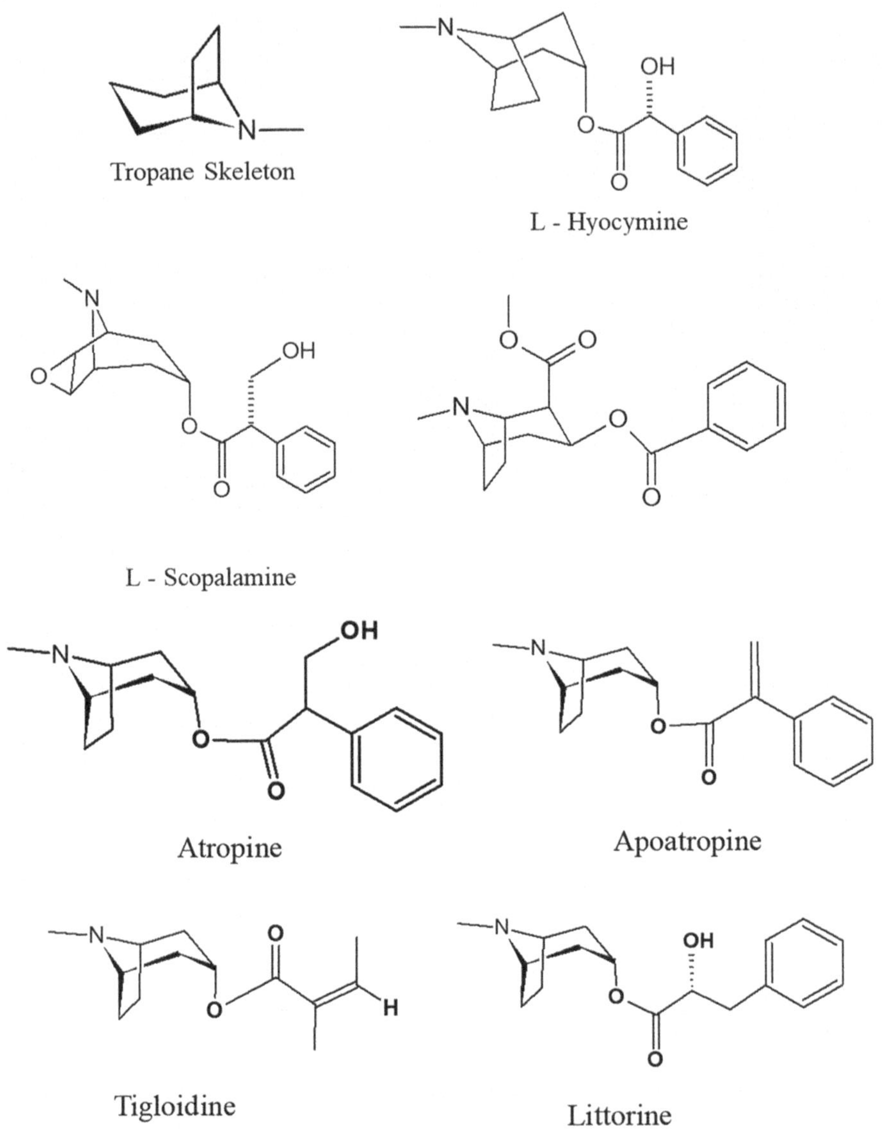

Fig. 3.2 Common tropane alkaloids. (Source: Huang et al. 2021; Kohnen-Johannsen and Kayser 2019; redrawn)

Anthocercis littorea Endl. mostly includes littorine and meteloidine (Cannon et al. 1969). *Cyphanthera anthocercidea* Haegi (*Anthocercis frondosa*) produces scopolamine and hyoscyamine, mostly nicotine (Evans and Ramsey 1981). *Anthocercis viscosa* R. Br. and *A. fasciculata* F. Muell. contain hyoscyamine (Cannon et al. 1969). Butyryl esters of tropine and 6β-hydroxytropine are produced by *Cyphanthera albicans* Miers (*Anthocercis albicans*), a member of the *Cyphanthera* genus. It was discovered that the alkaloids of *Cyphanthera tasmanica* Miers and *Cyphanthera odgersii* Haegi were compatible with those of other known *Cyphanthera* species (El-Imam et al. 1985). Hyoscyamine, norhyoscyamine, apoatropine, and scopolamine are found in three species of the *Anthotroche* genus: *A. myoporoides* C.A.Gardner, *A. pannosa* Endl., and *A. walcottii* F. Muell. (Evans and Ramsey 1981, 1983). The primary alkaloids in the two species of the *Symonanthus* genus are scopolamine and its derivative aposcopolamine. In contrast, the roots of *Symonanthus aromaticus* Haegi contain mono- and ditigloyl esters similar to those of *Datura* species (El-Imam et al. 1985).

3.6.2 *Tropane Alkaloids of* Atropa acuminata *Royle ex Miers*

Using the high-performance liquid chromatography (HPLC) technique, Fatemeh and Fatemeh (2011) assessed the tropane alkaloid content (atropine and scopolamine) in various sections of the cultivated *Atropa belladonna* L. and wild *Atropa acuminata* Royle ex Miers in Iran.

3.6.3 *Tropane Alkaloids of* Atropa belladonna

A quick and simple CE-electrospray interface (ESI)-TOF-MS method was created by Arraez-Roman et al. (2008) to separate several significant chemicals from *A. belladonna* leaf extract, including tropine, belladonnine, norhyoscyamine, apoatropine, hyoscyamine, 6β-hydroxyhyoscyamine, and scopolamine. Using sodium acetate-2-14C, Baralle and Gros (1969) reported the existence of cuscohygrine among other alkaloids. According to Van Haga (1954), cuscohygrine is a typical alkaloid ingredient of *A. belladonna.* Hyoscyamine is the main alkaloid found in *Atropa belladonna* L., along with apoatropine, scopolamine, and hyoscyamine N-oxides (Evans and Ramsey 1979). In addition, the extracts from leaves and roots contain two isomers, 3β-acetoxytropane and N-methylpyrrolidinylhygrine. Some secotropane alkaloids have been isolated from the roots as major alkaloids and have since been identified as (+)-physoperuvine, racemic physoperuvine, and (+)-N, N dimethyl-physoperuvinium salt (Evans and Ramsey 1979).

3.6.4 *Tropane Alkaloids of* Brugmansia suaveolens

Agrobacterium rhizogenes was used to infect the tips of the roots to create hairy root cultures of *Brugmansia suaveolens*. The polymerase chain reaction (PCR) analysis of the virC and rolC genes verified the effective transformation. The production of tropane alkaloids, including hyoscyamine, was investigated using hairy root cultures. To promote the manufacture of tropane alkaloids, the altered cells were incubated with possible elicitors such as salicylic acid, quercetin, and methyl-jasmonate. Tropane alkaloids' profiles and concentrations were examined using capillary GLC-MS.

About 18 tropane alkaloids (tropine, scopine, scopoline, pseudotropine, 3-acetoxy-6-hydroxytropane, 5 3a-acetoxy tropane, 3α-tigloyloxytropane, cuscohygrine, 3–3-tigloyloxy-6-hydroxytropane, hydroxy-6-(2-methyl butyryloxy)-tropane, apoatropine, 3-hydroxy-6-tigloyloxytropane, aposcopolamine, 3-tigloyloxy-6-(2-methylbutyryloxy)-tropane, 3α,6β-ditigloyloxytropane, hyoscyamine, 7β-hydroxyhyoscyamine, 6-β-hydroxyhyoscyamine) were identified. When compared to untreated controls, the alkaloid accumulation rose 25-fold when the cultures were treated with 200 μm of methyl jasmonate, reaching a level of 1 mg/g fresh weight (Zayed and Wink 2004).

3.6.5 Tropane Alkaloids of Datura *Genus*

3.6.5.1 *Datura ceratocaula* Ortega

Berkov et al. (2003) identified 36 alkaloids from the organs of *Datura ceratocaula* by GC/MS. Thirty-four, namely, pseudotropine, tropine, methylecgonine, 3-acetoxytropane, N-methylpyrrolidinyl-hygrine A, 3-acetoxy-6-hydroxytropane, N-methylpyrrolidinyl-hygrine B, cuscohygrine, 3β-tigloyloxytropane, 3α-tigloyloxytropane, 3,6-diacetoxytropane, 3-hydroxy-6-(2'methylbutyryloxy)tropane, 3(α)-tigloyloxy-6-hydroxytropane, 3-hydroxy-6-tigloyloxy tropane, 3-tigloyl oxy-6-propionyloxy-7-hydroxy-tropane, 3(β)-tigloyloxy-6-hydroxytropane, phenylacetoxytropane, 3(β)-tigloyloxy-6,7-dihydroxytropane, 3(α)-tigloyloxy-6,7-dihydroxytropane, aponorhyoscyamine, apohyoscyamine, littorine, aposcopolamine, norhyoscyamine, hyoscyamine, 6-hydroxyapohyoscyamine, 3,6-ditigloyloxytropane, 3-(3′-formyloxytropoyloxy)tropane, scopolamine, β 3-(3′-acetoxytropoyloxy)tropane, 3-(2′-hydroxytropoyloxy)tropane, 7-hydroxyhyoscyamine, 3,6-ditigloyloxy-7-hydroxytropane, 6-hydroxyhyoscyamine, and 6-tigloyloxyhyoscyamine, were identified. Furthermore, a new tropane ester was tentatively identified as 3-(3′-formyl oxytropoyloxy) tropane (Fig. 3.3) on basis of its mass spectral fragmentation. Hyoscyamine was the main alkaloid in the plant organs (Berkov 2003).

3.6.5.2 *Datura ferox* L.

The study by Vitale et al. (1995) describes chromatographic techniques (HPLC, GC-MS) for identifying alkaloids in toxic *Datura ferox* seeds from Argentina, from which only quantitative analysis for scopolamine was previously known. Three other tropane alkaloids have been identified: aposcopolamine (apohyoscine),

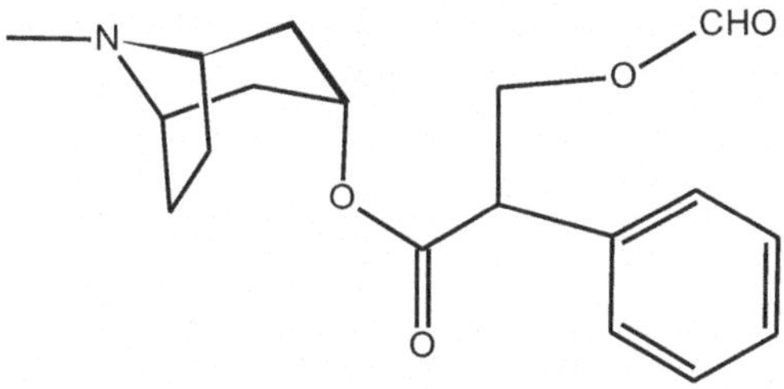

Fig. 3.3 3-(3′-formyl oxytropoyloxy) tropane

7β-hydroxy-6β-propenyloxy-3α-tropoyloxytropane, 3-phenylacetoxy-6β,7β-epoxytropane (3-phenylacetoxyscopine), traces of 7β-hydroxy-6β-isovaleroyloxy-3α-tigloyloxytropane, and a pyrrolidine alkaloid called hygrine. Based on their spectra, two novel structures are proposed: 7β-hydroxy-6β-propenyloxy-3α-tropoyloxytropane and 3-phenylacetoxy-6β,7β-epoxytropane (3-phenylacetoxyscopine) (Vitale et al. 1995).

3.6.5.3 *Datura innoxia* Mill

The alkaloid spectra of *Datura innoxia* plants grown in Egypt and Bulgaria were investigated through GC-MS analysis. Thirty-eight alkaloids (tropine, tropinone, pseudotropine, scopoline, scopine, 3-acetoxy-6-hydroxytropane, 3-acetoxytropane, 3a-tigloyloxytropane, 3-isobutyryloxy-6-hydroxytropane, cuscohygrine, 3-hydroxy-6-tigloyloxytropane, 3-tigloyloxy-6-hydroxytropane, 3-(2′-methylbutyryloxy)-6-hydroxytropane, 3-isovaleroyloxy-6-hydroxytropane, 3(α)-tigloyloxy-6-propionyloxy-7-hydroxy-tropane, 3-tigloyloxy-6-acetoxytropane, 3-tigloyloxy-6-propionyloxytropane, 3(β)-tigloyloxy-6-propionyloxy-7-hydroxy-tropane, 3-tigloyloxy-6,7-dihydroxytropane, 3-tigloyloxy-6-isobutyryloxytropane, 3-phenylacetoxytropane, 3,6-dihydroxy-7-tigloyloxytropane, 3-apotropoyloxytropane, norapoatropine, 3-tigloyloxy-6-methylbutyryloxytropane, aposcopolamine, 3-tigloyloxy-6-(2′-methylbutyryloxy)-tropane, 3(α),6-ditigloyloxytropane, hyoscyamine, 3-tigloyloxy-6-isovaleroyloxy-7-hydroxy-tropane, 3(β),6-ditigloyloxytropane, scopolamine, 3-(3′-acetoxytropoyloxy)-tropane, 7-hydroxyhyoscyamine, 3,6-ditigloyloxy-7-hydroxytropane, 3-tropoyloxy-6-isovaleroyloxytropane, 6-hydroxyhyoscyamine, and 6-tigloylhyoscyamine) were identified from the roots, leaves and fruits among these, five were new alkaloids (Berkov and Zayed 2004).

El Bazaoui et al. (2012) used GC/MS to identify 53 alkaloids in *Datura innoxia*'s roots, stems, leaves, flowers, and seeds. One nor-derivative, 3-phenylacetoxynortropane, for the genus *Datura*, and 17 of them are described for the first time for this species. Furthermore, based on their mass spectral data, four new tropane esters were tentatively identified as 3-(2′-phenylpropionyloxy)-6,7-epoxynortropane (dihydroaponorscopolamine), 6-acetoxy-6-propionyloxy-7-hydroxytropane, 3-acetoxy-6-epoxytropane (acetylscopine), and 6-acetoxy-6-propionyloxy-7-hydroxytropane. According to El Bazaoui et al.

(2012), hyoscyamine and scopolamine are the primary alkaloids found in the roots and aerial portions, respectively.

3.6.5.4 *Datura leichhardtii* Benth

It has been studied how the amounts of hyoscyamine and hyoscine in the aerial portions of *Datura leichhardtii* vary. According to Evans and Stevenson (1962), the alkaloids 7-hydroxy-3, 6-ditigloyloxytropane, 3α-tigloyloxytropane, and meteloidine were extracted from the roots.

3.6.5.5 *Datura metel* L.

Along with other withanolides, tropine (trigloyl esters), and pseudo-tropine, it includes many significant alkaloids, such as littorine, hyoscyamine, acetoxytropine, hyoscine, fastusine, valtropine, and fastusinine. Nonetheless, the nortropane alkaloids, or calystegines, that were separated from several *D. metel* species have exhibited glycosidase inhibitory action (Ghani 2003). Leaves include hyoscyamine, scopolamine, and atropine; roots contain atropine; and aerial portions contain scopolamine. Atropine, 3α-tigloyloxytropane, hyoscine, hyoscyamine, norhyoscine, tropine, and hyoscyamine from seeds are all abundant in *D. metel* roots (Afsharypuor et al. 1995) besides glycosides and terpenes (Islam et al. 2023; Cinelli and Jones 2021).

3.6.5.6 *Datura quercifolia* Kunth

Using *Agrobacterium rhizogenes* strain LBA 9402 to create hairy root cultures of *Datura quercifolia*, Dupraz et al. (1994) found eight tropane alkaloids in the hairy roots, with hyoscyamine being a significant component. The optimal medium for both growth and hyoscyamine accumulation has been shown to be Gamborg B5. After analyzing the impact of sucrose content, it was determined that a 5% concentration was optimal for both growth and alkaloid synthesis. The hyoscyamine concentration of the roots was 1.24% based on dry weight following 35 days of incubation in this medium (Dupraz et al. 1994).

3.6.5.7 *Datura stramonium* L.

It contains a variety of toxic tropane alkaloids such as atropine, hyoscyamine, and scopolamine (Gaire and Subedi 2013). It also contained hygrine, 6,7-dehydrotropine, cyclotropine, tropinone, tropine, pseudotropine, scopoline, and the other 67 tropane alkaloids methylecgonine, 3-(hydroxyacetoxy) tropane, 3-acetoxy-6-hydroxytropane, 3-acetoxytropane, 3,6-dihydroxytropane,

3-methylbutyryloxytropane, 3-hydroxy-6-acetoxytropane, 3,6-diacetoxytropane, 6,7-dehydro-3-tigloyloxytropane, and 3,7-dihydroxy-6-propionyloxytropane Tigloyloxytropane (3a) 36-tigloyloxytropane, 3-hydroxy-6-isobutyryloxytropane, 3-(2′-methylbutyryloxy) or 3-isovaleroyloxy-6-hydroxytropane.-6-hydroxytropane, 2-methylbutyryloxy-3-hydroxy-6-tropane, 3-tigloyloxy-6,7-epoxytropane, 3-hydroxy-6-methylbutyryloxytropane, 3-tigloyloxy-6-hydroxytropane, 6-tigloyloxytropane, 3a-hydroxy 3,7-dihydroxy-6-(2′-methylbutyryloxy) tropane, 3-tigloyloxy-6-acetoxytropane, and 3-b-hydroxy-66-tigloyloxytropane Apoatropine, 3,7-dehydro-3-apotropoyloxytropane, 3-tigloyloxy-6-propionyloxy-7-hydroxytropane, 3-phenylacetoxytropane, 3-(2-phenylpropionyloxy)tropane, 3-tigloyloxy-6-propionyloxytropane, 3-tigloyloxy-6-isobutyryloxytropne, and 3-tigloyloxy-6,7-dihydroxytropane,

The diversified alkaloids such as 3,7-dihydroxy-6-tigloyloxytropane 3-dihydroaposcopolamine, 3-phenylacetoxy-6,7-epoxytropane, 3-(2′-methylbutyryloxy) tropane-3-tigloyloxy-6- 6,7-dehydrohyoscyamine, 3-tigloyloxy-6-methylbutyryloxytropane, 3-tropane (3′-methoxytropoyloxy), apo-nor-scopolamine, aposcopolamine, 3-phenylacetoxy-6-hydroxytropane, littorine, 3-tigloyloxy-6-isobutyryloxy-7-hydroxytropane, hyoscyamine (atropine), ditigloyloxytropane (3,6) 6. Hydroxyatropine 3-isovaleroyloxy-7-hydroxytropane-3α-Tigloyloxy-6- Methylscopolamine, 3-tigloyloxy-6-isovaleroyloxy-7-hydroxytropane, scopolamine, 3-(3′-acetoxytropoyloxy) tropane, hydroxylittorine 4′ were identified from the capsules. By using GC/MS, the following compounds were found in the organs of *Datura stramonium*: 3,6-ditigloyloxy-7-hydroxytropane, 7-hydroxyhyoscyamine, 3-tropoyloxy-6-acetoxytropane, 6-hydroxyhyoscyamine, 3-tropoyloxy-6-isobutyryloxytropane, 3β-tropoyloxy-6 β-isovaleroyloxytropane, 3 β-tropoyloxy-68-isovaleroyloxytropane, and 3-tropoyloxy-6-tigloyloxytropane. There are nine novel tropane alkaloids: 3-tigloyloxy-6,7-epoxytropane, 6,7-dehydro-3-tigloyloxytropane, 3,7-dihydroxy-6-propionyloxytropane, 3,7-dihydroxy-6-(2′-methylbutyryloxy), 3-(3′-methoxytropoyloxy)tropane, 6,7-dehydroapoatropine, 5-hydroxytropane 3-tigloyloxy-6-isobutyryloxy, and 3-tropoyloxy-6-isobutyryloxytropane. The preliminary identification of 3β-tropoyloxy-6β-isovaleroyloxytropane were shown in Figs. 3.4, 3.5, and 3.6). For the first time, the alkaloids cyclotropine, dihydroaposcopolamine, 6,7-dehydrohyoscyamine, 6,7-dehydrotropine, and 4′-hydroxylittorine were identified for the genus *Datura*, and the family Solanaceae (Figs. 3.4, 3.5, and 3.6) (El Bazaoui et al. 2011).

3.6.5.8 Alkaloid Spectrum in Diploids and Tetraploids

By using gas chromatography/mass spectrometry to analyze the hairy root cultures of diploid and induced tetraploid plants of *Datura stramonium*, Berkov et al. (2003) discovered a new tropane ester 3-tigloyloxy-6-propionyloxy-7-hydroxytropane in addition to roughly 20 alkaloids for diploid and nine for tetraploid hairy root cultures. The predominant alkaloid in both diploid and tetraploid cultures was hyoscyamine. With the exception of apoatropine and hyoscyamine, the percentage

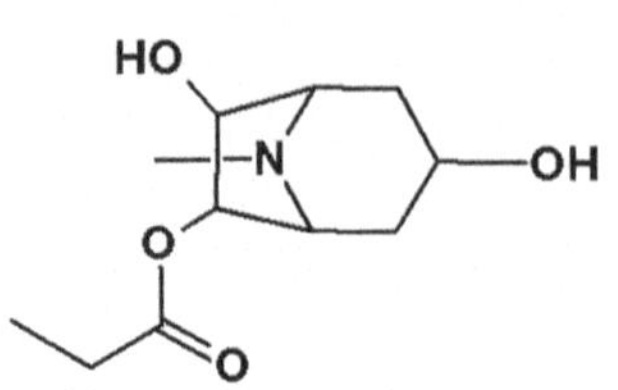

3,7-Dihydroxy-6-propionyloxy tropane

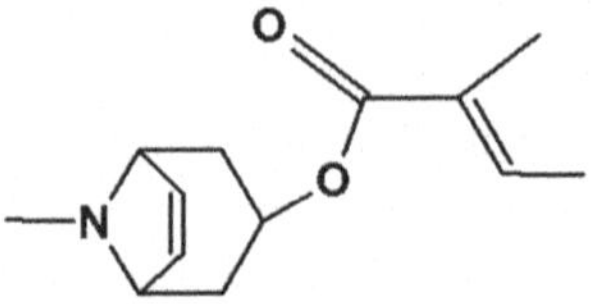

6,7-Dehydro-3-tigloyloxytropane

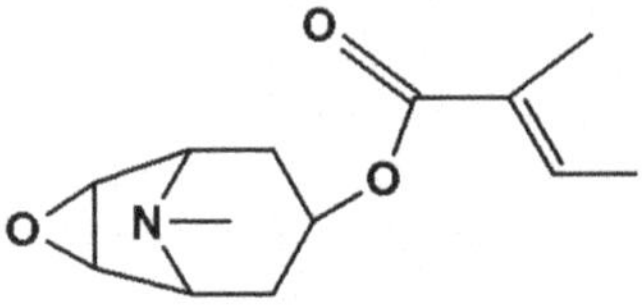

3-Tigloyloxy-6,7-epoxytropane

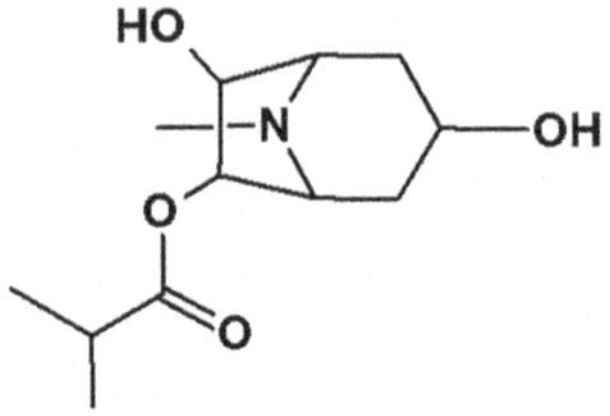

3,7-Dihydroxy-6-(2'-methyl butyryloxy)tropane

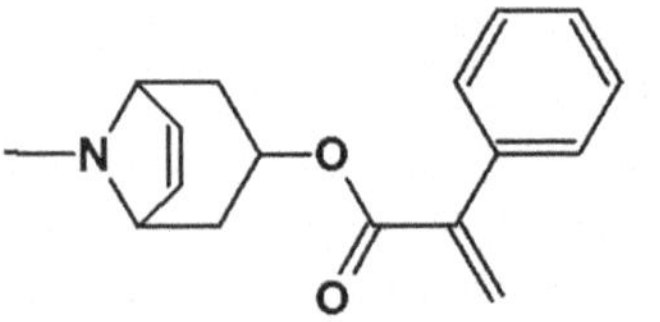

6,7-Dehydro-3-apotropoyloxytropane

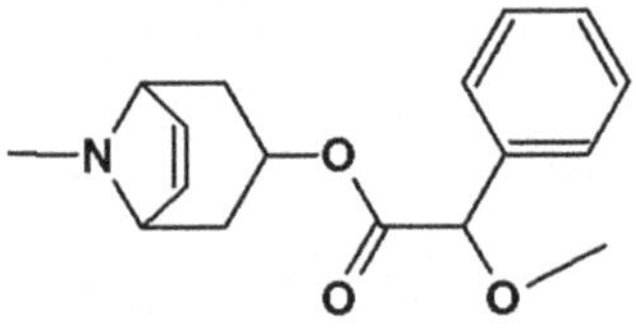

3-(3'-Methoxytropoyloxy)tropane

Fig. 3.4 Few Novel Tropane alkaloids identified from *Datura stramonium* capsules (Source: El Bazaoui et al. 2011, redrawn)

contributions of the alkaloids in the overall alkaloid combination of tetraploid hairy roots were larger than those of diploid hairy roots (Berkov et al. 2003).

3.6.6 Tropane Alkaloids of Duboisia leichhardtii *F. Muell*

Duboisia leichhardtii has yielded calystegines, including C1 (1β,2β,3β,4β,6β-pentahydroxynortropane) and C2 (1β,2α,3β,4α,6β-pentahydroxynortropane) (Nash et al. 1993; Kato et al. 1997).

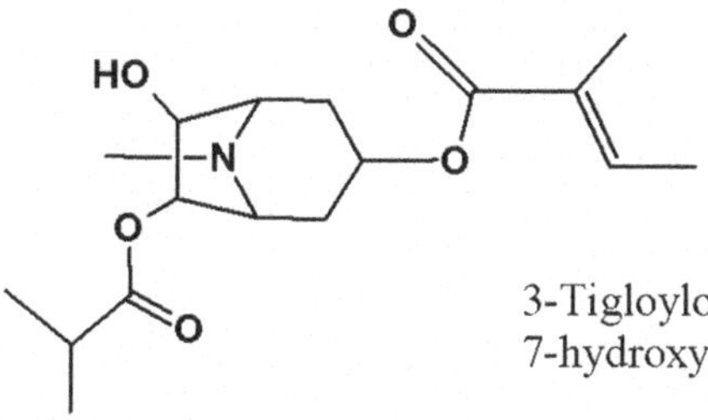

3-Tigloyloxy-6-isobutyryloxy-
7-hydroxytropane

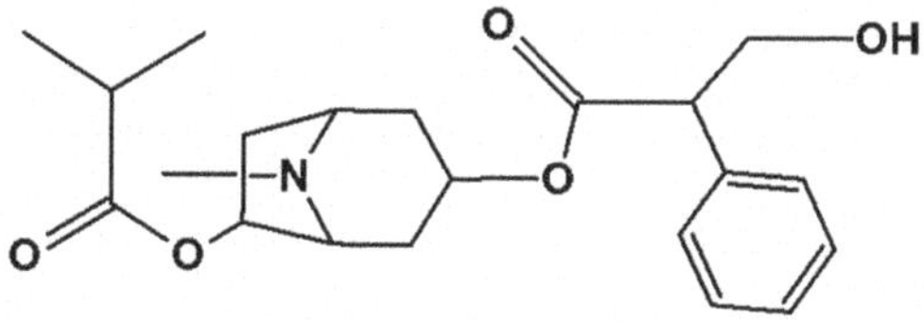

3-Tropoyloxy-6-isobutyryloxytropane

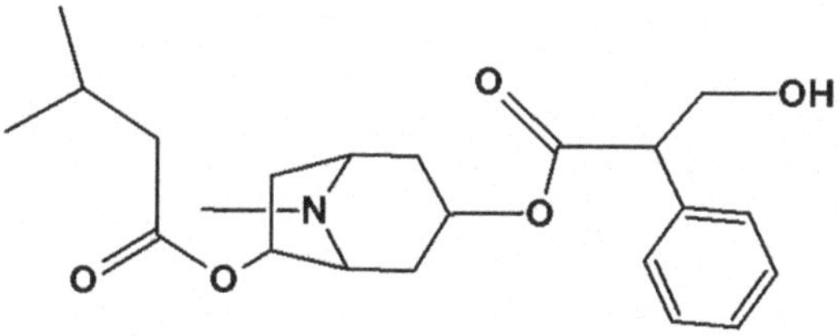

3α-Tropoyloxy-6α-isovaleroyloxy
tropane

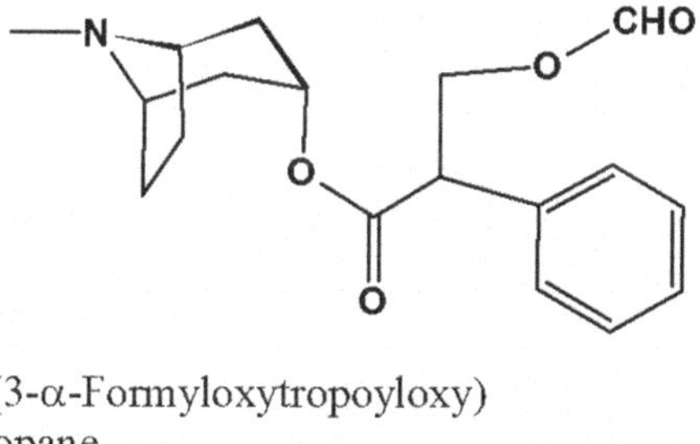

3-(3-α-Formyloxytropoyloxy)
Tropane

Fig. 3.5 Tropane alkaloids of *Datura stramonium*. (Source: El Bazaoui et al. 2011, redrawn)

3.6.7 Duboisia myoporoides *R. Br. (Corkwood Tree)*

The primary tropane alkaloids found in *Duboisia* include scopolamine, hyoscyamine, nicotine, and anabasine (Mortimer and Wilkinson 1957). The dried leaves of

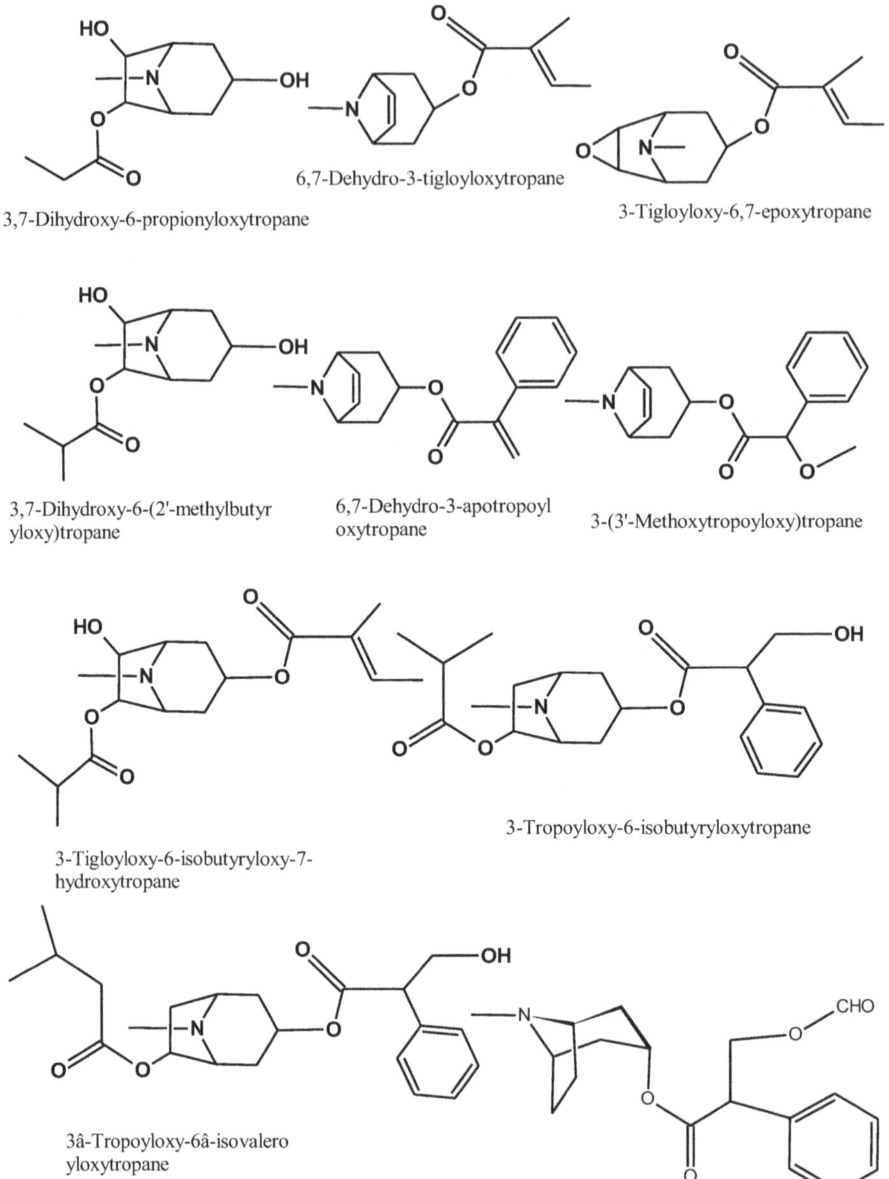

Fig. 3.6 Tropane alkaloids of *Datura stramonium*. (Source: El Bazaoui et al. 2011, redrawn)

the Solanaceae family members *Duboisia myoporoides*, *D. hopwoodii*, and *D. leichhardtii* make up *Duboisia*. The primary commercial source of atropine and scopolamine is thought to be *Duboisia myoporoides*. During extraction, the hyoscyamine it contains is changed into atropine. The medication also contains tigloidine, valtropine, norhyoscyamine, and tiglyoxytropine. Because atropine and scopolamine are exceedingly expensive to synthesize, a lot of dependence is put on their natural sources. Colorless crystals of atropine ($C_{17}H_{23}O_3N$) have no smell and a harsh taste. It dissolves in alcohol and chloroform (https://shorturl.at/wQtAP).

3.6.8 **Hyoscyamus niger** *L. (Black Henbane)*

The primary alkaloids of the root are L-hyoscyamine and scopolamine, along with apoatropine (atropamine) and cuscohygrine; tropine and scopolamine are more frequently found in leaves, while hyoscyamine, a small amount of hyoscine, and atropine are found in seeds. These tropane alkaloids have been identified from *H. niger* (Alizadeh et al. 2014).

3.6.9 **Hyoscyamus muticus L.** *(Egyptian Henbane)*

Hyoscyamus muticus seeds and stem or leaf midribs were used to create callus cultures. The culture thrives and endures throughout many cultural media. Every culture generated alkaloids. Hyoscyamine, scopolamine, and an unidentified alkaloid were detected by chromatographic analysis of the alkaloid composition (El-Olemy and Moustafa 1983). Sabri et al. (1973) evaluated the total alkaloidal contents in the various organs at the various phases of *Hyoscyamus desertorum* Täckh.'s development and reported their chromatographic pattern. One week after the flowers open is ideal for gathering the plant's organs, including the leaves, flowers, stems, roots, and fruits. It was discovered that the primary alkaloids in the plant were hyoscine and hyoscyamine. From 1.5 to 3 months of age, hyoscine is more common in radical leaves, ripe fruits, seeds, and leaves and in unopened and opened flowers. Hyoscyamine predominates in the stems, roots, and radical leaves that are between 3.5 and 5.5 months old. Unknown precursors took the place of hyoscyamine and hyoscine at specific developmental stages.

3.6.10 Tropane Alkaloids of the Genus **Mandragora**

In addition to the isomeric N-oxides of hyoscyamine and hyoscine, two N-oxides from the roots, stems with leaves, and fruits, *Mandragora officinarum* was reported to contain atropine, hyoscyamine, tropic acid, tropane, scopolamine (hyoscine), scopine, pseudohyoscyamine (norhyoscyamine), cuscohygrine, hyoscyamine, apoatropine, and belladonnines. Hanus et al. (2005) identified calystegines A3, B1, and B2 from the roots, leaves, spring leaves, flowers, and fruits. There have been reports of scopolamine, hyoscyamine, cuscohygrine, apoatropine, and belladonnines from the fresh and dried roots of *M. autumnalis* and *M. vernalis*, two species that are different from *M. officinarum*. Hanus et al. (2005) identified scopolamine, hyoscyamine, and apohyoscine from *M. turcomanica* Mizger's leaves, aerial parts, roots, rhizomes, fruits at various stages of growth, and seeds. A few tropane alkaloids, including norhyoscyamine, apoatropine, α- or β-belladonnine, (±)-3α-tigloyloxytropan-6β-ol, and tigloidine (3β-tigloyloxytropane), were confirmed and reported by TLC and MS.

3.6.11 **Physalis officinarum** *var.* **francheti** *(Mast.) R.J. Wang. (Syn.:* **Physalis alkekengi** *var.* **franchetii** *(Mast.) Makino)*

According to Asano et al. (1995), *Physalis alkekengi* var. *franchetii* (Solanaceae) contains calystegines (polyhydroxy-nor-tropanes with a bridgehead OH group) A3, A5, B1, B2, and B3. In contrast, tigloidine, 3α-tigloyloxytropane, cuscohygrine, hygrine, physoperuvine, tropine, and 3β-tigloyloxytropane were found in the roots of *Physalis alkekengi* L., according to a review by de Oliveira et al. (2011).

3.6.12 **Physalis peruviana** *L.*

The alkaloids tigloidine and 3-a-(tigloyloxy)-tropane was found in the roots of *Physalis peruviana* L. as well as in the leaves of the plant (Sahai and Ray 1980). GC-MS experiments on the root extracts of *Physalis peruviana* revealed the presence of tropane alkaloids, including hygrine, physoperuvine, trepine, 3 J3-acetoxytropane, N-methylpyrrolidinylhygrine A, N-methylpyrrolidinylhygrine B, 3α-tigloyloxytropane, and cuscohygrine (Kubwabo et al. 1993).

3.6.13 **Schizanthus tricolor** *Grau et Gronbach*

GC-MS was used to examine the alkaloid profile of *Schizanthus tricolor*'s aerial parts. Seventeen tropane alkaloids were detected and reported, such as hygrine, hygroline A or B, hygroline A or B, tropinone, 3α-tropine, 3α-acetoxytropane, 3α-senecioyloxytropane, 3α-hydroxy-6β-angeloyloxytropane, 3α-senecioyloxy-6β-hydroxytropane, 3α-hydroxy-6β-senecioyloxytropane, 3α-hydroxy-6β-tigloyloxytropane, 3α-methylitaconyloxytropane, 3α-methylmesaconyloxytropane, 3α-methylitaconyloxy-6β-angeloyloxytropane, 3α-methylmesaconyloxy-6β-angeloyloxytropane (Figs. 3.7 and 3.8), and Isomer of 337 Da. Among these

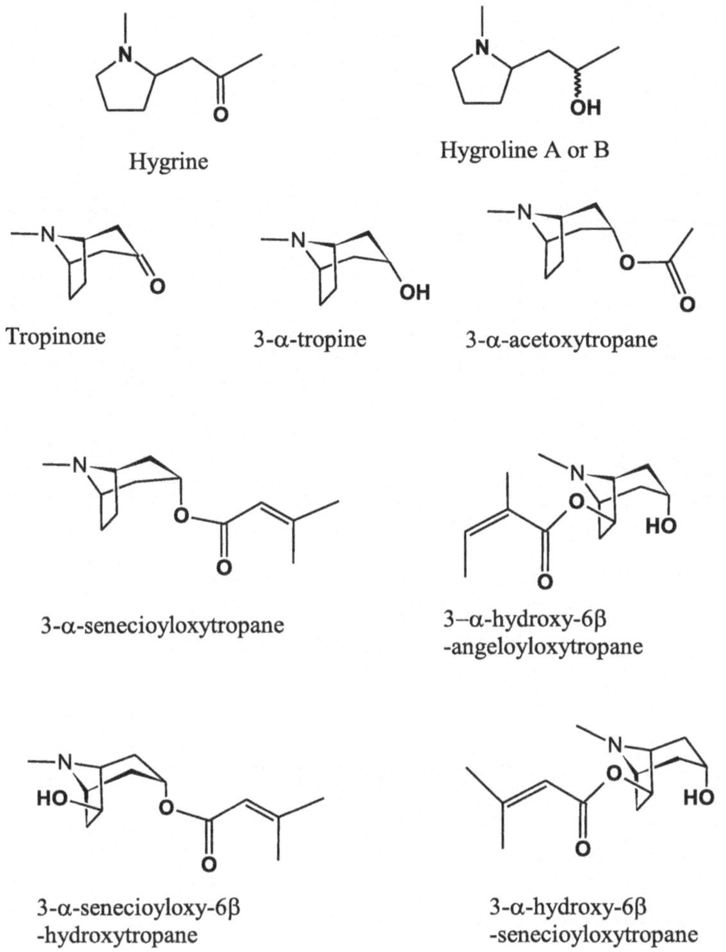

Fig. 3.7 Tropane alkaloids of *Schizanthus tricolor*. (Source: Humam et al. 2007, redrawn)

3α-methylitaconyloxytropane has been reported as a new alkaloid by Humam et al. (2007).

3.6.14 Schizanthus grahamii *Gillies*

Tropane alkaloids identified in *S. grahamii* include 3α-hydroxy-6β-angeloyloxytropane, 3α-senecioyloxy-6β-hydroxytropane, 3α-hydroxy-6β-senecioyloxytropane, and 3α-hydroxy-6β-tigloyloxytropane (Bieri et al. 2006).

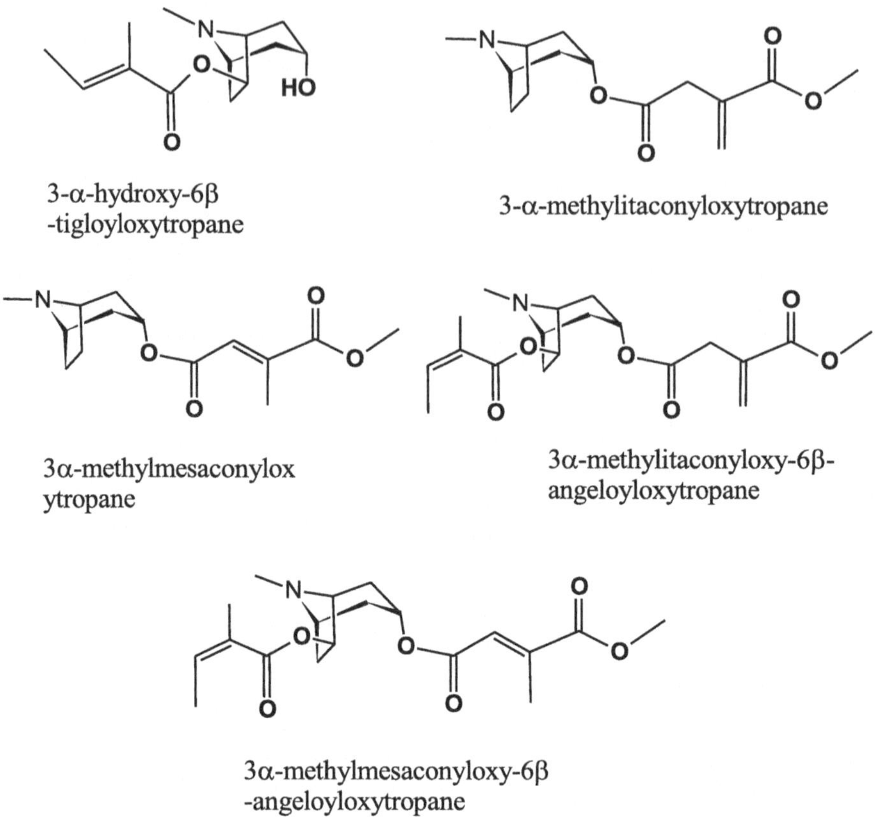

Fig. 3.8 Tropane alkaloids of *Schizanthus tricolor*. (Source: Humam et al. 2007, redrawn)

3.6.15 Solanum torvum *Sw*

Two steroidal-glycosylated tropane-based alkaloids (Solasodine and Solasonine) were characterized from the leaves of *Solanum torvum* by Perez-Amador et al. (2007).

3.7 Calystegines

Plant alkaloids known as calystegines have been found in a variety of plant species, including the Solanaceae family's culinary plants (sweet peppers, eggplants, potatoes, and tomatoes). The polyhydroxylated nortropane alkaloids known as calystegines (CA) were initially discovered in the roots of *Atropa belladonna*, *Convolvulus arvensis*, and *Calystegia sepium* (Tepfer et al. 1988). The plant kingdom, which includes the Solanaceae, Convolvulaceae, Moraceae, and Brassicaceae, has now been discovered to contain CA in large quantities (Bekkouche et al. 2001; Schimming et al. 1998; Brock et al. 2006). From a chemical perspective, calystegines are a class of alkaloids that share a ring system of polyhydroxylated nortropane. Groups A, B, and C, which include three, four, and five hydroxyl groups, are the three groups into which the calystegines are separated based on the increased degree of hydroxylation (Goldmann et al. 2001).

A nortropane ring system, a tertiary hydroxyl group at the bicyclic ring bridgehead, and two to four additional hydroxyl groups at various locations are the three structural characteristics that CAs share (Andersson 2002). The CAs are classified into three categories, A, B, and C, according to the amount of hydroxyl groups they contain. These types have three, four, or five hydroxyl groups, respectively (Goldmann et al. 1996). CA A3, B1, and B2 structures were the first to be clarified (Goldmann et al. 1990). CAs A3 and B2 were discovered in potato tubers, and B2 was also discovered in some cultivars' leaves (Nash et al. 1993). Furthermore, Solanaceous plants have also been shown to contain dihydroxylated nortropanes (Asano et al. 2001). Additionally, several CAs depart from the basic structural characteristics listed above. One such is CA N1, which has an amino substituent on C-1 of the nortropane molecule rather than a hydroxyl group. The two N-methylated calystegines, N-methyl-calystegine B2 and N-methyl-calystegine C1, which were both extracted from *Lycium chinense* roots, are further rare types of CA (Asano et al. 1997). Although calystegines and tropane alkaloids have a similar ring structure, they are not as bioactive as tropane alkaloids because they do not have the nitrogen-associated methyl group (Fig. 3.9) (Arif 2013).

Along with other alkaloids, several calystegines have been found in toxic plants linked to plant-induced disorders that affect animals. Certain calystegines are known to block glycosidase, just as other polyhydroxylated cyclic alkaloids. Through the buildup of partly catabolized saccharides in the lysosomes, inhibition of certain glycosidases can cause cellular toxicity (Stegelmeier et al. 2013). Originally believed to be found exclusively in roots, CA has since been found in almost every section of plants. It is still unknown, nevertheless, if CA are produced in every tissue or if they are carried from the roots to the plant's aerial portion. Additionally, CA has been found in leaves, sprouts, and tubers (peel, meat), with the largest quantities seen in sprouts (Drager et al. 1994; Drager 1995). There was a significant difference in the amounts of CA in the meat and the peel, with the peels having greater levels (Friedman et al. 2003). Asano et al. (1997) examined a number of consumable fruits and vegetables to determine whether calystegines A3, B1, B2, and C1 were present. While tomatoes and potatoes had comparatively lower amounts, aubergines (up to 80 mg/kg fresh weight), sweet peppers (53 mg/kg fresh weight), and sweet potatoes (up to 44 mg/kg fresh weight) had higher overall values. Higher amounts of B2, the individual calystegine, were found in the majority of the samples (up to 73 mg/kg fresh weight in aubergines).

Friedman et al. (2003) found that the amounts of calystegines A3 and B2 in several potato cultivars ranged from 5.4 to 68.1 mg/kg (fresh weight), with B2 being more prevalent in the majority of the samples. Mulder et al. (2016) recently examined the levels of six calystegines (A3, A5, B1, B2, B3, and B4) in 308 samples of

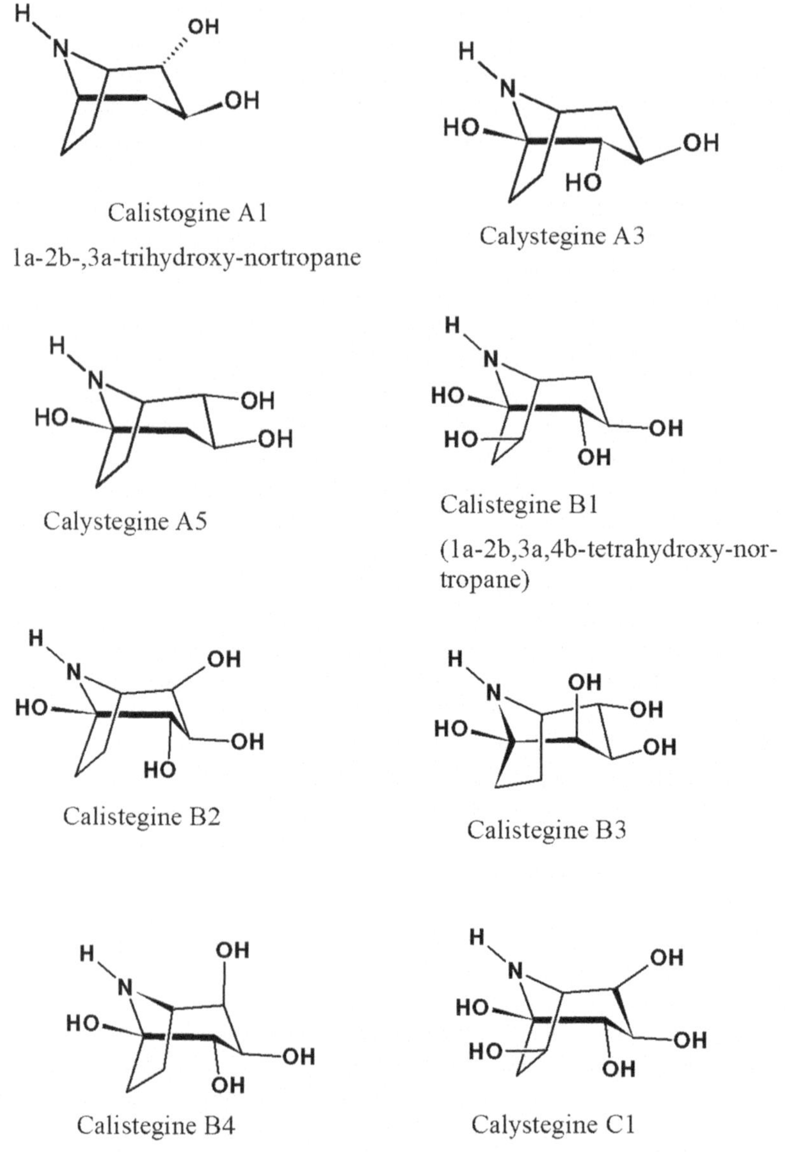

Fig. 3.9 Chemical structure of different types of calystegines in potato. (Source: Arif 2013, redrawn)

potato products (297 fresh samples and 11 processed samples), 90 aubergine samples, and 6 bell pepper samples that were gathered from different EU Member States (EU MSs). All of the potato product samples and over 93% of the aubergine and bell pepper samples contained calystegines.

With a mean concentration of 164.0 mg total calystegines/kg and a maximum of 507.3 mg, fresh potatoes exhibited the highest levels. Processed potato products and aubergines followed, with mean and maximum levels of 95.6 and 207.7 mg total calystegines/kg and 21.1 and 181.5 mg total calystegines/kg, respectively. The calystegines that were most often found in potato products at the individual level were A3, B2, and B4. A3 also had the highest mean and maximum amounts (about 100 and 400 mg/kg, respectively). Bell peppers and aubergines displayed a distinct pattern, with B2 being the chemical assessed at higher mean and maximum levels (about 15 and 125 mg/kg, respectively) and A3, B1, and B2 being the most often identified compounds. Two samples of sweet potatoes were subjected to further investigation, and the primary calystegines in those samples were determined to be B1 and B2 (EFSA et al. 2019).

3.8 Tropane Alkaloids of Other Than Solanaceae Plant Resources

3.8.1 *Brassicaceae*

Some calystegines, including calystegine A, the 3-hydroxybenzoate ester of tropine and cochlearine, and a new group of hydroxylated nortropane alkaloids, are found in Brassicaceae or Cruciferae. The final one is the first tropane alkaloid found in *Cochlearia arctica* Schlecht and *Cochlearia officinalis* L. (Jenett-Siems et al. 2005). The alcohol moiety of cochlearine and pseudotropine was detected in trace amounts in the majority of *Cochlearia* tropine species. High levels of calystegines were detected in the tissues of *Cochlearia officinalis* L. Though calystegines A3 and B2 predominate in Solanaceae, B1 and B2 in Convolvulaceae, and A3 and B2 in Erythroxylaceae, Calystegine A5 is the main component found in it (Jenett-Siems et al. 2005).

3.8.2 *Convolvulaceae*

The presence of one to six polyhydroxy alkaloids of the nortropane type (calystegines) in 62 species from 22 genera across all tribes, with the exception of the unusual parasitic Cuscuteae, was first identified in 189 Convolvulaceous species from 29 genera (12 tribes). The trihydroxynortropane A3 (38%) and the pentahydroxylated congener-C1 (26%) are the next most common substances, after the tetrahydroxylated alkaloids B2 and calystegine B1 (90% and 68%), respectively (Schimming et al. 2005). The minor frequency of calystegines A5, B3, and B5, which were similarly isolated from *Scopolia japonica* Maxim. (Solanaceae) (Asano et al. 1996), ranged from 16% to 20%. Only *Stictocardia mojangensis* has been found to contain calystegine A6, which is distinguished by the absence of a hydroxy group at C-3 (Schimming et al. 2005).

Convolvulus lineatus L. and *C. pseudocantabricus* Schrenk have yielded the nortropan-3α-yl veratrate convolvine and its N-methyl derivative convolamine, as well as closely related alkaloids with a veratroyl or vanilloyl moiety as the acyl component. Convolvine and convolamine were also isolated from *C. subhirsutus* L. and *C. krauseanus* Regel & Schmalh. Additionally, a number of tropine esters containing methoxy substituted benzoic acids of tropane alkaloids are found in *C. krauseanus* Regel & Schmalh and *C. subhirsutus* L. These include convolvine (3α-veratroyloxynortropane), convolidine (3α-vanilloyloxynortropane), convolidine, convolicine, convosine, subhirsine, convolvidine, convolamine N-oxide, convolacine (Aripova and Abdilalimov 1993; Aripova et al. 1977, 1983), confoline (3α-veratroyl-N-formylnortropine) (Sharova et al. 1980), and convolamine-N-oxide (Aripova 1985).

The confolidine ((±)3a-vanillyl-Nformylnortropane), a novel tropane alkaloid, was recently isolated from the aerial portion of *C. subhirsutus* L. Similarly, convoline, convolamine, and convolidine are found in *Evolvulus sericeus* Sw. Erycibelline (2β,7β-dihydroxynortropane) and baogongten C have been isolated from *Erycibe* species, *E. elliptilimba* Merr. et Chun, and *E. hainanensis* Merr. (Lu et al. 1986). *Erycibe obtusifolia* Benth. is the source of baogongten A, baogongten B, and (2β-hydroxy-6β-acetoxynortropane). Calystegine (calystegin) B1 (1β,2α,3β,6α-tetrahydroxynortropane), calystegine B2 (1β,2α,3β,4α-tetrahydroxynortropane), and calystegine A3 (1β,2α,3β-trihydroxynortropane) are found in the roots of *C. arvensis* L. and transformed root cultures of *Calystegia sepium* R. Br. Additionally, calystegines A1, A2, and A4 have been isolated from the roots of *C. sepium* R. Br. (Brock et al. 2006).

While the roots of *Convolvulus sabatius* Viv. ssp. *mauritanicus* contain consabatine (2-deoxyconsiculine) and consiculine, the roots of *C. siculus* L. contain merresectine B, a tropan-3α-ol esters, and consiculine. 3α-[1,4-dihydroxy-3-(3-methyl-but-2-enyl)-cyclohex-2-enecarboxy]tropane (4-dihydroconsabatine) is isolated from the roots of *Convolvulus sabatius* ssp. *sabatius* (Jenett-Siems et al. 2005). The β-isomer of phyllalbine was recently discovered in *C. cneorum* L., whose epigeal vegetative portions also contain concneorine, a 3β-vanilloyloxytropane (Asano et al. 1994).

Merresectines A–D, 3α-(4-methoxybenzoyloxy)-nortropane (merresectine A), 3α-kurameroyloxytropane (merresectine B), 3α-nervogenoyloxytropane (merresectine C), and 3α-[4-(β-D-glucopyranosyloxy)-3-methoxy-5-(3-methyl-2-butenyl) benzoyloxy]tropane (β-D-glucoside of merresectine D) were all isolated from *Merremia dissecta* Hallier f. and *M. guerichii* A. Meeuse. Additionally, *M. dissecta* Hallier f. has yielded the new 3α,6β-di-(4-methoxybenzoyloxy)-tropane (merredissine). Six species of *Merremia* have been found to have the 4-β-D-glucoside of merresectine E (Jenett-Siems et al. 2005). Tropane alkaloids of Convolvulaceae are shown in Figs. 3.10 and 3.11.

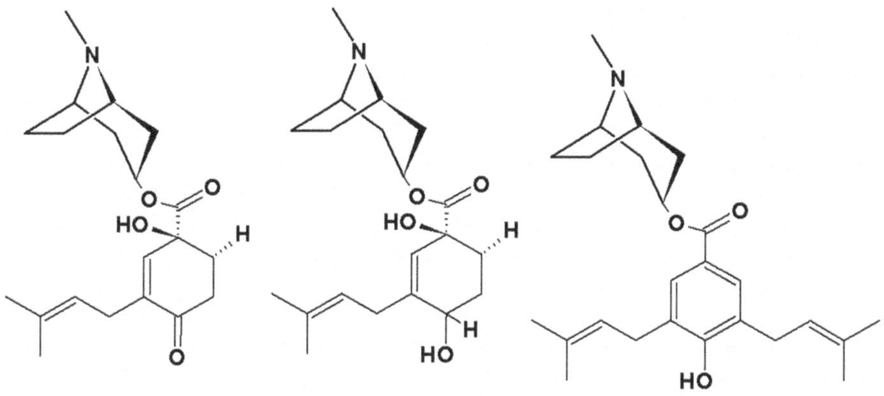

Consabatine 4'-dihydroconsabatine Merresectine A

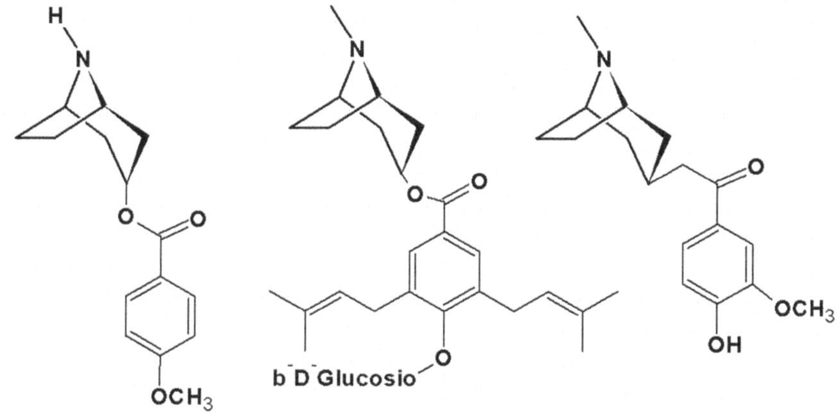

Merresctine A Merresectine B Concneorine

Fig. 3.10 Tropane alkaloids of Convolvulaceae. (Jenett-Siems et al. 2005; redrawn)

3.8.3 Erythroxylaceae

The genus *Erythroxylum* is a fascinating group of plants that includes around 250 distinct species, collectively known as the coca plant family. This genus thrives in tropical regions, with a notable concentration in South America, particularly in the Andean countries of Peru and Bolivia. The coca bush (*Erythroxylum coca*) holds a rich history, having been recognized and utilized for at least 5000 years for its therapeutic and psychoactive properties. Erythroxylaceae, the family to which this genus belongs, can be found in the Andean region, across the African tropics, and parts of Southern Asia. This widespread distribution highlights the plant family's adaptability to various ecosystems. The *Erythroxylum* species are particularly notable for

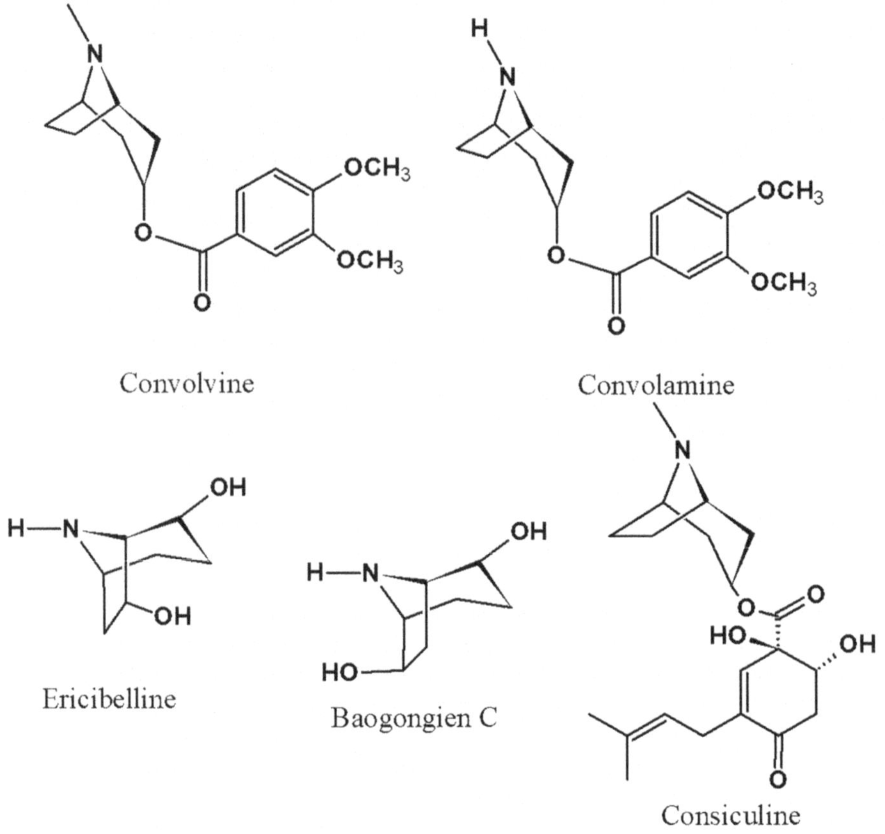

Fig. 3.11 Tropane alkaloids of Convolvulaceae. (Jenett-Siems et al. 2005; redrawn)

Fig. 3.12 Hygrine and hygroline from *Erythroxylum* species

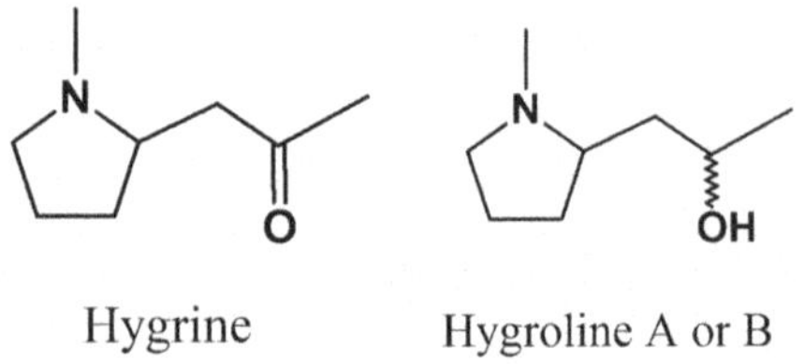

producing a diverse array of L-ornithine-derived alkaloids, which exhibit various biochemical effects. Among these species, three are renowned for containing significant alkaloids: *Erythroxylum coca*, *Erythroxylum truxillense*, and *Erythroxylum novogranatense*. These plants are known to produce several important alkaloids, including cocaine, ecgonine, and cinnamyl cocaine, as well as others like α-truxilline, truxilline, methylergonovine, tropine, hygrine, hygroline, and cuscohygrine (Fig. 3.12) (Seneca 2007).

The following discussion will delve deeper into the various types of tropane alkaloids, exploring their unique features and roles within different *Erythroxylum* species and shedding light on their significance in historical and modern contexts.

3.8.3.1 *Erythroxylum caatingae* Plowman

In addition to the known tropane alkaloids 3α,6β-dibenzoyloxytropane and 6β-benzoyloxy-3α-[(3,4,5-trimethoxybenzoyl)oxy], three tropane alkaloids, 1–3, were isolated from *Erythroxylum caatingae*, including 6β-benzoyloxy-3α-[(4-hydroxy-3,5-dimethoxybenzoyl)oxy]tropane, a novel tropane alkaloid (de Oliveira et al. 2011).

3.8.3.2 *Erythroxylum coca* Lam

Coca was first cultivated and used several thousand years ago BC. Albert Niemann isolated an active ingredient from *Erythroxylum coca* in 1860 and called it "cocaine": methyl (1R, 2R, 3S, 5S)-3-(benzoyloxy)-8-methyl cocaine. The coca shrub's leaves contain a lot of −8-azabicyclo[3.2.1]octane-2-carboxylate, which is the first local anesthetic to come from a particular plant alkaloid (Chandrika and Karunarathna 2022). Particularly well-known are the phytoconstituents of tropane alkaloids (TAs), including cocaine, found in the representative genus *Erythroxylum* of the Erythroxylaceae family. *E. coca* and *E. novogranatense* are well-known plant sources of cocaine. Species of this genus are mostly found in tropical and subtropical locations, such as South America, South Africa, Southeast Asia, and Australia (Hegnauer 1981). Several alkaloids, including methylecgonidine, tropine, 3alpha-acetoxytropane, ecgonine methyl ester, cuscohygrine, N-norbenzoyltropine, benzoyltropine, hexanoylecgonine methyl ester, cocaine, cis-cinnamoylcocaine, and trans-cinnamoylcocaine, were identified and described by alkaloid extracts from the seeds of *Erythroxylum coca* var. *coca* cultivated in Bolivia's Chapare Valley (Casale et al. 2005).

3.8.3.3 *Erythroxylum hypericifolium* Lam

Tropan-3α-yl-3-hydroxy-phenylacetate, (+)-tropane-3α,6p-diol 3-phenylacetate, tropane-3 α,6p-d iol 3-phenylacetate 6-acetate, tropane-3 α, 6fJ,7p-triol 3-p-phenylacetate, and nortropan-3α-yl phenylacetate were all isolated from the root-bark of *E. hypericifolium*. Hygrine was identified as a prominent component among 13 basic chemicals in the stem bark of *E. hypericifolium*. Furthermore, alkaloids from the root bark were first identified, including 3-acetoxy-6-phenylacetoxytropane, 6-acetoxy-3α-benzoyloxytropane, and 3-α-phenylacetoxynortropan-6β-ol (Fig. 3.13) (Al-Said et al. 1986, 1989a, b).

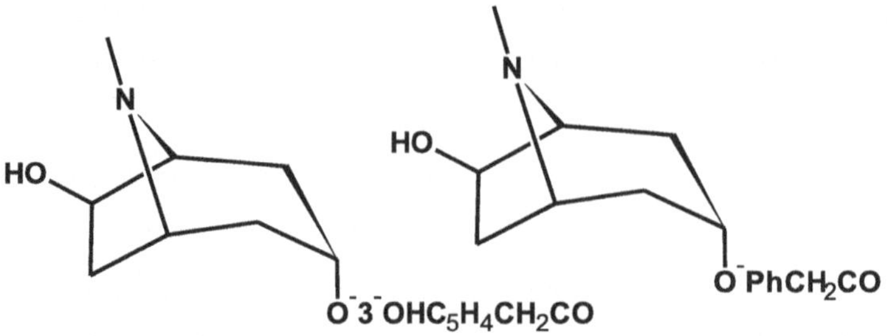

Tropan-3-α-yl-Hydroxyphenyl acetate

Tropan-3-α-yl-Phenylacetate

Fig. 3.13 Tropane alkaloids of *E. hypericifolium*. (Source: Al-Said et al. 1986; redrawn)

Fig. 3.14 Benzoyl esters tropan-3β-ol of *Erythroxylum macrocarpum*. (Source: David et al. 2019; redrawn)

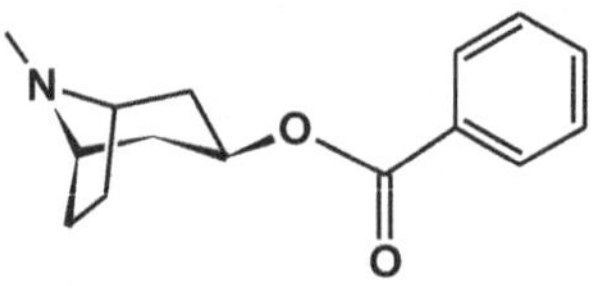

3.8.3.4 *Erythroxylum macrocarpum* O.E. Schulz

Benzoyl esters of tropan-3α-ol, tropan-3β-ol, and tropan-3α, 6β-diol, together with their nor-derivates, are the primary alkaloids of the plant and are responsible for some of its therapeutic properties (Fig. 3.14) (David et al. 2019).

3.8.3.5 *Erythroxylum monogynum* Roxb

By using GC and GC-MS, 26 tropane alkaloids were found in the root bark of *Erythroxylum monogynum*; 20 of these alkaloids had never before been recognized as root bark components. Three novel alkaloids were discovered as 3α-isobutyryloxynortropane, 3α-(4-methylvaleroyloxy)tropane, and tentatively 3β-phenylacetoxytropane. The primary base was 3α-(3′,4′,5′-trimethoxybenzoyloxy)tropane (Christen et al. 1995).

3.8.3.6 *Erythroxylum moonii* Hochr

Mooniine A and Mooniine B, two novel dimeric tropane alkaloids, were discovered in *E. moonii* leaves. Mass spectrometry and multidimensional NMR spectroscopy were used to identify the structures of these alkaloids (Fig. 3.15) (Rahman et al. 1998).

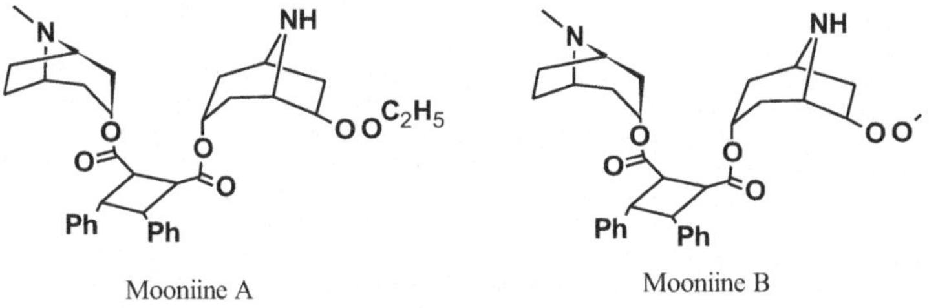

Fig. 3.15 Dimeric tropane alkaloids of *Erythroxylum moonii.* (Source: Rahman et al. 1998)

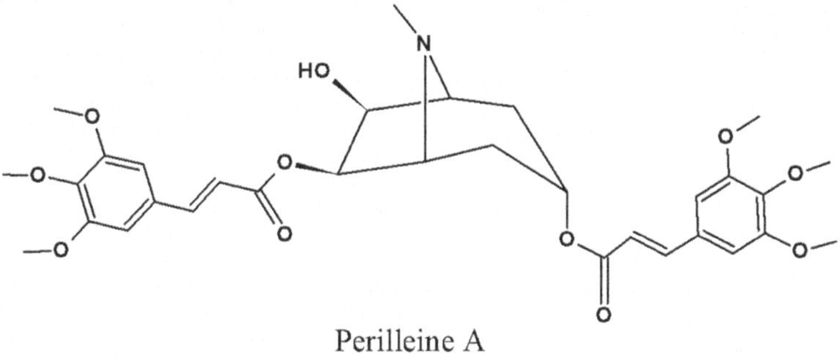

Fig. 3.16 Perilleine A. (Source: Chin et al. 2006; redrawn)

3.8.3.7 *Erythroxylum pervillei* Baill

Four tropane aromatic ester alkaloids, namely, pervilleines G and H, cis-pervilleines B and F, and pervilleine A, were found as minor ingredients in the stem bark extract of *E. pervillei* (from southern Madagascar) (Fig. 3.16) (Chin et al. 2006).

3.8.3.8 *Erythroxylum pungens* O. E. Shulz

The mass spectra of *E. pungens* stem bark, root bark, and leaf extracts revealed around 11 recognized tropane alkaloids, and 3-(2-methylbutyryloxy)tropan-6,7-diol and 3-(2-methylbutyryloxy)nortropan-6,7-diol were separated and described using mass spectrometry (Pereira et al. 2018).

3.8.3.9 *Erythroxylum rimosum* O. E. Schulz

A methanol extract of *E. rimosum* leaves yielded a new tropane alkaloid called 7β-acetoxy-3β,6β-dibenzoyloxytropane, as well as known compounds like quercetin, kaempferol-3-O-α-L-arabinofuranoside, quercetin-3-O-α-arabinofuranoside, quercetin-3-O-β-arabinopyranoside, quercetin-3-O-β-arabinopyranoside,

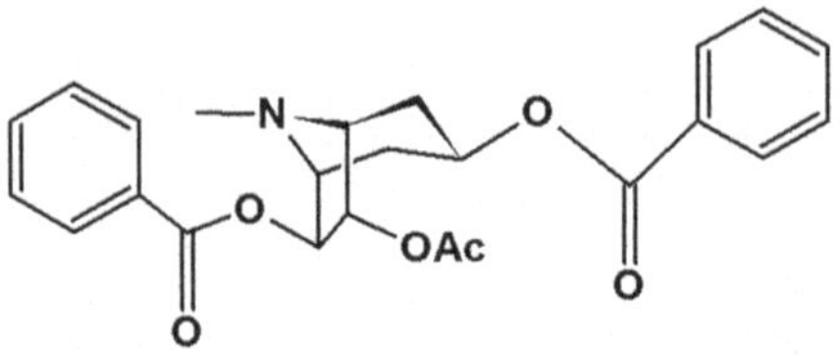

7β-acetoxy-3α,6α-dibenzoyloxytropane

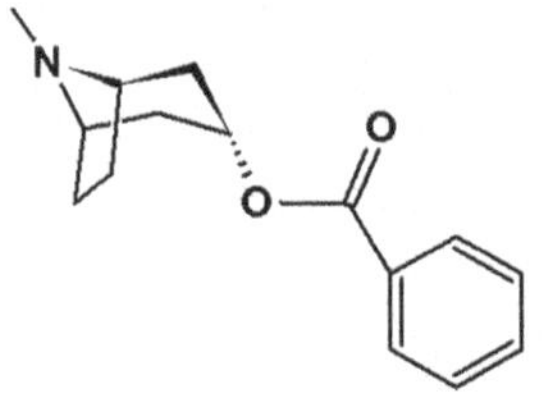

Benzoyl esters Tropan-3α-ol

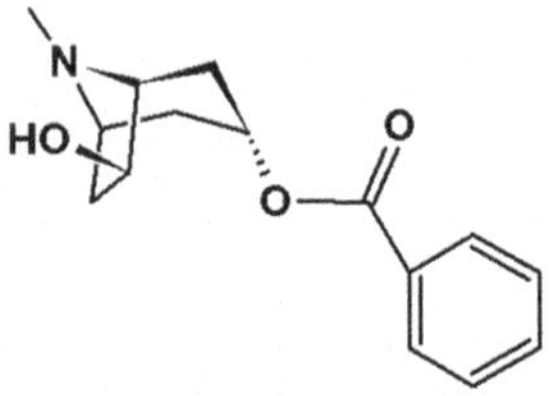

Benzoyl esters Tropan 6β-diol

Fig. 3.17 Tropane alkaloids of *Erythroxylum rimosum*. (Ribeiro et al. 2013; redrawn)

quercetin-3-O-β-glucopyranoside, β-sitosterol, α-amyrin, β-amyrin, and their ester derivatives (Fig. 3.17) (Ribeiro et al. 2013).

3.8.3.10 *Erythroxylum vacciniifolium* Mart

The bark of *Erythroxylum vacciniifolium*, a native Brazil, known locally as Catuaba and utilized as an aphrodisiac in traditional medicine, yielded two novel tropane alkaloid N-oxides that were replaced by a methylpyrrole moiety. Queiroza et al. (2009) characterized two tropane alkaloids, namely, 3α,7α-dihydroxy-6β-[(1-methyl-1H-pyrrol-2-yl)carbonyloxy] tropane N-oxide and 6β-hydroxy-4α-[(1-methyl-1H-pyrrol-2 yl)carbonyloxy] tropane N-oxide, which they called vaccinine B N-oxide. Eight novel tropane alkaloids, including catuabine D, 7-α-hydroxycatuabine D, catuabine E, 7-α-hydroxycatuabine E, 7-α-acetylcatuabine E, catuabine F, 7-α-hydroxycatuabine F, and catuabine G, as well as catuabines A, B, and C, were identified from the bark of *E. vacciniifolium* (Fig. 3.18) (Zanolari et al. 2003).

3.8.3.11 *Erythroxylum zeylanicum* O.E. Schulz

The Sri Lankan indigenous plant *Erythroxylum zeylanicum* (Erythroxylaceae) yielded six tropane alkaloids, which were then structurally clarified using NMR and MS studies. Along with known tropane alkaloids in other *Erythroxylum* species, three of them—erythrozeylanines A

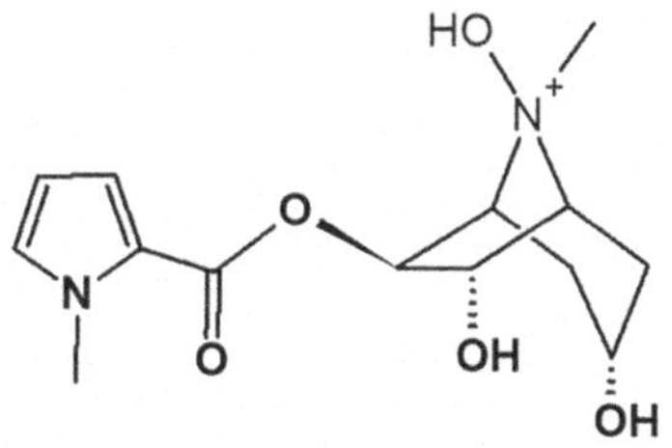

7α-hydroxycatuabine H N-oxide

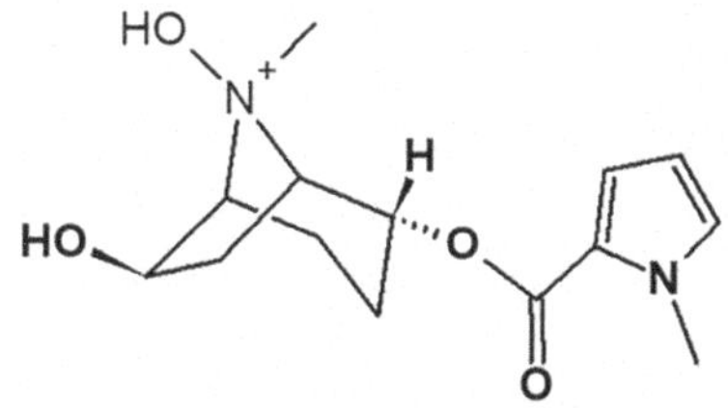

6β-hydroxy--α[(1-methyl-1H-pyrrol-2-yl) carbonyloxy]tropane N-oxide

(Vaccinine B N-oxide)

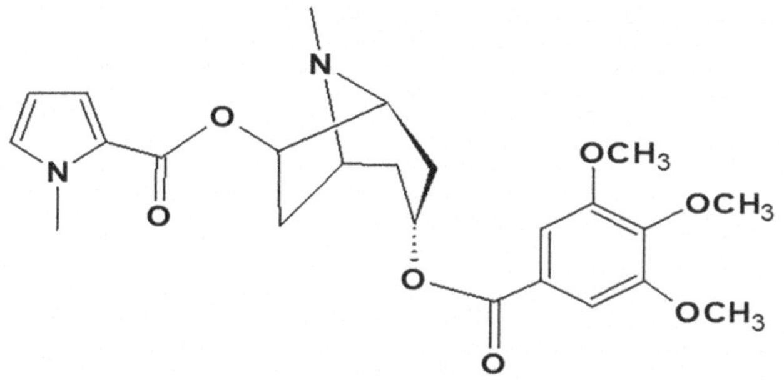

Catuabin A

Fig. 3.18 Tropane alkaloids from the bark of *Erythroxylum vacciniifolium*. (Source: Zanolari et al. 2003; redrawn)

[1R,3R,5S,6R-6-acetoxy-3-(3′,4′,5′-trimethoxybenzoyloxy)tropane], B [cis-3β-(cinnamoyloxy)tropane], and C [cis-6β-acetoxy-3α-(cinnamoyloxy)tropane]—are novel (Bringmann et al. 2000).

3.8.3.12 Tropane Alkaloids Application in Chemotaxonomy of Genus *Erythroxylum*

Erythroxylum argentinum, *E. cumanense*, *E. glaucum*, *E. cataractarum*, *E. mamacoca*, and *E. shatona* total alkaloids vary from 0.06% to 0.20% (El-Imam et al. 1985). Benzoic and phenylacetic acids combine with alkamines, primarily tropan-3-ols, tropan-3,6-diols, and their derivatives, to create esters in the six *Erythroxylum* spp. of *Archerythroxylum*. The most common alkaloid is 3β-benzoyloxytropane; other species contain hydrocuscohygrine, cuscohygrine, and nortropacocaine (3β-benzoyloxynortropane), which is recovered from *E. mamacoca*.

Along with nortropacocaine (3β-benzoyloxynortropane) isolated from *E. mamacoca*, 3β-benzoyloxytropane is the most often occurring alkaloid. Certain species

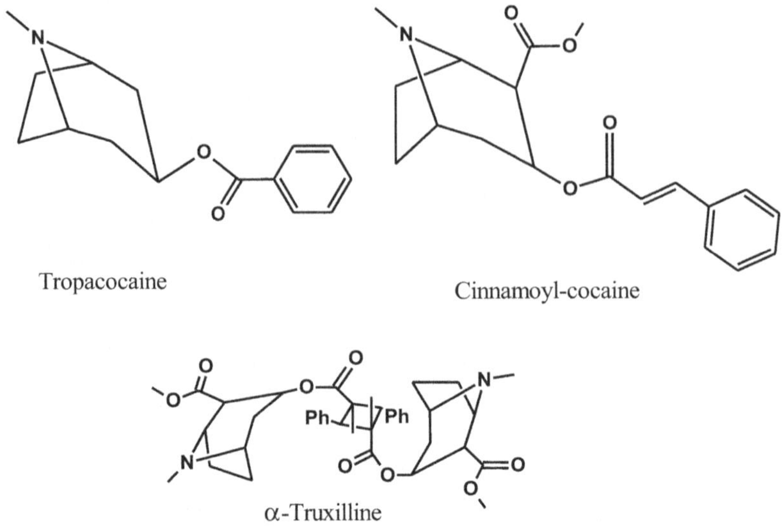

Fig. 3.19 Tropane alkaloids. (Redrawn)

also include hydrocuscohygrine and cuscohygrine. *E. cataractarum* Spruce ex Peyr's primary alkaloid, cuscohygrine, is distinct from other members of the genus. Tropacocaine (Fig. 3.19), which is not found in the sections *Macrocalyx* and *Rhabdophyllum*, is isolated from *E. ulei* O. E. Schulz (section Leptogramme), *E. mamacoca* Mart., and *E. argentinum* O. E. Schulz (section Archerythroxylon) (El-Imam et al. 1985). The cultivated cocaine-producing species *E. coca*, *E. novogranatense*, and its variant *truxillense* also contain dihydrocuscohygrine and cuscohygrine. *E. coca* Lam. and *E. novogranatense* var. *truxillense* Morris contain cocaine and the cis and trans isomers of cinnamoyl cocaine (Rivier 1981).

The α-truxilline and β-truxilline were the most prevalent dimeric methylecgonine (methyl ester of ecgonine, the basic moiety of cocaine) and truxillic acid (1,3-diphenylcyclobutane-2,4-dicarboxylic acids, dimers of cinnamic acid) esters found in Bolivian coca leaf (Moore et al. 1987). *E. novogranatense* Morris var. *novogranatense* and var. *truxillense* contained 1-hydroxytropacocaine, a novel alkaloid (Moore et al. 1994).

Both *E. cumanense* Kunth and *E. glaucum* O. E. Schulz (in the *Archerythroxylum* section) contain the base 6β-benzoyloxytropan-3α-ol. *E. cumanense* Kunth roots contain trimethoxybenzoyl esters. The comparable spectrum of alkaloids, primarily benzoyl esters of tropan-3α-ol, tropan-3β-ol, and tropan-3α,6β-diol together with its nor-derivatives, are found in *E. macrocarpum* O. E. Schulz and *E. sideroxyloides* Lam. of the Packylobus section. According to Al-Said et al. (1986), three alkaloids have been identified: 3α-benzoyloxytropan-6β-ol (sourced from E. sideroxyloides Lam.), 3α-benzoyloxynortropane, and 3β-benzoyloxynortropan-6β-ol (both species). The bases 3α-(3-hyroxyphenylacetoxy)-tropane, (+)-3α-phenylacetoxytropan-6β-ol,

6β-acetoxy-3α-phenylacetoxytro-pane, 3α-phenylacetoxytropan-6β,7β-diol, 3α-phenylacetoxynortropane, and 3α-phenylacetoxytropane are found in the root barks of *E. pervillei* Baill. and *E. hypericifolium* Lam. (Al-Said et al. 1986).

3,3′-truxilloyloxy-6′-acetoxyditropane, 3,3′-truxilloyl-6hydroxy-6′-acetoxyditropane, and 3,3′-truxilloyloxy-6′-hydroxyditropane were the names given to these. Similar to the stem bark of *E. hypericifolium* Lam., the major alkaloids, 3α-phenylacetoxynortropan-6β-ol, 6β-acetoxy-3α-benzoyloxytropane, 3-acet oxy-6-phenylacetoxytropane, and hygrine, are the predominant esters of phenylacetic acid in the stem bark (Al-Said et al. 1989a, b). The root bark alkaloids of *E. zambesiacum* N. Robson (section Melanocladus) are also complex. Three α-(3,4,5-trimethyoxybenzoyloxy) are the most significant: 3α-(3,4,5-trimethoxybenzoyloxy)nortropane3α-(3,4,5-trimethoxybenzoyloxy) and -tropan-6β-ol6β-benzoyloxytropan-3α,7β-diol, 6β-benzoyloxy, and nortropan-6β-ol(3,4,5-trimethoxycinnamoyloxy)-3α-7β-acetoxy-6β-benzoyloxy-3α-(3,4,5-trimethoxycinnamoyloxy)-tropane and 7β-tropan-7β-ol. Among the other minor bases found are 3α-(3,4,5-trimethoxybenzoyloxy), 3α-(3,4,5-trimethoxycinnamoyloxy)-tropane-tropane, 3α-(3,4,5-trimethoxybenzoyloxy), 3α-phenylacetoxytropan-6β-ol6β-benzoyloxytropan-3α-ol, 7β-diol, 6β-benzoyloxy, and tropan-6β-3α-(3,4,5-trimethoxycinnaomoyloxy) tropane (El-Imam et al. 1987). In *E. zambesiacum* primary alkaloid is 3α-(3′,4′,5′-trimethoxybenzoyl)-oxytropane besides three novel alkaloids, 6-isovaleryloxytropan-3-ol, 3-(2-methyl butyryloxy), and 6-β-benzoyloxytropan-3-one (Christen et al. 1993).

Mooniine A and mooniine B are two more dimeric tropane alkaloids that have been identified from *E. moonii* Hochr., albeit in extremely low yield (approximately 0.003%) (Rahman et al. 1998). They are carboxylic dibasic acid diesters of tropine and nortropine, most likely due to the biogenesis of tropic acids. The tropane alkaloid spectrum of the genus *Erythroxylum* is very similar to that of the Solanaceae, except for tropylesters (Evans 1979; El-Imam et al. 1987). Tigloyl esters of meteloidine, 3α-tigloyloxytropan-6β-ol, 3α-tigloyloxynortropan-6β-ol, and an unidentified base that was provisionally determined to be 7-hydroxy-6-tigloyloxynortropan-3-yl, 2-hydroxy-3-phenylpropionate are found in *E. australe* F. Muell (section Coelocarpus). In addition to meteloidine, hydroxy-3-phenyl-propionic acid, a member of the solanaceous alkaloid littorine and a close relative of tropic acid, provides another connection to the Solanaceae family (Griffin 1978).

Additionally, calystegines were found in the genus *Erythroxylum*. The dried leaves of *E. novogranatense* Morris var. *novogranatense* have a total calystegine content of 0.2%. Calystegines were found in 46 species of *Erythroxylum* herbarium, which were primarily composed of leaf tissue. In the majority of species, the predominant calystegines were A3 and B2. The idea of shared biosynthesis stages of these alkaloids in *Erythroxylum* is supported by the simultaneous presence of calystegines, cocaine, other alkaloids with a 3α-hydroxy- or 3β-hydroxytropane structure, and nicotine (Brock et al. 2005). Lastly, *E. pervillei* Baill was recently used to extract pervilleine A, a new tropane alkaloid (Mi et al. 2001).

3.8.4 *Euphorbiaceae*

Tropane alkaloids were isolated only from *Peripentadenia mearsii* (C.T. White) L. S. Sm. in this family. They are 3β-benzoyloxytropane, 2α-benzoyloxy-3β-hydroxytropane, 3α-acetoxy-6β-hydroxytropane, and tropacocaine (Johns et al. 1971).

3.8.5 *Moraceae*

The tropane alkaloids found in *Morus alba* L. make it noteworthy among its family. In particular, the penthydroxylated nortropane calystegine C1 (Asano et al. 1994) was extracted from its root bark, whereas calystegines A8 and A9 are found in fruits (Kusano et al. 2002).

3.8.6 *Proteaceae*

Bellendine (Fig. 3.20), an uncommon tropane alkaloid, was the first alkaloid from the Proteaceae to be produced by *Bellendena montana* R. Br. (Bick et al. 1971). *Darlingia darlingiana* L. A. S. Johnson was used to identify similar alkaloids, including darlingine, a pyranotropane base (Ralph et al. 1979a). Darlingine,

Fig. 3.20 Tropane alkaloids of Proteaceae. (Ralph et al. 1979b; redrawn)

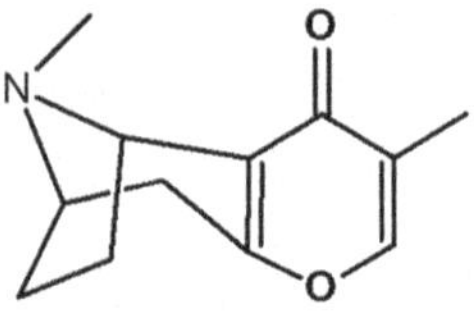

Bellendine

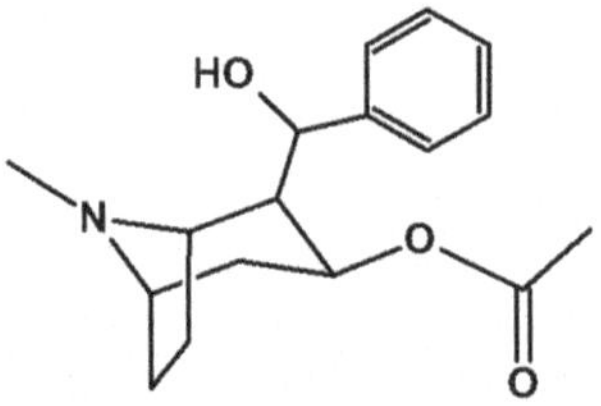

Knightinol

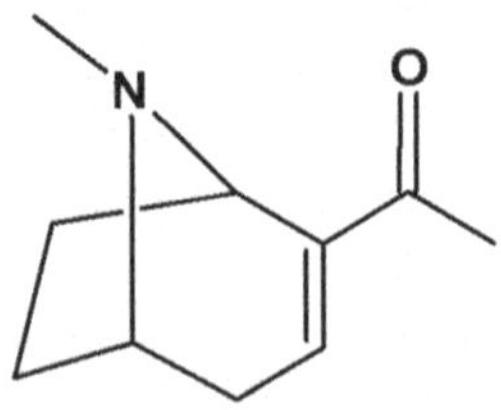

Ferruginine

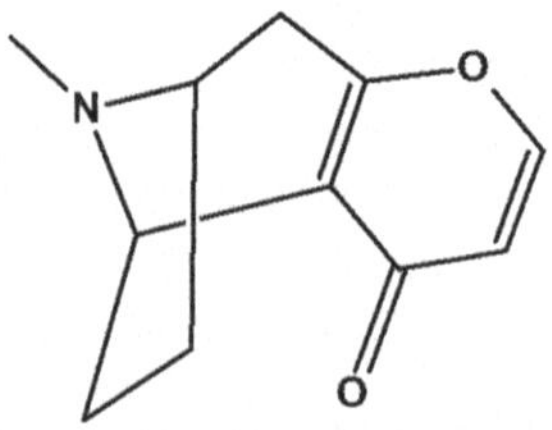

Strobiline

ferruginine, and 3α-benzoyloxy-2α-hydroxybenzyltropine are all found in *D. ferruginea* J. F. Bailey (Ralph et al. 1979b). Bellendine, isobellendine, and darlingine are found in *B. montana* R. Br. (Bick et al. 1979). The genus *Knightia*, *K. deplanchei* Vieill ex Brongn. & Gris, and *K. strobilina* R. Br. were the sources of alkaloids of the 2-benzyltropine type (Lounasmaa et al. 1975). D-strobiline is the primary alkaloid of *K. strobilina* R. Br., along with dihydrostrobiline, strobamine B, chalcostrobamine, strobolamine, knightalbinol, knightolamine, 3α-cinnamoyloxytropan-6β-ol, 3α-acetoxy-2α-acetoxybenzyltropane (acetylknightionl), 3α-acetoxy-2α-acetoxybenzyltropan-6β-ol (knightoline), 6β-benzoyloxytropan-3α-ol, and 2α-hydroxybenzyl-3α-acetoxytropane (knightinol). The first naturally occurring tropene base, 3α-(p-hydroxybenzoyloxy)-trop-6-ene, and 6β-acetoxy-3α-tigloyloxytropane, are found in *Agastachys odorata* R. Br. (Ralph et al. 1979b). The alkaloids 3α-cinnamoyloxytropan-6β-ol, which were isolated as the (+) base from *Knightia* spp. leaves, are also noteworthy (Lounasmaa et al. 1980).

Erythroxylum hypericifolium Lam. was the source of the pseudotropine ester. In *Erythroxylum* species, the benzoyl analogue is extensively dispersed. Along with cinnamate tropane dimers and esters of truxillic acid, the alkaloids 3α-cinnamoyloxytropane, 3α,6β-dicinnamoyloxytropane (similar to the dibenzoyl ester found in *E. cuneatum* Kurz (El-Imam et al. 1988), 3-cinnamoyloxytropan-6-ol (stereochemistry not established), 6β-acetoxy-3α-cinnamoyloxytropane, and 6-phenylacetoxytropan-3-ol were reported from *Crossostylis* spp. (Medina et al. 1983).

3.8.7 Rhizophoraceae

Other dithiolane esters of acetic, propionic, n-butyric, isobutyric, isovaleric, and benzoic acids are found in *Bruguiera exaristata* Ding Hou, whereas brugine (tropine 1,2-dithiolane-3-carboxylate) was isolated from *B. sexangula* Poir. (Loder and Russel 1966). Along with tropine, 3α-benzoyl-oxytropane, and two esters of tropine, tropine cinnamate and ferulate, brucine has also been detected in *Crossostylis* species (*C. biflora* Forst., *C. multiflora* Brongn. & Gris, and *C. sebertii* Brongn. & Gris) (Fig. 3.21). The tropane alkaloids tropan-3α-yl cinnamate and tropan-3α-yl benzoate are found in *C. sebertii* (Pancher) Pierre Arbuste, tropan-3α-yl-ferulate, 3α-tropanol, and brugine, in *C. multiflora* Brongn. & Gris, while hygrine and tropanone are found in *C. biflora* J.R.Forst. & G. Forst. (Medina et al. 1983).

3.9 Conclusion

The tropane alkaloids have highly diversified structural and functional properties with a huge number of natural combinations. Especially they have a protective role within the plants against the gazing animals and natural enemies. However, few of

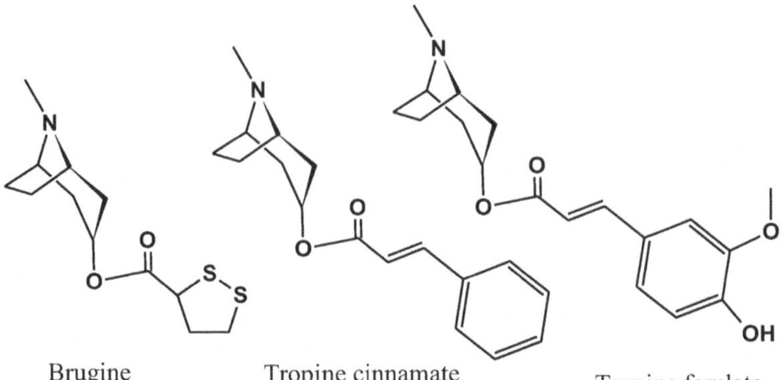

Fig. 3.21 Tropane alkaloids. (Johns et al. 1971; Medina et al. 1983 redrawn)

them are poison to animals as well as human beings with their psychoactive properties. Moreover, a few tropane alkaloids play an important role in the prevention of certain painful diseases, as well as sedatives if used under the guidance of a pharmacist. Research has to be focused on the modification of certain tropane alkaloids based on their functional group versus pharmacology versus safety issues. The process of biosynthesis of tropane alkaloids in these plants, as well as their enhancement through the interference of biotechnological methods, will be discussed in the later chapters, which may promote the sustainable utilization of beneficial alkaloids, excluding the toxic substances in view of bio-safety issues.

References

Afsharypuor S, Mostajeran A, Mokhtary R (1995) Variation of scopolamine and atropine in different parts of *Datura metel* during development. Planta Med 61:383–384

Alizadeh A, Moshiri M, Alizadeh J, Balali-Mood M (2014) Black henbane and its toxicity—a descriptive review. Avicenna J Phytomed 4(5):297–311

Al-Said MS, Evans WC, Grout RJ (1986) Alkaloids of the genus *Erythroxylum*. Part 5. E. *hypericifolium* Lam. root-bark. J Chem Soc 6:957–959

Al-Said MS, Evans WC, Grout RJ (1989a) Alkaloids of the genus *Erythroxylum*. Part 10. Alkaloids of *Erythroxylum hypericifolium* leaves. Phytochemistry 28(11):3211–3215

Al-Said MS, Evans WC, Grout RJ (1989b) Alkaloids of *Erythroxylum hypericifolium* stem bark. Phytochemistry 28(2):671–673. https://doi.org/10.1016/0031-9422(89)80086-X

Alsamarrai SHA (2019) Synthesis of tropane derivatives. IntechOpen. https://doi.org/10.5772/intechopen.83382

Andersson HC (2002) Calystegine alkaloids in Solanaceous food plants. Nordic Council of Ministers, Copenhagen

Arif U (2013) Effect of wounding and light exposure on sterol, glycoalkaloid, and calystegine levels in potato plants (*Solanum tuberosum* L. group *tuberosum*). Doctoral thesis. Swedish University of Agricultural Sciences, Uppsala. https://doi.org/10.13140/RG.2.2.35878.52800

Aripova SF (1985) N-oxide of convolamine from *Convolvulus krauseanus*. Khimiya Prirodnykh Soedinenii 2:275

Aripova SF, Abdilalimov O (1993) Convolacine? A new alkaloid from *Convolvulus subhirsutus*. Chem Nat Compd 29(1):74–75. https://doi.org/10.1007/bf00631024

Aripova SF, Malikov VM, Yunusov SY (1977) Convolidine-new alkaloid from *Convolvulus krauseanus*. Khimiya Prirodnykh Soedinenii 2:290–291

Aripova SF, Sharova EG, Yunusov SY (1983) Convoline – a new alkaloid from *Convolvulus krauseanus*. Khimiya Prirodnykh Soedinenii 6:749–751

Arraez-Roman D, Zurek G, Babmann C, Antonio SC, Alberto FG (2008) Characterization of *Atropa belladonna* L. compounds by capillary electrophoresis-electrospray ionization-time of flight-mass spectrometry and capillary electrophoresis-electrospray ionization-ion trap-mass spectrometry. Electrophoresis 29(10):2112–2116

Asano N, Oseki K, Tomioka E, Kizu H, Matsui K (1994) N-containing sugars from *Morus alba* and their glycosidase inhibitory activities. Carbohydr Res 259(2):243–255

Asano N, Kato A, Oseki K, Kizu H, Matsui K (1995) Calystegins of *Physalis alkekengi* var. *franchetii* (Solanaceae). Eur J Biochem 229:369–376

Asano N, Kato A, Kizu H, Matsui K, Watson AA, Nash RJ (1996) Calystegine B4, a novel trehalase inhibitor from *Scopolia japonica*. Carbohydr Res 293(2):195–204

Asano N, Kato A, Matsui K, Watson AA, Nash RJ, Molyneux RJ, Hackett L, Topping J, Winchester B (1997) The effects of calystegines isolated from edible fruits and vegetables on mammalian liver-glycosidases. Glycobiology 7:1085–1088

Asano N, Yokoyama K, Sakurai M, Ikeda K, Kizu H, Kato A, Arisawa M, Höke D, Dräger B, Watson AA (2001) Dihydroxynortropane alkaloids from calystegine-producing plants. Phytochemistry 57(5):721–726

Baralle FE, Gros EG (1969) Biosynthesis of cuscohygrine in *Atropa belladonna* from sodium acetate-2-14C*. Phytochemistry 8(5):853–855

Bekkouche K, Daali Y, Cherkaoui S, Veuthey JL, Christen P (2001) Calystegine distribution in some Solanaceous species. Phytochemistry 58(3):455–462

Berkov S (2003) Alkaloids of *Datura ceratocaula*. Z Naturforsch C J Biosci 58(7–8):455–458. https://doi.org/10.1515/znc-2003-7-801

Berkov S, Zayed R (2004) Comparison of tropane alkaloid spectra between *Datura innoxia* grown in Egypt and Bulgaria. Z Naturforsch C J Biosci 59(3–4):184–186. https://doi.org/10.1515/znc-2004-3-409

Berkov S, Pavlov A, Kovatcheva P, Stanimirova P, Philipov S (2003) Alkaloid spectrum in diploid and tetraploid hairy root cultures of *Datura stramonium*. Z Naturforsch C J Biosci 58(1–2):42–46. https://doi.org/10.1515/znc-2003-1-207

Bick IRC, Bremer JB, Gillard JW (1971) Methyl(p-hydroxybenzoyl)acetate and an alkaloid, bellendine, from *Bellendena montana*. Phytochemistry 10(2):475–477

Bick IRC, Gillard JW, Huck-Meng L (1979) Alkaloids of *Bellendena montana*. Austral J Chem 32(8):1827–1840

Bieri S, Munoz O, Veuthey JL, Christen P (2006) Analysis of isomeric tropane alkaloids from *Schizanthus grahamii* by very fast gas chromatography. J Sep Sci 29:96–102

Bringmann G, Christian G, Jorg M, Lalith MD, Gunathilake P, Wickramasinghe A (2000) Tropane alkaloids from *Erythroxylum zeylanicum* O.E. Schulz (Erythroxylaceae). Phytochemistry 53(3):409–416. https://doi.org/10.1016/S0031-9422(99)00561-0

Brock A, Bieri S, Christen P, Drager B (2005) Calystegines in wild and cultivated *Erythroxylum* species. Phytochemistry 66(11):1231–1240

Brock A, Herzfeld T, Paschke R, Koch M, Dräger B (2006) Brassicaceae contain nortropane alkaloids. Phytochemistry 67(18):2050–2057

Cannon JR, Joshi KR, Meehan GV, Williams J (1969) Tropane alkaloids from three western Australian *Anthocercis* species. Austral J Chem 22(1):221–227

Casale JF, Toske SG, Colley VL (2005) Alkaloid content of the seeds from *Erythroxylum coca* var. *coca*. J Forensic Sci 50:1402–1406. https://doi.org/10.1520/JFS2005121

Chandrika UG, Karunarathna U (2022) Chapter 5 – Anaesthetics and analgesic activities of herbal medicine: review of the possible mechanism of action, in Features and assessments of pain, anaesthesia, and analgesia, editor(s): Rajendram R, Patel VB, Preedy VR, Martin CR. Academic, 47–56. doi:https://doi.org/10.1016/B978-0-12-818988-7.00003-0

Chin YW, Jones WP, Waybright TJ, McCloud TG, Rasoanaivo P, Cragg GM, Cassady JM, Kinghorn AD (2006) Tropane aromatic ester alkaloids from a large-scale re-collection of *Erythroxylum pervillei* stem bark obtained in Madagascar. J Nat Prod 69(3):414–417. https://doi.org/10.1021/np050366v

Christen P, Roberts MF, Phillipson JD, Evans WC (1993) Alkaloids of the genus *Erythroxylum*. Part 11. Alkaloids of *Erythroxylum zambesiacum* stem-bark. Phytochemistry 34(4):1147–1151

Christen P, Roberts MF, Phillipson William JD, Evans C (1995) Alkaloids of *Erythroxylum monogynum* root-bark. Phytochemistry 38(4):1053–1056. https://doi.org/10.1016/0031-9422(94)00777-Q

Cinelli MA, Jones AD (2021) Alkaloids of the genus *Datura*: review of a rich resource for natural product discovery. Molecules 26(9):2629. https://doi.org/10.3390/molecules26092629

David AR, Saenz E, Jara-Muñoz OA et al (2019) *Erythroxylum* in focus: an interdisciplinary review of an overlooked genus. Molecules 24:3788. https://doi.org/10.3390/molecules24203788

de Oliveira SL, Tavares JF, Branco MV, Lucena HF, Barbosa-Filho JM, de F Agra M, do Nascimento SC, dos S Aguiar J, da Silva TG, de Simone CA, de Araújo-Júnior JX, da Silva MS (2011) Tropane alkaloids from *Erythroxylum caatingae* Plowman. Chem Biodivers 8(1):155–165. https://doi.org/10.1002/cbdv.200900400

Drager B (1995) Identification and quantification of calystegines, polyhydroxyl nortropane alkaloids. Phytochem Anal 6(1):31–37

Drager B, Funck C, Hohler A, Mrachatz G, Nahrstedt A, Portsteffen A, Schaal A, Schmidt R (1994) Calystegines as a new group of tropane alkaloids in Solanaceae. Plant Cell Tissue Organ Cult 38(2):235–240

Du N, Zhou W, Jin H, Liu Y, Zhou H, Liang X (2019) Characterization of tropane and cinnamamide alkaloids from *Scopolia tangutica* by high-performance liquid chromatography with quadrupole time-of-flight tandem mass spectrometry. J Sep Sci 42:1163–1173

Dupraz J, Christen P, Ilias K (1994) Tropane alkaloids in transformed roots of *Datura quercifolia*. Planta Med 60(2):158–162. https://doi.org/10.1055/s-2006-959441

EFSA (2013) Panel on contaminants in the food chain (CONTAM). Scientific opinion on tropane alkaloids in food and feed. EFSA J 11:3386

EFSA, Binaglia M, Baert K, Schutte M, Rositsa S (2019) Scientific report: overview of available toxicity data for calystegines. EFSA J 17(1):e05574. https://doi.org/10.2903/j.efsa.2019.5574

El Bazaoui A, Bellimam MA, Soulaymani A (2011) Nine new tropane alkaloids from *Datura stramonium* L. identified by GC/MS. Fitoterapia 82(2):193–197. https://doi.org/10.1016/j.fitote.2010.09.010

El Bazaoui A, Bellimam MA, Soulaymani A (2012) Tropane alkaloids of *Datura innoxia* from Morocco. Z Naturforsch C J Biosci 67(1–2):8–14. https://doi.org/10.1515/znc-2012-1-202

El-Imam YMA, Evans WC, Plowman T (1985) Alkaloids of the genus *Erythroxylum*. Part 4. Alkaloids of some South American *Erythroxylum* species. Phytochemistry 24(10):2285–2289

El-Imam YMA, Evans WC, Grout RJ, Ramsey KPA (1987) Alkaloids of the genus *Erythroxylum*. Part 7. Alkaloids of *Erythroxylum zambesiacum* root-bark. Phytochemistry 26(8):2385–2389

El-Imam YMA, Evans WC, Grout RJ (1988) Alkaloids of the genus *Erythroxylum*. Part 8. Alkaloids of *Erythroxylum cuneatum*, *E. ecarinatum* and *E. australe*. Phytochemistry 27(7):2181–2184

El-Olemy MM, Moustafa SMI (1983) Alkaloids of *Hyoscyamus muticus* tissue cultures. Bull Pharmaceut Sci (Assiut University) 6(1):196–211

Evans WC (1979) Tropane alkaloids of the Solanaceae. Linnean Soc Symp 7:241–254

Evans WC, Ramsey KPA (1979) Alkaloids of the Solananceae tribe Anthocercideae. J Pharm Pharmacol 31:9P

Evans WC, Ramsey KPA (1981) Tropane alkaloids from *Anthocercis* and *Anthotroche*. Phytochemistry 20(3):497–499

Evans WC, Ramsey KPA (1983) Alkaloids of the Solananceae tribe Anthocercideae. Phytochemistry 22(10):2219–2225

Evans WC, Stevenson NA (1962) Studies on *Datura leichhardtii* Muell. ex Benth. J Pharm Pharmacol 14(1, Suppl):107T–110T. https://doi.org/10.1111/j.2042-7158.1962.tb10542.x

Evans WC, Treagust PG (1973) Alkaloids of *Datura pruinosa*. Phytochemistry 12(8):2077–2078

Fatemeh A, Fatemeh S (2011) Tropane alkaloids of *Atropa belladonna* L. and *Atropa acuminata* Royle ex Miers plants. J Med Plants Res 5(29):6515–6522. https://doi.org/10.5897/JMPR11.482

Friedman M, Roitman JN, Kozukue N (2003) Glycoalkaloid and calystegine contents of eight potato cultivars. J Agric Food Chem 51:2964–2973

Gaire BP, Subedi L (2013) A review on the pharmacological and toxicological aspects of *Datura stramonium* L. J Integr Med 11(2):73–79. https://doi.org/10.3736/jintegrmed2013016

Ghani A (2003) Medicinal plants of Bangladesh with chemical constituents and uses. Asiatic Society of Bangladesh, Dhaka, pp 42–48

Goldmann A, Milat ML, Ducrot PH, Lallemand JY, Maille M, Lepingle A, Charpin I, Tepfer D (1990) Tropane derivatives from *Calystegia sepium*. Phytochemistry 29(7):2125–2127

Goldmann A, Message B, Tepfer D, Molyneux RJ, Duclos O, Boyer FD, Pan Y, Elbein AD (1996) Biological activities of the nortropane alkaloid, calystegine B2, and analogs: structure-function relationships. J Nat Prod 59(12):1137–1142

Goldmann A, Asano N, Yokoyama K, Sakurai M, Ikeda K, Kizu H, Kato A, Arisawa M, Hoke D, Drager B, Watson AA (2001) Dihydroxynortropane alkaloids from calystegine-producing plants. Phytochemistry 57(5):721–726

Gonzalez-Gomez L, Morante-Zarcero S, Pereira JAM, Câmara JS, Sierra I (2022) Improved analytical approach for determination of tropane alkaloids in leafy vegetables based on μ-QuEChERS combined with HPLC-MS/MS. Toxins 14:650

Griffin WJ (1978) A phytochemical investigation of *Erythroxylon australe* F. Muell. Austral J Chem 31(5):1161–1165

Griffin WJ, Lin GD (2000) Chemotaxonomy and geographical distribution of tropane alkaloid producing plants. Phytochemistry 53:623–637

Grynkiewicz G, Gadzikowska M (2008) Tropane alkaloids as medicinally useful natural products and their synthetic derivatives as new drugs. Pharmacol Rep 60:439–463

Gumus ZP (2023) Assessment of toxic pyrrolizidine and tropane alkaloids in herbal teas and culinary herbs using LC-Q-ToF/MS. Food Secur 12:3572. https://doi.org/10.3390/foods12193572

Hanus OL, Tomas R, Jaroslav S, Dembitsky VM (2005) Substances isolated from *Mandragora* species. Phytochemistry 66:2408–2417. https://doi.org/10.1016/j.phytochem.2005.07.016

Hegnauer R (1981) Chemotaxonomy of Erythroxylaceae (including some ethnobotanical notes on old world species). J Ethnopharmacol 3:279–292

Huang JP, Wang Y, Tian T et al (2021) Tropane alkaloid biosynthesis: a centennial review. Nat Prod Rep 38:1634–1658

Humam M, Muñoz O, Christen P, Hostettmann K (2007) Tropane alkaloids of the aerial parts of *Schizanthus tricolor*. Nat Prod Commun 2(7):743–747

Islam T, Ara I, Islam T et al (2023) Ethnobotanical uses and phytochemical, biological, and toxicological profiles of *Datura metel* L.: a review. Current research. Toxicology 4:100106. https://doi.org/10.1016/j.crtox.2023.100106

Jenett-Siems K, Weigl R, Böhm A, Mann P et al (2005) Chemotaxonomy of the pantropical genus *Merremia* (Convolvulaceae) based on the distribution of tropane alkaloids. Phytochemistry 66(12):1448–1464

Johns SR, Lamberton JA, Sioumis AA (1971) New tropane alkaloids, (+)-(3R, 6R)-3α-acetoxy-6β-hydroxytropane and (+)-2α-benzoyloxy-3β-hydroxynortropane, from *Peripentadenia mearsii* (Euphorbiaceae). Austral J Chem 24(11):2399–2403

Kato A, Asano N, Kizu H, Matsui K, Suzuki S, Arisawa M (1997) Calystegine alkaloids from *Duboisia leichhardtii*. Phytochemistry 45(2):425–429. https://doi.org/10.1016/S0031-9422(96)00865-5

Kaur S, Pandey NS (2020) Phytochemistry and pharmacological properties of *Datura stramonium*: an analysis. Indian J Pure Appl Biosci 8(3):92–105

Klein LM, Gabler AM, Rychlik M, Gottschalk C, Kaltner F (2022) A sensitive LC–MS/MS method for isomer separation and quantitative determination of 51 pyrrolizidine alkaloids and two tropane alkaloids in cow's milk. Anal Bioanal Chem 414:8107–8124

Kohnen-Johannsen KL, Kayser O (2019) Tropane alkaloids: chemistry, pharmacology, biosynthesis and production. Molecules 24(4):796. https://doi.org/10.3390/molecules24040796

Kowalczyk E, Kwiatek K (2022) Scopolamine and atropine in feeds–determination with liquid chromatography mass spectrometry. Food Addit Contam Part A 39:977–989

Kubwabo C, Rollmann B, Tilquin B (1993) Analysis of alkaloids from *Physalis peruviana* by capillary GC, capillary GC-MS, and GC-FTIR. Planta Med 59(2):161–163. https://doi.org/10.1055/s-2006-959634

Kusano G, Orihara S, Tsukamoto D, Shibano M, Coskun M, Guvenc A, Erdurak CS (2002) Five new nortropane alkaloids and six new amino acids from the fruit of *Morus alba* Linne growing in Turkey. Chem Pharm Bull 50(2):185–192

Leon N, Miralles P, Yusà V, Coscolla C (2022) A green analytical method for the simultaneous determination of 30 tropane and pyrrolizidine alkaloids and their N-oxides in teas and herbs for infusions by LC-Q-Orbitrap HRMS. J Chromatogr A 1666:462835

Loder JW, Russel GB (1966) Tropine 1,2-dithiolane-3-carboxylate, a new alkaloid from *Bruguiera sexangula*. Tetrahedron Lett 51:6327–6329

Lounasmaa M, Tamminen T (1993) The tropane alkaloids: chemistry and biology. In: Cordell GA (ed) The alkaloids, vol 44. Academic Press, New York, pp 1–114

Lounasmaa M, Wovkulich PM, Wenkert E (1975) Carbon-13 nuclear magnetic resonance spectroscopy of naturally occurring substances. XXXVI. Structures of some *Knightia deplanchei* alkaloids. J Org Chem 40(25):694–697

Lounasmaa M, Pusset J, Sevenet T (1980) The new Caledonian plants. Part 57. 2-benzyl pseudotropanes and dihydropyranotropanes, alkaloids of *Knightia strobilina*. Phytochemistry 19(5):953–955

Lu Y, Yao T, Chen Z (1986) Studies on the constituents of *Erycibe elliptilimba*. Acta Pharm Sin 21(11):829–835

Marin-Saez J, Romero-González R, Garrido Frenich A (2019) Reliable determination of tropane alkaloids in cereal based baby foods coupling on-line spe to mass spectrometry avoiding chromatographic step. Food Chem 275:746–753

Martinello M, Manzinello C, Gallina A, Mutinelli F (2022) In-house validation and application of UHPLC-MS/MS method for the quantification of pyrrolizidine and tropane alkaloids in commercial honey bee-collected pollen, teas and herbal infusions purchased on Italian market in 2019–2020 referring to recent European Union regulations. Int J Food Sci Technol 57:7505–7516

Medina GDH, Pusset J, Pusset M, Husson HP (1983) Tropane alkaloids from *Crossostylis* sp. J Nat Prod 46(3):398–400

Mi Q, Cui B, Silva GL, Lantvit D, Lim E, Chai H, You M, Hollingshead MG, Mayo JG, Kinghorn AD, Pezzuto JM (2001) Pervilleine A, a novel tropane alkaloid that reverses the multidrug-resistance phenotype. Cancer Res 61(10):4030–4037

Moore JM, Cooper DA, Lurie IS, Theodore CM, Carr S, Harper C, Yeh J (1987) Capillary gas chromatographic-electron capture detection of coca leaf-related impurities in illicit cocaine: 2,4-diphenylcyclobutane-1,3-dicarboxylic acids, 1,4-diphenylcyclobutane-2,3-dicarboxylic acids and their alkaloidal precursors, the truxillines. J Chromatogr 410(2):297–318

Moore JM, Hayes PA, Cooper DA, Casale JF, Lydon J (1994) 1-Hydroxytropacocaine: an abundant alkaloid of *Erythroxylum novogranatense* var. *novogranatense* and var. *truxillense*. Phytochemistry 36(2):357–360

Mortimer PI, Wilkinson S (1957) The occurrence of nicotine, anabasine, and isopelletierine in *Duboisia myoporoides*. J Chem Soc:3967–3970. https://pubs.rsc.org/en/Content/ArticleLanding/1957/JR/JR9570003967

Mulder PPJ, De Nijs M, Castellari M, Hortos M, MacDonald S, Crews C, Hajslova J, Stranska M (2016) Occurrence of tropane alkaloids in food. EFSA Support Publ 13(12):EN-1140. https://doi.org/10.2903/sp.efsa.2016.EN-1140

Nash RJ, Rothschild M, Porter EA, Watson A, Waigh R, Waterman PG (1993) Calystegines in *Solanum* and *Datura* species and the death's-head hawk-moth (*Acherontia atropos*). Phytochemistry 34(5):1281–1283

Pereira GM, Moreira LGL, Neto TDSN et al (2018) Isolation, spectral characterization, molecular docking, and cytotoxic activity of alkaloids from *Erythroxylum pungens* O. E. Shulz. Phytochemistry 155:12–18

Perez-Amador MC, Ocotero VM, Castaneda JMG, Esquinca ARG (2007) Alkaloids in *Solanum torvum* Sw (Solanaceae). Int J Exp Bot 76:39–45

Queiroza EF, Zanolarib B, Marstonb A, Guiletc D, Burgenerb L, Paulod MQ, Hostettmann K (2009) Two new tropane alkaloids from the bark of *Erythroxylum vacciniifolium* Mart. (Erythroxylaceae). Nat Prod Commun 4(10):1337–1340

Qureshi M, Sidhartha DR (2024) Belladonna alkaloids. In: Wexler P (ed) Encyclopedia of toxicology, 4th edn. Academic Press, Amsterdam, pp 933–938

Rahman AU, Khattak KF, Nighat F, Shabbir M, Hemalal KD, Tillekeratne LM (1998) Dimeric tropane alkaloids from *Erythroxylum moonii*. Phytochemistry 48(2):377–383. https://doi.org/10.1016/S0031-9422(97)01079-0

Ralph I, Bick C, Gillard JW, Leow HM (1979a) Alkaloids of *Darlingia darlingiana*. Austral J Chem 32(11):2523–2536

Ralph I, Bick C, Gillard JW, Leow HM (1979b) Alkaloids of *Darlingia ferruginea*. Austral J Chem 32(11):2537–2543

Ribeiro EM, Luciano SL, Jorge MD, do Vale AE, Lucia MXL, Juceni PD (2013) A new tropane alkaloid and other constituents of *Erythroxylum rimosum* (Erythroxylaceae). Phytochem Lett 6(2):232–235. https://doi.org/10.1016/j.phytol.2013.02.008

Rivier L (1981) Analysis of alkaloids in leaves of cultivated *Erythroxylum* and characterization of alkaline substances used during coca chewing. J Ethnopharmacol 3(2–3):313–335

Rocio M, Carrera A, Meijler MM, Janda KD (2004) Cocaine pharmacology and current pharmacotherapies for its abuse. Bioorg Med Chem 12:5019–5030

Romera-Torres A, Romero-González R, Martínez Vidal JL, Garrido Frenich A (2020) Comprehensive tropane alkaloids analysis and retrospective screening of contaminants in honey samples using liquid chromatography-high resolution mass spectrometry (Orbitrap). Food Res Int 133:109–130

Sabri M, El-Masry S, Khafagy SM (1973) Phytochemical investigation of *Hyoscyamus desertorum*. Planta Med 23(1):4–9

Sahai M, Ray AB (1980) Secotropane alkaloids of *Physalis peruviana*. J Org Chem 45:3265–3268

Schimming T, Tofern B, Mann P, Richter A, Jenett-Siems K, Drager B, Asano N, Gupta MP, Correa MD, Eich E (1998) Distribution and taxonomic significance of calystegines in the Convolvulaceae. Phytochemistry 49(7):1989–1995

Schimming T, Jenett-Siems K, Mann P, Tofern-Reblin B, Milson J, Johnson RJ, Deroin T, Austin DF, Eich E (2005) Calystegines as chemotaxonomic markers in the Convolvulaceae. Phytochemistry 66(4):469–480

Seneca (2007) Chapter 1 – Definition, typology and occurrence of alkaloids. In: Aniszewski T (ed) Alkaloids – secrets of life. Elsevier, pp 1–59. https://doi.org/10.1016/B978-044452736-3/50003-9

Sharova EG, Aripova SF, Yunusov SY (1980) Alkaloids of *Convolvulus subhirsutus*. Khimiya Prirodnykh Soedinenii 5:672–676

Srinivasan P, Smolke CD (2019) Engineering a microbial biosynthesis platform for de novo production of tropane alkaloids. Nat Commun 10:3634. https://doi.org/10.1038/s41467-019-11588-w

Stegelmeier BL, Field R, Panter KE et al (2013) Selected poisonous plants affecting animal and human health. In: Haschek WM, Rousseaux CG, Walling MA (eds) Haschek and Rousseaux's handbook of toxicologic pathology. Academic Press, New York, pp 1259–1314

Tepfer D, Goldmann A, Pamboukdjian N, Maille M, Lepingle A, Chevalier D, Dénarié J, Rosenberg C (1988) A plasmid of *Rhizobium meliloti* 41 encodes catabolism of two compounds from root exudate of *Calystegia sepium*. J Bacteriol 170(3):1153–1161

Van Haga RP (1954) Cuscohygrine, a normal constituent alkaloid of *Atropa belladonna*. Nature 174(4435):833–834

Versilovskis A, Mulder PPJ, Pereboom-de FDPKH, de Stoppelaar J, de Nijs M (2020) Simultaneous quantification of ergot and tropane alkaloids in bread in The Netherlands by LC-MS/MS. Food Addit Contam Part B Surveill 13:215–223

Vitale AA, Acher A, Pomilio AB (1995) Alkaloids of *Datura ferox* from Argentina. J Ethnopharmacol 49(2):81–89. https://doi.org/10.1016/0378-8741(95)90035-7

Weil AT (1981) The therapeutic value of coca in contemporary medicine. J Ethnopharmacol 3:367–376

World Health Organization (2015) WHO model list of essential medicines—19th list. WHO, Geneva

Zanolari B, Guilet D, Marston A, Queiroz EF, Paulo Mde Q, Hostettmann K (2003) Tropane alkaloids from the bark of *Erythroxylum vacciniifolium*. J Nat Prod 66(4):497–502. https://doi.org/10.1021/np020512m

Zayed R, Wink M (2004) Induction of tropane alkaloid formation in transformed root cultures of *Brugmansia suaveolens* (Solanaceae). Z Naturforsch 59c:863D867

Zhang F, Qiu F, Zeng J et al (2023) Revealing evolution of tropane alkaloid biosynthesis by analyzing two genomes in the Solanaceae family. Nat Commun 14:1446. https://doi.org/10.1038/s41467-023-37133-4

Chapter 4
Biosynthesis of Tropane Alkaloids

Lepakshi Md. Bhakshu and T. Pullaiah

Abstract The tropane alkaloids have shown a highly diversified class in the alkaloids, which possess complexity in their structure and pronounced biological effects. The main objective of the present review is to disclose and elaborate on the mechanisms of biosynthesis in the plant system and its storage localization. The biosynthesis mechanism provides a route to engage scientists to design and synthesize nontoxic and highly efficient medicines to cope with the demand. The substrates for biosynthesis are ornithine, phenylalanine arginine, etc., and various tropane alkaloids derived from the presence of specific enzymes. The common biosynthetic path and certain specific tropane alkaloids have been discussed.

Keywords Biosynthesis · Tropane alkaloids · Solanaceae · Erythroxylaceae · Cocaine · Calystegines · Atropine · Scopolamine · Hyoscyamine · Anticholinergic

4.1 Introduction

All plants that manufacture tropane alkaloids follow similar metabolic routes, which begin with putrescine, made from the polyamine pool that contains either arginine or ornithine (Hashimoto and Yamada 1994). The enzyme that transforms putrescine into N-methylputrescine, the first stage in the production of alkaloids, is called putrescine N-methyltransferase (PMT) (EC 2.1.1.53) (Hashimoto and Yamada 1994). Many diverse pathways are displayed by various species from this point on, including those that lead to the production of spermine, spermidine, and other alkaloids such as nicotine and cocaine (Kohnen-Johannsen and Kayser 2019). An essential enzyme in the biosynthesis of alkaloids, tropinone reductase I (TRI) (EC

L. Md. Bhakshu (✉)
Dr. YSR Government Degree College, Vedurukuppam, Andhra Pradesh, India

T. Pullaiah
Department of Botany, Sri Krishnadevaraya University, Anantapur, Andhra Pradesh, India

T. Pullaiah et al. (eds.), *Tropane Alkaloids*,
https://doi.org/10.1007/978-981-96-5805-3_4

1.1.1.206), is in charge of converting tropinone to tropine when the N-methyl-Δ1-pyrrolinium cation is spontaneously produced (Drager 2006).

Littorine is created when the enzyme littorine synthase combines with tropine and phenyllactyl-glucose, which is produced from phenylalanine (Qiu et al. 2020). Hyoscyamine is produced by reactions with many recently identified enzymes (Huang et al. 2021; Li et al. 2006). The final significant change is the intramolecular epoxide that H6H produces (EC 1.14.11.11) (Hashimoto and Yamada 1986). Figure 4.1 provides a general summary of TA biosynthesis. Jirschitzka et al. (2012) have conducted a detailed investigation into the biosynthetic and evolutionary characteristics of the Solanaceae and Erythroxylaceae, and they have also explained the evolutionary trend that is presently being reviewed.

Kramer et al. (2023) examined the biosynthetic route for polymorphic loci in *Hyoscyamus niger* accessions from Russia, Eastern Europe, and the United States. Annotated and sequenced was the new allele. The F2 population of a monohybrid cross and the leaf and root tissues of the accessions were analyzed for tropane alkaloid profiles. Improved TA production in *H. niger* and related species was discussed after a chemotype × genotype investigation.

To identify elite groups and individuals with higher TA concentrations and high scopolamine-to-hyoscyamine ratios, Kramer et al. (2023) set out to measure the tropane alkaloid concentrations and ratios of the available germplasm resources of *H. niger* from North America, Eastern Europe, and Russia. Three loci involved in TA biosynthesis were evaluated to identify gene polymorphisms: the branch pointing to the tropane alkaloids hyoscyamine and scopolamine (TRI); the final conversion from hyoscyamine to scopolamine (H6H); and the first committed step to alkaloid biosynthesis (PMT) (Fig. 4.1). Additionally, the relationship between the inheritance behavior in the segregating population and the existence of allelic forms with tropane alkaloid concentrations and ratios was evaluated.

4.2 Synthesis of Hyoscyamine (Figs. 4.1 and 4.2)

The main alkaloid of solanaceous plants, hyoscyamine, is the ester of tropine with levorotatory {l(−)} or (S)-tropic acid. It is thought to be produced biogenetically by skeletal rearrangement of phenyl lactic acid, which is obtained from L-phenylalanine. Recently, there has been debate about the 1,2-carboxy group change, which is commonly acknowledged by contemporary treatises on alkaloid biosynthesis, in the metabolic pathway from tropine phenyl lactic ester, littorine, to tropic acid ester, and hyoscyamine (Fig. 4.2) (Patterson and O'Hagan 2002).

6-Hydroxytropine, also called anisodamine, is produced by the enzyme hyoscyamine 6-hydroxylase, also known as 6-(R)hydroxylase (H6H). It is a precursor of scopine, another alkamine that is a basic component of hyoscine and has an extra epoxide ring. Benzoic, cinnamic, tiglic, truxillic, isovaleric, methylbutyric, and other acids can be esterified at position C-3 with minor tropane alkaloids, which

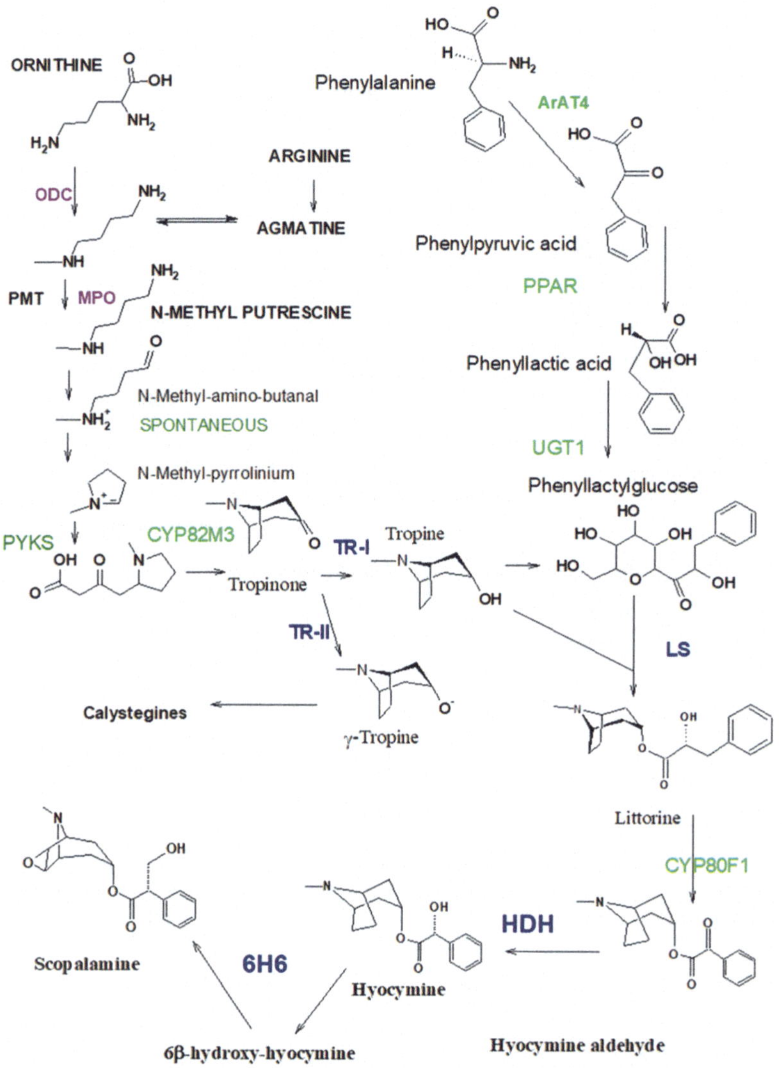

Fig. 4.1 The production of tropane alkaloids in *Hyoscyamus niger*. *ADC* Arginine decarboxylase, *ArAT4* aromatic amino acid aminotransferase, *CYP80F1* littorine mutase, *CYP82M3* tropinone synthase, *H6H* hyoscyamine 6β-hydroxylase, *HDH* hyoscyamine dehydrogenase, *LS* littorine synthase, *MPO* N-oxidase, *ODC* ornithine decarboxylase, *PPT* putrescine N-methyltransferase, *PPAR* phenylpyruvic acid reductase, *PYKS* pyrrolidine ketide synthase, *TRI* tropinone reductase I and II, *UGT1* phenyllactate UDP-glycosyltransferase. The enzymes' gene sequences are bolded. (Source: Redrawn from Kramer et al. 2023)

comprise tropine, nortropine, or hydroxylated tropines as an alkamine component. Except for anisodamine (3-tropoyl-6-hydroxytropine), none of these secondary metabolites are currently being developed as medicines (Poupko et al. 2006).

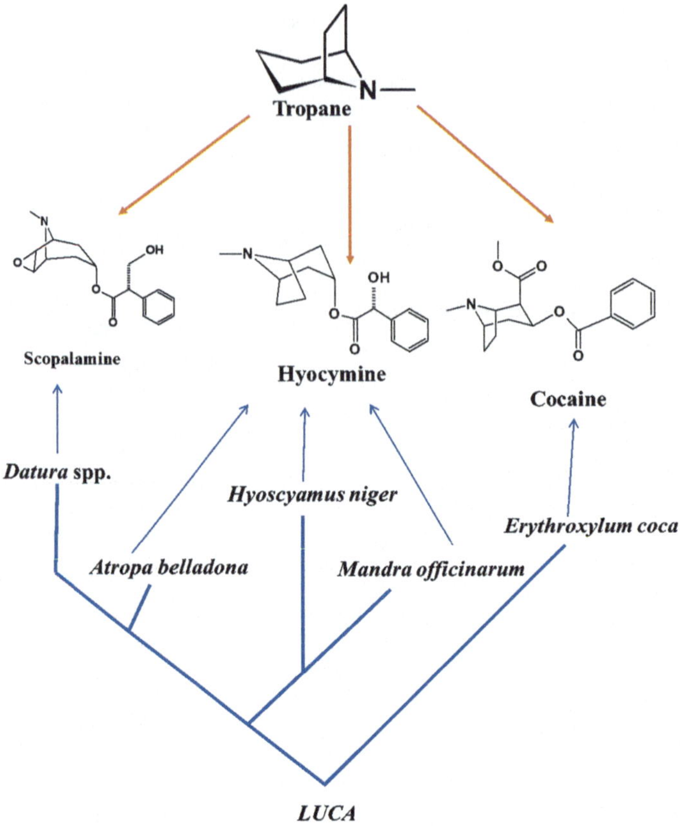

Fig. 4.2 Tropane alkaloids' phylogeny and chemical structure showed how they were synthesized and found in five different plant species. (Redrawn Source: https://commons.wikimedia.org/w/index.php?curid=115787548, by Mplanine, their own work, CC BY-SA 4.0)

4.3 Early Steps in TA Biosynthesis: A United Way

The precursor biosynthesis pathway for the various classes of tropane alkaloids (TAs), including cocaine, scopolamine/hyoscyamine, and calystegines, is similar (Fig. 4.3), starting with the amino acids L-ornithine (Orn) and L-arginine (Arg). In plants, glutamate (Glu), an amino acid that is directly related to the absorption of nitrogen, is used to generate Orn and Arg. Through the glutamine synthetase-glutamate synthase (GS-GOGAT) pathway, ammonia—which can be produced from nitrate or absorbed from the soil—is integrated into Glu. In a number of polyamines (PA) routes, glu serves as the precursor. Because of their pleiotropic activities, the regulation of PAs is extremely complicated and not entirely understood (Agostinelli 2014). Additionally, the concentration of PAs in plants might serve as an indication of different types of abiotic stress.

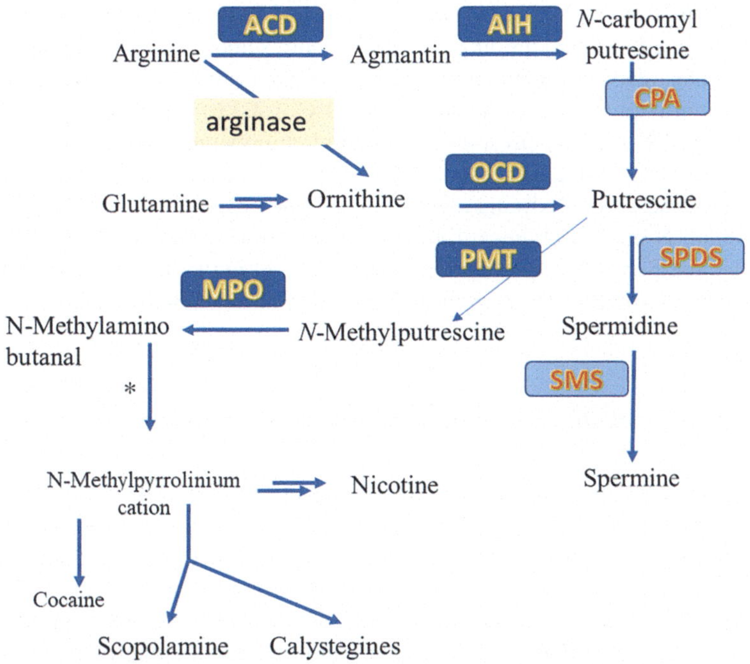

Fig. 4.3 ACD stands for arginine decarboxylase, AIH for agmatine deiminase, OCD for ornithine decarboxylase, CPA for N-carbamoylputrescine amidase, PMT for putrescine N-methyltransferase, SPDS for spermidine synthase, SMS for spermine synthase, and MPO for N-methylputrescine oxidase. The joint biosynthesis is shown with * denoting spontaneous cyclization. Nicotinic acid, or more specifically its reactive derivative 2,5-dihydropyrindine, is produced by condensation with the N-methyl-D1-pyrrolinium cation. (Kohnen-Johannsen and Kayser 2019; redrawn)

The enzymes ADC (arginine decarboxylase; EC 4.1.1.19), AIH (agmatine deiminase; EC 3.5.3.12), and CPA (N-carbamoylputrescine amidase; EC 3.5.1.53) catalyze the three-step reaction that forms putrescine (1,4-diaminobutane) from the amino acids Orn and Arg. Orn is decarboxylated by the enzyme ODC (ornithine decarboxylase; EC 4.1.1.17), while Arg goes through a three-step reaction that includes decarboxylation, hydrolysis of the imine functionality of guanidine, and hydrolysis of urea. To investigate the nature of these two pathways to putrescine production, the activities of ADC and ODC were inhibited in *Datura* plants using the particular irreversible inhibitors DL-α-difluoromethyl arginine and DL-α-difluoromethylornithine, respectively. These tests showed that the ADC was more active than the ODC and that the two pathways do not function independently of one another (Robins et al. 1991a, b).

Tetramethylenediamine, or putrescine, is an intermediary in a number of metabolic processes (Fig. 4.3). A spermidine synthase (SPDS; EC. 2.5.1.16) catalyzed reaction with putrescine and S-adenosyl methioninamine (decarboxylated

S-adenosyl methionine) as substrates can convert it to spermidine. Using SAM (S-adenosyl methionine), the enzyme PMT (putrescine N-methyltransferase; EC 2.1.1.53) may also methylate putrescine to N-methylputrescine (Biastoff et al. 2009). N-methylputrescine oxidase (MPO; EC 1.4.3.6) catalyzes the oxidative deamination of N-methylputrescine to 4-methylaminobutanal, the subsequent stage in TA biosynthesis (Mizusaki et al. 1972). Copper is necessary as a cofactor for this diamine oxidase. N-methylputrescine spontaneously cyclizes to generate a central intermediate, N-methyl-D1-pyrrolinium. This reaction is known chemically as an intramolecular Schiff base production. A branchpoint in the biosynthesis of TA and nicotine is the N-methyl-D1-pyrrolinium cation (Courdavault 2010).

4.4 Hyoscyamine and Scopolamine Biosynthesis

The subsequent stages of the scopolamine biosynthesis (Fig. 4.4), which starts with N-methyl-D1-pyrrolinium, were long unclear. The production of the intermediate 4-(1-methyl-2-pyrrolidinyl)-3-oxobutanoic acid is catalyzed by an unusual polyketide synthase from *Atropa belladonna*, according to Bedewitz et al. (2018). A cytochrome P450 enzyme called AbCYP82M3 catalyzes the malonyl-coenzyme A-mediated decarboxylative condensation that converts this intermediate to tropinone. Two stereospecific reductases, tropinone reductase I (TR-I; EC 1.1.1.206) and tropinone reductase II (TR-II; EC 1.1.1.236), use tropinone as a substrate (Hashimoto et al. 1992). Tropinone is reduced to pseudotropine (3β-tropanol) by TR-II, whereas its reduction to tropine (3α-tropanol) is catalyzed by TR-I. Tropine is utilized to make scopolamine, whereas pseudotropine is the precursor of calystegine biosynthesis. The third ring intermediate is delivered to littorine by the condensation of tropine with activated (R)-phenyllactate (phenyllactyl-CoA). Phenylalanine, an intermediate in the shikimate pathway, is converted to phenylpyruvate to produce phenyllactate. Bedewitz et al. (2014) identified the coding sequence of a unique aromatic amino acid aminotransferase (ArAT) that co-expresses with genes involved in tropane alkaloid production. By lowering phenyllactate levels, ArAT4 silencing in *A. belladonna* interfered with the production of scopolamine. A newly identified phenylpyruvic acid reductase (AbPPAR) catalyzes the subsequent step, reducing of ketone function.

This reductase was found in the endodermis and the root pericycle, and it also showed cell-specific expression (Qiu et al. 2018). An enzyme associated with the cocaine synthase may be involved in the production of littorine, even though no enzymatic activity has been reported (Robins et al. 1991a, b). The mutase/monooxygenase (CYP80F1; EC 1.6.2.4) reorganizes littorine to hyoscyamine aldehyde, which is then reduced to the equivalent alcohol hyoscyamine (Li et al. 2006). The enzyme H6H (hyoscyamine 6 β-hydroxylase; EC 1.14.11.11) is responsible for the conversion of hyoscyamine. According to Hashimoto and Yamada (1985), The

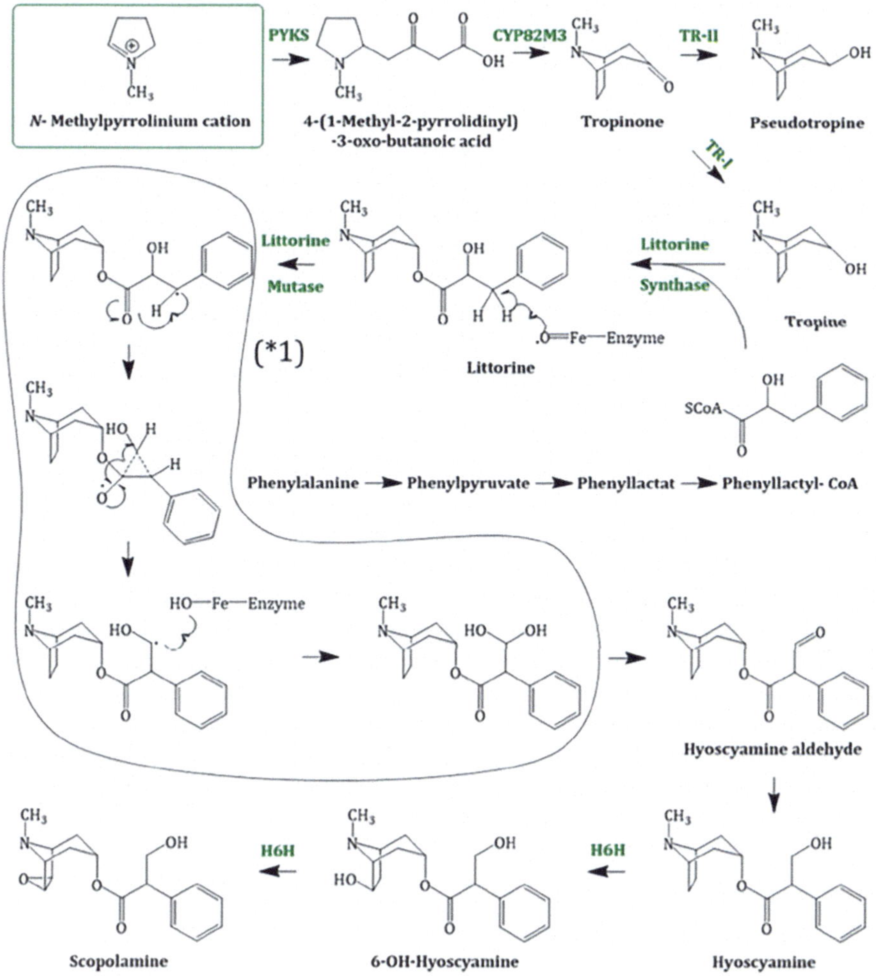

Fig. 4.4 Scopolamine biosynthesis. **1* suggested mechanism of littorine rearrangement, *H6H* hyoscyamine 6β-hydroxylase, *PYKS* polyketide synthase, *CYP82M3* cytochrome P450 enzyme, *TR-I/II* tropinone reductase I/II; littorine synthase (sequence unknown); littorine mutase/monooxygenase (CYP80F1). (Source: Kohnen-Johannsen and Kayser 2019)

dioxygenase H6H is reliant on 2-oxoglutarate. It catalyzes the hydroxylation of hyoscyamine to 6-hydroxy hyoscyamine and the epoxidation of 6-hydroxy hyoscyamine to scopolamine. Compared to the rate-limiting epoxidase activity, the bifunctional dioxygenase shows a significant hydroxylase activity (Hashimoto et al. 1993).

4.5 Localization and Organization of Scopolamine Biosynthesis in Plants

The structure and location distribution of TA biosynthesis are varied and intricate. TA production occurs in the roots of Solanaceae plants, after which the alkaloids are carried to the aerial sections for storage. It is believed that the TAs are transported through the xylem; however, there is little information known on the transit or the transport form. In the root tissue pericycle, where expression of the genes pmt. in *A. belladonna* (Suzuki et al. 1999) and h6h in *Hyoscyamus niger* (Kanegae et al. 1994) were found, cell-specific compartmentalization of scopolamine production was previously noted. However, in *H. niger*, the enzyme TR-I is found in the endodermis and adjacent cortical cells (Hashimoto et al. 1992; Nakajima and Hashimoto 1999). In potatoes, the cortex and phloem parenchyma of roots and stolons were shown to contain TR-II, a protein that supplies pseudotropine for calystegine production; in tuber spouts, the protein was found in companion cells. Although its role in potatoes is still unknown, TR-I was also found in tuber tissue protein extracts, but in amounts too small to allow localization to individual cells (Kaiser et al. 2006). Additionally, the initial stage of nicotine production (described above) is catalyzed by the enzyme PMT. PMT is expressed in the endodermis, outer cortex, and root xylem of *Nicotiana sylvestris*, a plant that produces nicotine (Shoji et al. 2000).

Plant biosynthesis compartmentalization might make future attempts at heterologous production in single-celled microbial systems (described below) more challenging. Since eukaryotic cells, such as yeasts or microalgae, have been utilized for the successful metabolic engineering of complex compounds and display organelle compartmentalization, it is possible that these cells would make good host organisms for their production (Lauersen 2018).

4.6 Cocaine Biosynthesis

Beyond its branch point with common intermediates shared with other TAs, cocaine biosynthesis (Fig. 4.5) is still being studied and is not entirely understood. Two distinct pathways for the production of cocaine have been proposed in the literature. The standard hypothesis states that the bridgehead carbon atoms C-1 and C-2 of methylecgonine come from acetoacetate and N-methyl-D1-pyrrolinium cations, respectively. However, this idea could not be confirmed by feeding tests using the labeled N-methyl-D1-pyrrolinium cation, which were inconclusive in planta. Therefore, it was proposed that the gradual addition of C2 units to the ecgonine skeleton, which is produced from acetate, was compatible with the observed regiochemistry of incorporation of the labeled N-methyl-D1-pyrrolinium cation into cocaine (Leete et al. 1990).

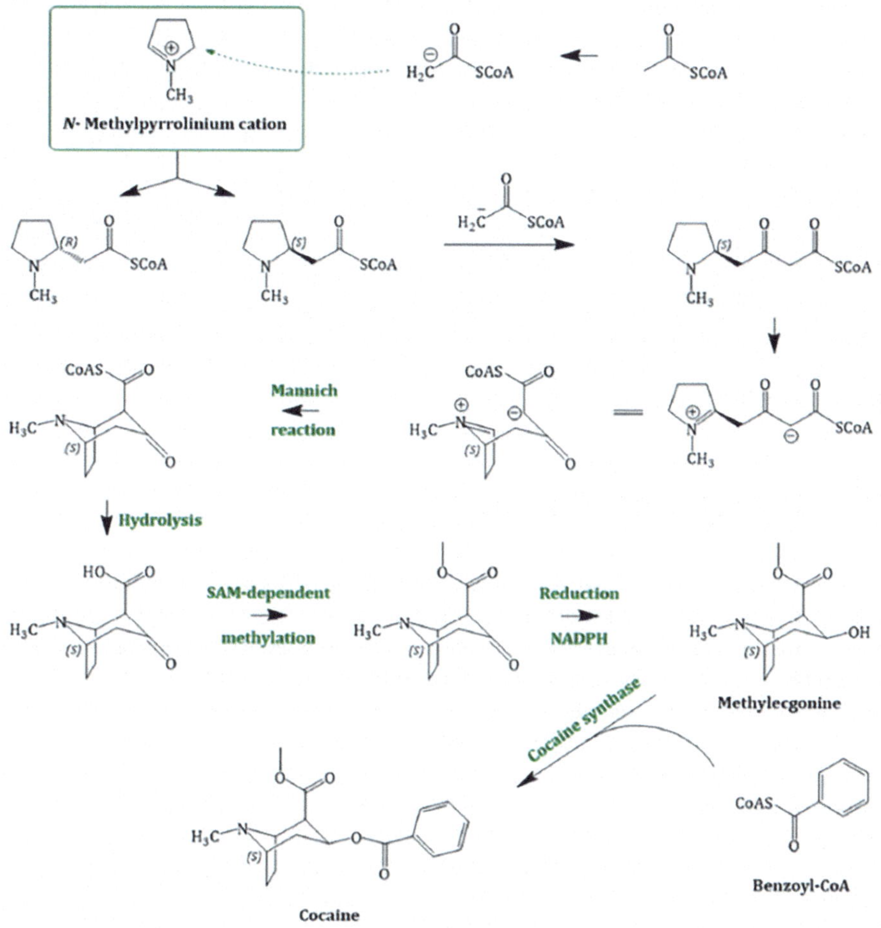

Fig. 4.5 Cocaine biosynthesis: Less is known about the enzymatic role in the production of cocaine, which begins with the intermediate N-methyl-Δ1-pyrrolinium cation. In an intramolecular Mannich reaction, the iminium cation combines with two acetyl-CoA moieties to form an intermediate that cyclizes. Methyl-ecgonine is produced via hydrolysis, methylation, and reduction. The final stage in the route, the condensation of methylecgonine with benzoyl-CoA, is catalyzed by the cocaine synthase. (Source: Kohnen-Johannsen and Kayser 2019)

As a result, this theory suggests N-methyl-2-pyrrolidineacetic acid as a novel biosynthetic intermediary for cocaine. Despite being found in a number of plants, this chemical has not yet been successfully incorporated into any ester or thioester forms (Hemscheidt and Spenser 1992). Chemically, the initial acetyl-CoA moiety reaction is thought to be a Mannich-like reaction using the enolate anion; Claisen condensation is responsible for the side-chain extension (Dewick 2002). Through an intramolecular Mannich reaction, the (S)-enantiomer cyclizes and creates the bicyclic structure of the cocaine tropane ring skeleton (Hoye et al. 2000). After the CoA-ester is hydrolyzed, methylecgonine (2-carbomethoxy-3β-tropine) is produced via SAM-dependent methylation and reduction. Benzoyl-CoA, which is generated from L-phenylalanine (Leete et al. 1998), condenses with methylecgonine to produce cocaine.

Schmidt et al. (2015) identified an enzyme they called the cocaine synthase that catalyzes this process. This synthase, which is a member of the BAHD family (D'Auria 2006), catalyzes the transfer of CoA-activated acyl thioesters to acceptor molecules containing oxygen or nitrogen. The two primary structural differences between cocaine and scopolamine are the preservation of acetoacetate's C-1 carboxyl group, which is later methylated, and the distinct stereochemistry of the C-3 hydroxy groups.

4.7 Calystegine Biosynthesis

Similar to the cocaine and hyoscyamine/scopolamine pathways, the specific mechanisms behind calystegine production remain unknown. It is believed that tropinone is involved, and TR-II catalyzes the reduction of tropinone to pseudotropine. There is currently no more information known about the biosynthesis (Scholl et al. 2001). There have been no documented attempts to clarify the hydroxylation or demethylation of these molecules so far. This could be because calystegines were only discovered in 1990 (Drager 2004), and there is less interest in using them for medicinal purposes than there is for scopolamine or cocaine. The calystegine skeleton (8-azabicyclo [3.2.1] octane) is polyhydroxylated rather than N-methylated like cocaine and scopolamine are. Depending on how many hydroxy groups they contain, these chemicals are divided into three subgroups (A), four subgroups (B), and five subgroups (C) (Goldmann et al. 1996).

4.8 L-Ornithine-Based Biosynthesis of Tropane Alkaloids

It is easy to observe from the chemical structure of the tropane alkaloids that all of the molecules have the tropanic core, and a potential shared biosynthesis pathway has been suggested, beginning with the amino acid L-Ornithine. This chemical undergoes a number of metabolic processes before being transformed into hygrine, which is the source of α-tropine, β-tropine, ecgonine, and cuscohygrine. These hygrines serve as the building blocks for producing all known tropane alkaloids (Fig. 4.6; De Simone et al. 2008).

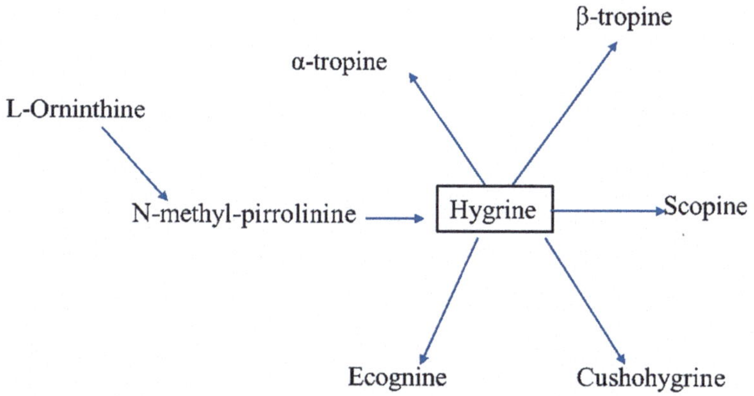

Fig. 4.6 Biosynthesis of tropane alkaloids from L-Ornithine. (Source: De Simone et al. 2008; redrawn)

4.9 Conclusion

Research on tropane alkaloids' production is crucial for comprehending the many metabolic mechanisms that might be helpful in controlling the pathway. By redesigning its metabolic intermediates and specific enzymes involved in its creation, the biosynthesis technique aids in the removal of harmful alkaloids or their accessory groups. Through enhanced biotechnological procedures, it is also feasible to regulate and manage the synthesis of modified cells. It may be possible to produce the necessary amounts of tropane alkaloids by feeding the cells the precursor chemicals. The review aids in comprehending biosynthesis and target molecules for long-term use in various pharmaceutical applications while optimizing yields.

References

Agostinelli E (2014) Polyamines and transglutaminases: biological, clinical and biotechnological perspectives. Amino Acids 46:475–485

Bedewitz MA, Gongora-Castillo E, Uebler JB et al (2014) A root-expressed L-phenylalanine:4 hydroxyphenylpyruvate aminotransferase is required for tropane alkaloid biosynthesis in *Atropa belladonna*. Plant Cell 26:3745–3762

Bedewitz MA, Jones AD, D'Auria JC, Barry CS (2018) Tropinone synthesis via an atypical polyketide synthase and P450-mediated cyclization. Nat Commun 9:5281

Biastoff S, Brandt W, Drager B (2009) Putrescine N-methyltransferase – the start for alkaloids. Phytochemistry 70:1708–1718

Courdavault V (2010) Biosynthesis and regulation of alkaloids. In: Plant developmental biology, vol 655. Humana Press, Totowa, pp 139–160

D'Auria JC (2006) Acyltransferases in plants: a good time to be BAHD. Curr Opin Plant Biol 9:331–340

De Simone R, Margarucci L, De Feo V (2008) Tropane alkaloids: an overview. Pharmacologyonline 1:70–89

Dewick PM (2002) Medicinal natural products: a biosynthetic approach (2nd ed.), John Wiley & Sons, Ltd, ISBN: 0471496405 (Hardback), p. 514

Drager B (2004) Chemistry and biology of calystegines. Nat Prod Rep 21:211–223. https://doi.org/10.1039/B300289F

Drager B (2006) Tropinone reductases, enzymes at the branch point of tropane alkaloid metabolism. Phytochemistry 67:327–337. https://doi.org/10.1016/j.phytochem.2005.12.001

Goldmann A, Message B, Tepfer D, Molyneux RJ, Duclos O (1996) Biological activities of the nortropane alkaloid, calystegine B 2 and analogs: structure—function relationships. J Nat Prod 59:1137–1142

Hashimoto T, Yamada Y (1985) Hyoscyamine 6b-hydroxylase, a 2-oxoglutarate-dependent dioxygenase, in alkaloid producing root cultures. Plant Physiol 81:619–625

Hashimoto T, Yamada Y (1986) Hyoscyamine 6β-hydroxylase, a 2-oxoglutarate-dependent dioxygenase, in alkaloid-producing root cultures. Plant Physiol 81:619–625. https://doi.org/10.1104/pp.81.2.619

Hashimoto T, Yamada Y (1994) Alkaloid biogenesis: molecular aspects. Annu Rev Plant Physiol Plant Mol Biol 45:257–285. https://doi.org/10.1146/annurev.pp.45.060194.001353

Hashimoto T, Nakajima K, Ongena G, Yasuyuki Y (1992) Two tropinone reductases with distinct stereospecificities from cultured roots of *Hyoscyamus niger*. Plant Physiol 100:836–845

Hashimoto T, Matsuda J, Yamada Y (1993) Two-step epoxidation of hyoscyamine to scopolamine is catalysed by bifunctional hyoscyamine-6 β-hydroxylase. FEBS Lett 329:35–39

Hemscheidt T, Spenser ID (1992) Biosynthesis of 6β-hydroxytropine in *Datura stramonium*: nonregiospecific incorporation of [1,2-13C2]acetate. J Am Chem Soc 114:5472–5473

Hoye TR, Bjorklund JA, Koltun DO, Renner MK (2000) N-methylputrescine oxidation during cocaine biosynthesis: study of prochiral methylene hydrogen discrimination using the remote isotope method. Org Lett 2:3–5

Huang JP, Wang YJ, Tian T, Li W, Yan Y, Sheng-Xiong H (2021) Tropane alkaloid biosynthesis: a centennial review. Nat Prod Rep 38:1634–1658

Jirschitzka J, Schmidt GW, Reichelt M, Schneider B, Gershenzon J, D'Auria JC (2012) Plant tropane alkaloid biosynthesis evolved independently in the Solanaceae and Erythroxylaceae. Proc Natl Acad Sci 109(26):10304–10309. https://doi.org/10.1073/pnas.1200473109

Kaiser H, Richter U, Keiner R, Brabant A, Hause B, Draeger B (2006) Immuno-localisation of two tropinone reductases in potato (*Solanum tuberosum* L.) root, stolon and tuber sprouts. Planta 225:127–137

Kanegae T, Kajiya H, Amano Y, Hashimoto T, Yamada Y (1994) Species-dependent expression of the hyoscyamine 6 beta-hydroxylase gene in the pericycle. Plant Physiol 105:483–490

Kohnen-Johannsen KL, Kayser O (2019) Tropane alkaloids: chemistry, pharmacology, biosynthesis and production. Molecules 24:796. https://doi.org/10.3390/molecules24040796

Kramer L, Jayanty S, Reckhow DA, Sathuvalli V (2023) Identification of tropane alkaloid chemotypes and genotypes in *Hyoscyamus Niger* L. J Am Soc Hortic Sci 148(6):304–314. https://doi.org/10.21273/JASHS05309-23

Lauersen KJ (2018) Eukaryotic microalgae as hosts for light-driven heterologous isoprenoid production. Planta 249:155–180

Leete E, Endo T, Yamada Y (1990) Biosynthesis of nicotine and scopolamine in a root culture of *Duboisia leichhardtii*. Phytochemistry 29:1847–1851

Leete E, Bjorklund JA, Sung HK (1998) The biosynthesis of the benzoyl moiety of cocaine. Phytochemistry 27:2553–2556

Li R, Reed DW, Liu E, Nowak J, Pelcher LE, Page JE, Covello PS (2006) Functional genomic analysis of alkaloid biosynthesis in *Hyoscyamus Niger* reveals a cytochrome P450 involved in littorine rearrangement. Chem Biol 13:513–520

Mizusaki S, Tanabe Y, Noguchi M, Tamaki E (1972) N-methylputrescine oxidase from tobacco roots. Phytochemistry 11:2757–2762

Nakajima K, Hashimoto T (1999) Two tropinone reductases, that catalyze opposite stereospecific reductions in tropane alkaloid biosynthesis, are localized in plant root with different cell-specific patterns. Plant Cell Physiol 40:1099–1107

Patterson S, O'Hagan D (2002) Biosynthetic studies on the tropane alkaloid hyoscyamine in *Datura stramonium*; hyoscyamine is stable to in vivo oxidation and is not derived from littorine via a vicinal interchange process. Phytochemistry 61:323–329. https://doi.org/10.1016/S0031-9422(02)00200-5

Poupko JM, Baskin SI, Moore E (2006) The pharmacological properties of anisodamine. J Appl Toxicol 27:116–121

Qiu F, Yang C, Yuan L, Xiang D, Lan X, Chen M, Liao Z (2018) A phenylpyruvic acid reductase is required for biosynthesis of tropane alkaloids. Org Lett 20:7807–7810

Qiu F, Zeng J, Wang J, Huang J-P, Zhou W, Yang C, Lan X, Chen M, Huang S-X, Kai G, Liao Z (2020) Functional genomics analysis reveals two novel genes required for littorine biosynthesis. New Phytol 225:1906–1914 https://doi.org/10.1111/nph.16317

Robins RJ, Bachmann P, Robinson T, Rhodes MJ, Yamada Y (1991a) The formation of 3 alpha- and 3 beta acetoxytropanes by *Datura stramonium* transformed root cultures involves two acetyl-CoA-dependent acyltransferases. FEBS Lett 292:293–297

Robins RJ, Parr AJ, Walton NJ (1991b) Studies on the biosynthesis of tropane alkaloids in *Datura stramonium* L. transformed root cultures. Planta 183:196–201

Schmidt GW, Jirschitzka J, Porta T et al (2015) The last step in cocaine biosynthesis is catalyzed by a BAHD acyltransferase. Plant Physiol 167:89–101

Scholl Y, Höke D, Drager B (2001) Calystegines in *Calystegia sepium* derive from the tropane alkaloid pathway. Phytochemistry 58:883–889

Shoji T, Yamada Y, Hashimoto T (2000) Jasmonate induction of putrescine N-methyltransferase genes in the root of *Nicotiana sylvestris*. Plant Cell Physiol 41:831–839

Suzuki K, Yamada Y, Hashimoto T (1999) Expression of *Atropa belladonna* putrescine N-methyltransferase gene in root pericycle. Plant Cell Physiol 40:289–297

Chapter 5
Pharmacology of Tropane Alkaloids

Aziza Mahrous Amer and Mohamed M. Amer

Abstract Tropane alkaloids are a class of naturally occurring compounds primarily derived from plants in the Solanaceae family, such as *Atropa belladonna* (deadly nightshade), *Hyoscyamus niger* (henbane), and *Datura* species. These plants contain high concentrations of tropane alkaloids, synthesized from the amino acid ornithine. Tropane alkaloids primarily act as anticholinergic agents, blocking the action of acetylcholine at muscarinic receptors in the nervous system. This inhibition decreases parasympathetic nervous system activity, leading to increased heart rate, muscle relaxation, and reduced secretions. It is a competitive antagonist of the muscarinic acetylcholine receptors, used as a medication to treat various conditions. Derivatives of tropane alkaloids include atropine, used for treating bradycardia and as an antidote for organophosphate poisoning, and scopolamine, effective for motion sickness and as a sedative and homatropine, used in ophthalmology for pupil dilation. Tropane alkaloids have various medical applications including ophthalmology, pupil dilation for examinations and treatment of eye conditions; cardiology, management of bradycardia and cardiac arrest; anesthesia, reducing salivation and respiratory secretions during surgery; gastroenterology, alleviating gastrointestinal spasms and hypermotility; antidote, effective against poisoning from organophosphate pesticides and nerve agents; and neurology, treatment for certain movement disorders and muscle rigidity. Beyond medical contexts, atropine is sometimes used in veterinary medicine and research. Its effects on the nervous system make it valuable for studying neurotransmitter pathways and developing other pharmacological agents. Overall, atropine remains a critical drug in both emergency and routine medical practice due to its diverse therapeutic applications.

Keywords Tropane alkaloids · Mode of action · Atropine · Scopolamine · Pharmacology

A. M. Amer (✉) · M. M. Amer
Faculty of Veterinary Medicine, Cairo University, Giza, Egypt
e-mail: aziza_amer@cu.edu.eg; profdramer@cu.edu.eg

T. Pullaiah et al. (eds.), *Tropane Alkaloids*,
https://doi.org/10.1007/978-981-96-5805-3_5

Abbreviations

ADME	Pharmacokinetics investigates the absorption, distribution, metabolism, and excretion
Arg	L-arginine
ASCs	Adipose-derived stromal stem cells
AUC	Area under the curve and
BC	Before Christ
CL	Clearance
Cmax	Peak plasma concentration = maximum concentration
CNS	Central nervous system
COPD	Chronic obstructive pulmonary disease
EEG	Electroencephalogram
ER	Endoplasmic reticulum
ERCP	Endoscopic retrograde cholangiopancreatography
FDA	Food and Drug Administration
GC-MS	Gas chromatography-mass spectrometry
GPCRs	G-protein-coupled receptors
HSV	Herpes simplex virus
HT	Hydroxytryptamine
JEV	Japanese encephalitis virus
LD_{50}	Lethal dose for 50% deaths
mAChR	Muscarinic acetylcholine receptor
NTAs	Nortropane alkaloids
Orn	Acids L-ornithine
PD	Parkinson's disease
PNS	Peripheral nervous system
TA	Tropane alkaloid
Tmax	Time to reach maximum concentration
TTS-patch	Transdermal therapy system patch
Vd	Volume of distribution
WHO	World Health Organization

5.1 Introduction

As defined by Vallance and Smart (2006), pharmacology includes the comprehensive study of drugs and medications. This field delves into various aspects such as these substances' origin, composition, pharmacodynamics, pharmacokinetics, therapeutic applications, and toxicology. It primarily focuses on understanding the intricate interactions between chemicals and the human body, influencing normal and abnormal biological functions. Consequently, any substances possessing medicinal properties are classified as pharmaceuticals (Merriam-Webster 2023).

Pharmacology studies the composition, effects, origins, and production methods of drugs, including their design and interactions. It investigates the molecular and cellular mechanisms of drug actions within various organs and systems, covering signal transduction, molecular diagnostics, and drug therapy. The field is divided into pharmacodynamics, which examines how drugs affect living organisms, and pharmacokinetics, which looks at how organisms impact drugs—specifically their absorption, distribution, metabolism, and excretion (ADME). Pharmacology also encompasses herbalism and natural substances, with remedies derived from natural sources documented in pharmacopoeias since ancient times.

Pharmacometabolomics analyzes metabolites produced by the body during treatment (Kaddurah-Daouk et al. 2008; Kaddurah-Daouk and Weinshilboum 2014). Pharmacomicrobiomics focuses on how microbiome variations affect drug disposition, action, and toxicity (Rizkallah et al. 2010). Lastly, pharmacogenetics studies the epigenetic factors influencing individual responses to treatments (Gomez and Ingelman-Sundberg 2009).

5.2 History of Plant-Based Pharmacology

Plants have long been utilized as traditional remedies for addressing various health issues. Different components of natural plant products have been explored as potential foundations for the design, discovery, and development of new drugs. Recently, there has been a focus on identifying molecular targets based on proteins, as well as the necessity for novel chemical variations in plant-based screening and the study of their metabolites (Chaachouay and Zidane 2024). Natural products are essential for meeting this demand. As medicinal plants continue to be a significant source of new therapeutic options, challenges such as developing effective screening methods, increasing the production of bioactive compounds, and obtaining plant materials still need to be addressed (Vaou et al. 2021).

The early connection between humans and natural resources in maintaining human health and well-being was established through the history of herbal medicine. According to Reis (2001) and Chaachouay and Zidane (2024), archaeological discoveries indicate the utilization of herbal treatments dating back to ancient times, as early as the Paleolithic era approximately 60,000 years ago (Aboelsoud 2010; Sam 2019; Chaachouay and Zidane 2024).

Around 5000 years ago, an ancient Sumerian civilization left behind clay slabs containing valuable information about the usage of medicinal plants. These slabs contained 12 recipes for drug preparations, utilizing over 200 different plants (Sumner 2000). Similarly, ancient Mesopotamia, dating back to approximately 2600 BC, recorded their knowledge of medicinal plants on clay tablets inscribed with cuneiform script (Scurlock 2014). The utilization of medicinal plants can be traced back to early civilizations, as exemplified by the Chinese emperor Shen Nung around 2500 BC. The properties of more than 300 medicinal herbs enrich the historical narrative of Chinese Materia medica (Zhu 1998). The use of medicinal plants as therapeutic agents has also been recorded in various cultures, including those of ancient Egypt, Mesopotamia, and Europe. Opium, recognized as one of the earliest and most potent drugs known to humanity, was employed thousands of years ago (Hassan 2015). Secondary metabolites are natural substances synthesized by plants that do not play a direct role in vital processes such as growth, development, or reproduction (Thirumurugan et al. 2018). However, these metabolites often fulfill ecological functions, including defense against herbivores, attracting pollinators, or competing with other plant species (Pagare et al. 2015; Chaachouay and Zidane 2023).

For numerous years, these compounds have been utilized in traditional medicine for their potent analgesic, antimalarial, and stimulant properties (Ravina 2011). Traditional pharmacopoeias, which are rooted in the knowledge and practices of Indigenous communities, primarily rely on plant-based substances, particularly those abundant in secondary metabolites (Chaachouay et al. 2023). Herbal remedies and formulations based on traditional knowledge have been transmitted through generations and continue to play a significant role in healthcare practices across various societies (Alamgir 2018; Ahmed et al. 2024; Chaachouay and Zidane 2024).

Various secondary plant metabolites have been utilized for their medicinal properties and are regarded as crucial components of traditional pharmacopoeias. Alkaloids, which are nitrogen-containing compounds, are frequently present in various plant species and exhibit numerous pharmacological effects. For example, morphine is derived from the opium poppy, quinine is obtained from the cinchona tree, and caffeine is extracted from coffee beans (Chaachouay and Zidane 2023, 2024).

Tropane alkaloids, including atropine, cocaine, and scopolamine, are nitrogenous organic compounds that are commonly found in plants belonging to the Solanaceae family, such as *Atropa* and *Scopolia*. These alkaloids have been used as medicinal substances for centuries and are among the oldest known remedies. They exhibit a wide range of pharmacological effects on the central nervous, respiratory, and gastrointestinal systems. Additionally, they can act as antidotes against poisoning (Heinrich et al. 2021). Tropane alkaloids also possess anticholinergic properties (Shim et al. 2022). This chapter aims to gather information on the properties of tropane alkaloids from various reliable sources (Sweta and Lakshmi 2015).

5.3 History of Tropane Alkaloids

Herbal sources have acquired a variety of natural medicinal products derived from plants. Alkaloids represent a class of nitrogen-containing metabolites found in numerous plant species, known for their beneficial pharmacological and physiological effects. These products, such as atropine, cocaine, ephedrine, colchicine, caffeine, digitoxin, morphine, quinine, scopolamine, theobromine, and Taxol, are now primarily produced through economic synthesis methods (Chaachouay and Zidane 2024).

Atropine derivatives are derived from a naturally occurring metabolite. The primary metabolic process of atropine occurs through enzymatic hydrolysis in the liver. Key metabolites from this process include noratropine, atropin-n-oxide, tropine, and tropic acid. Organophosphate pesticides can hinder the metabolism of atropine. It is estimated that between 13% and 50% of atropine is excreted in its unchanged form through urine (Van der Meer et al. 1986). Compared to other bioactive compounds sourced from plants, synthetic forms of atropine are widely employed in approved pharmaceuticals (Grynkiewicz and Gadzikowska 2008; Chaachouay and Zidane 2024).

Atropine and scopolamine, which are present in plants belonging to the Solanaceae family, have been historically utilized in ancient medicinal systems for the treatment of mental disorders, skin ailments, and tumors (Wink 1998; Shim et al. 2022). The first tropane alkaloid to be isolated from *Atropa belladonna* leaves and roots was atropine, accomplished by Mein in 1832; however, the results were not published. A year later, Geiger and Hesse published the data on atropine isolation from *Atropa belladonna* and *Hyoscyamus niger*. Scopolamine, on the other hand, was isolated from *Scopolia japonica* 5 years after the isolation of atropine (Shim et al. 2022). Cocaine, which was isolated by Gaedcke in 1855 from *Erythroxylon coca* leaves, is renowned for its local anesthetic effects and its significant role in the treatment of postnatal depression and morphine overdose (Shim et al. 2022).

5.4 Types of Tropane Alkaloids

There exist two primary categories of tropane alkaloids (TAs), namely, the esters of 3-tropanole (tropine) and 3-tropanole (pseudotropine), with the former being the more prevalent variant. TAs can be classified into three distinct groups: those derived from Solanaceae plants, such as hyoscyamine and scopolamine; coca alkaloids, including cocaine sourced from *Erythroxylum coca*; and the newly identified calystegines group, which consists of polyhydroxylated nortropane alkaloids (NTAs) predominantly found in the families Convolvulaceae, Solanaceae, Moraceae, Erythroxylaceae, and Brassicaceae (Kohnen-Johannsen and Kayser 2019). The tropane skeleton, illustrated in Fig. 5.1, constitutes the fundamental structural framework for TAs, which can be derived from these three principal categories. To date, approximately 200 distinct TAs have been recognized (Gadzikowska and Grynkiewicz 2002; Kohnen-Johannsen and Kayser 2019).

5.5 Tropane Alkaloids and Pandemics COVID-19

The identification of key interactions and binding affinities between SARS-CoV-2 papain-like protease and tropane alkaloids from *Schizanthus porrigens* was accomplished through the utilization of Autodock Vina (Alfaro et al. 2020). In recent

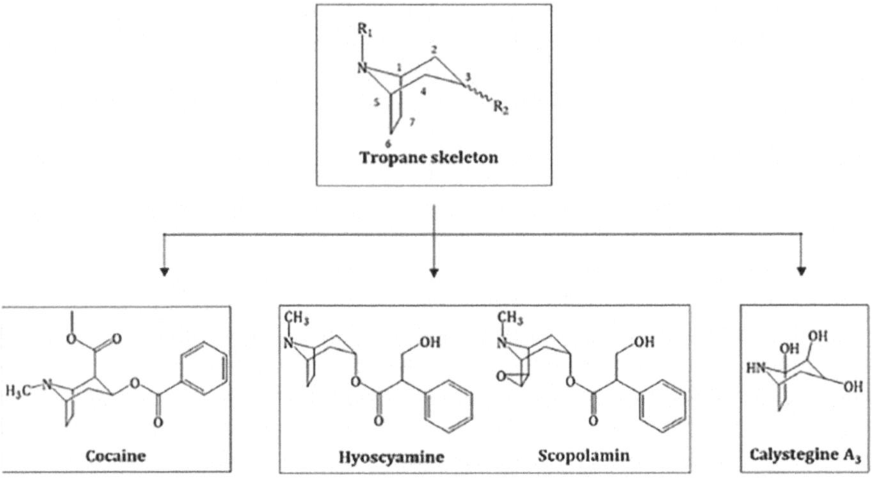

Fig. 5.1 Structure of the tropane skeleton and the three major groups of TAs derived from this skeleton. (After Kohnen-Johannsen and Kayser 2019)

times, there has been a growing interest in phytocomposites, prompting researchers to turn to computational chemistry for assistance in investigating various plant activities, particularly focusing on *Schizanthus porrigens* as a rich source of four tropane alkaloids with molecular docking and ADME properties (Kohnen-Johannsen and Kayser 2019; Alfaro et al. 2020).

5.6 Pharmacology of Tropane Alkaloids

Alkaloids are biomolecules that contain a nitrogen atom within a heterocyclic ring. They are named for their alkaline-like basic properties. Alkaloids that consist of at least a bicyclic amine with a pyrrolidine and a piperidine ring, along with a common nitrogen atom, are commonly referred to as tropane alkaloids. Tropane alkaloids are highly bioactive and have significant pharmaceutical value, which is why they are widely utilized as active ingredients in numerous drug formulations (Michelon et al. 2021).

Tropane alkaloids (Tas) comprise approximately 200 alkaloids characterized by a typical bicyclic tropane ring in their chemical makeup (Lounasmaa and Tamminen 1993; Shim et al. 2022). These alkaloids are commonly found in various plant families such as Convolvulaceae, Erythroxylaceae, Euphorbiaceae, Cruciferae, Proteaceae, Rhizophoraceae, and Solanaceae (Griffin and Lin 2000). The highest concentrations of TAs are typically found in the Solanaceae family, including plants like *Atropa belladonna*, *Datura*, *Hyoscyamus niger*, *Mandragora officinarum*, *Duboisia*, and *Scopolia lurida*, Erythroxylaceae (*Erythroxylum coca*) (Kohnen-Johannsen and Kayser 2019; Zeng et al. 2016). Common derivatives of tropane alkaloids include tropane, tropine, tropanone, scopolamine, cocaine, and pseudotropine (Sweta and Lakshmi 2015). The main bioactive compounds obtained from these plants are atropine, hyoscyamine, and hyoscine (scopolamine). These plants are considered poisonous if ingested orally and can have significant harmful effects. High doses of their extracts or compounds may lead to delirium, confusion, and severe hallucinations (Diaz 2010; Shim et al. 2022).

Tropane alkaloids, including atropine, cocaine, and scopolamine, have been used as medicinal substances for centuries and are among the oldest known remedies. They exhibit a wide range of pharmacological effects on the central nervous, respiratory, and gastrointestinal systems. Additionally, they can act as antidotes against poisoning and possess anticholinergic properties.

Despite the tropane ring structure shared by all TAs, there are notable differences in the pharmacological effects of these compounds. While cocaine and hyoscyamine/scopolamine can induce hallucination and psychoactive effects in a dose-dependent manner by crossing the blood-brain barrier, calystegines lack these effects due to their polarity, hydrophilicity, and inability to penetrate this barrier (Kohnen-Johannsen and Kayser 2019).

The Solanaceae family contains secondary metabolites that have a common tropane skeleton. These metabolites can be divided into two classes: tropane derivatives and ecgonine derivatives. Tropane alkaloids, which are derived from this skeleton, have been used as medicine for centuries. Atropine, obtained from *Atropa belladonna* L., and hyoscyamine, obtained from *Hyoscyamus niger* L., are examples of common tropane alkaloids. Poisonous plants from the Solanaceae family, such as *Atropa*, *Brugmansia*, *Datura*, and *Duboisia*, have been known since ancient times and have been used in folk medicine by various ethnic groups (Sneader 2005; Sweta and Lakshmi 2015). On the other hand, cocaine, derived from two *Erythroxylon* species, has limited medical use due to its highly addictive properties (Carrera et al. 2004; Sweta and Lakshmi 2015).

5.7 Pharmacological Action of Tropane Alkaloids

In clinical pharmacology, the crucial aspect for pharmacological action is stereoselectivity as it directly impacts the binding affinity of the molecule to the receptor (Fig. 5.2). Studies have shown that the S-(−)-hyoscyamine is approximately 30–300 times more potent than the R-(+)-isomer. However, due to its instability, the S-(−)-isomer rapidly converts to + hyoscyamine (+ atropine). As atropine is a racemic mixture and more stable, it is commonly utilized in clinical practice instead of its isomers (Gyermek 2002; Shim et al. 2022).

TAs such as atropine, scopolamine, and hyoscyamine can be given through oral, intravenous, intramuscular, endotracheal, and transdermal routes (McLendon and Preuss 2024). Oral administration is more convenient for patients than transdermal applications, as it ensures rapid absorption and pharmacological effects. Both atropine and scopolamine are readily absorbed from the gastrointestinal tract, although

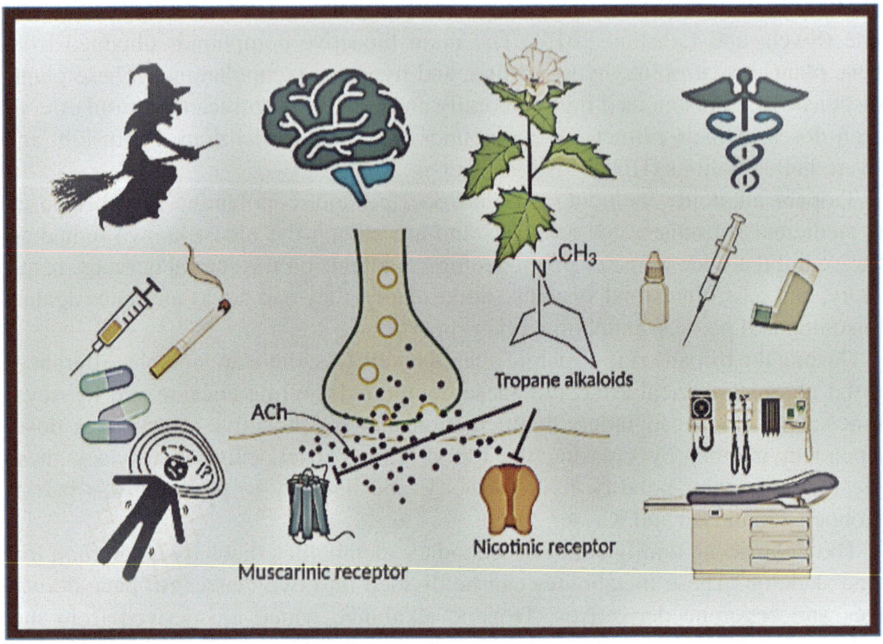

Fig. 5.2 Tropane alkaloids binding receptors. (After Shim et al. 2022)

their pharmacokinetics may differ among patients (Mirakhur 1978). However, transdermal delivery is preferred for scopolamine due to its ability to minimize drug variabilities and side effects, making it more patient-friendly (Shaoul et al. 2012).

Pharmacokinetics of tropane alkaloids: TAs are absorbed in the gastrointestinal tract, quickly distributed throughout the body, and primarily eliminated through the renal system (EFSA 2011). Due to its short half-life in plasma and adverse effects that vary with dosage, scopolamine is mainly restricted to transdermal application (Renner et al. 2005). Following absorption, scopolamine undergoes a substantial first-pass effect, as only a small percentage (2.6%) is excreted in the urine in its pharmacologically active state (Kanto et al. 1989). Cytochrome P450 enzymes play a crucial role in the metabolism of scopolamine through oxidative demethylation. The intake of grapefruit juice, which inhibits CYP3A, has been found to extend the time to reach maximum concentration (tmax) and increase the AUC0–24 h value of scopolamine metabolization (Ebert et al. 2000). Furthermore, scopolamine and its apo- and nor-metabolites undergo glucuronidation or sulfation during phase II metabolism for elimination through urine. It is worth noting that scopolamine and hyoscyamine are not retained in the human body, and there is no evidence of genotoxicity or chronic toxicity. Adverse effects from prolonged exposure have not been documented (EFSA 2011).

5.8 Atropine

Pharmacodynamics: Atropine, as atropine sulfate, is an FDA-approved medication used as the first line of treatment for bradycardia, organophosphate poisoning, and anti-sialagogue/anti-vagal action (More et al. 2015).

Pharmacokinetics: The peak plasma levels of atropine vary depending on the route of administration, ranging from a few minutes (intramuscular: 13 min) to several hours (oral: 1 h; aerosol: 1.5–4 h) (Brown and Taylor 2003). Sex-based differences in the pharmacokinetics of atropine have been observed, with females showing a 15% higher AUC (area under the curve) and Cmax (peak plasma concentration) than males. Additionally, females have a shorter half-life of atropine (approximately 20 min) than males. Atropine binds to 14–22% plasma proteins and is metabolized in the liver, with almost 90% of the drug being excreted through urine within 24 h (John et al. 2018). Between 12 and 16 min, there was a noticeable correlation between the maximum plasma concentration of atropine and an increased pulse rate. However, no correlation was observed between the time of maximum response and the concentration of atropine in the plasma. Age is also a significant factor in determining the kinetics of atropine, as children and the elderly are more sensitive to its effects. In the past, rectal administration of atropine (at a dosage of 0.01 mg/kg) was the preferred method for delivering the drug to children. This method helped slow down systemic absorption rate and resulted in better preoperative sedation with fewer side effects (Fuchs and Schwabe 1990).

Atropine has a delayed onset and lower peak plasma concentration when given rectally (15 min, 0.7 ng/mL) compared to intramuscularly (5 min, 2.4 ng/mL) (Olsson et al. 1983). For deep sedation in pediatric patients, a combination of 8 mg/kg ketamine, 0.5 mg/kg midazolam, and 0.02 mg/kg atropine is more effective (Wang et al. 2010). It is vital to monitor atropine concentration, as 0.3 mg/L in the peripheral blood can be fatal (Ricard et al. 2012).

Species exhibit significant differences in the metabolism of atropine, primarily through glucuronidation and N-demethylation reactions (Kalser 1971). While rabbits have been found to metabolize atropine into tropic acid and tropine through serum carboxylesterase, this activity has not been observed in humans and monkeys (Harrison et al. 2006). The main metabolite (29%) in humans is tropine (Hinderling et al. 1985). In urine, the metabolites of atropine include (+)-hyoscyamine (followed by nortropine at 24%), atropine oxide (15%), tropic acid (3%), and tropine (2%). However, no conjugated products of atropine have been detected (Shim et al. 2022).

5.9 Therapeutic Uses of Atropine and Atropine Derivatives

Hyoscine is utilized in endoscopic retrograde cholangiopancreatography (ERCP) to relax smooth muscles (Salmanroghani et al. 2020). Atropine effectively treats nocturnal asthma by enhancing the mucociliary function of the lungs. It also impacts the vagal tone of diabetic patients and managed tremors in monkeys with PD (Gomez-Mancilla et al. 1991). Benztropine, developed by USL Pharma, is a tropane-based dopamine inhibitor approved by the FDA as adjunctive therapy for PD. It acts as a selective M1 mAChR antagonist, inhibits dopamine uptake, and reduces serotonin and norepinephrine levels. Therefore, it is used in the symptomatic treatment of PD (Schlagmann and Remien 1986) and cocaine addiction (Rothman et al. 2008). Merck Darmstadt introduced homatropine, another tropane derivative, as a mydriatic drug in 1883 (Sneader 2005). It binds to muscarinic receptors in the stomach and atria with similar affinity (Gilani and Cobbin 1987). Although homatropine itself is not FDA approved, a combination of homatropine methyl bromide and hydrocodone bitartrate has been approved as an antitussive agent.

Atropine effectively inhibits excessive fluid production, reversing bronchorrhea and bronchoconstriction (Takemura et al. 2013; Hulse et al. 2014). It prevents leukocyte migration to infection sites by blocking chemotaxis and chemokines (Razani-Boroujerdi et al. 2008) and significantly reduces neutrophil influx in the lungs (McQueen et al. 2007). Additionally, intraperitoneal atropine is as effective as intraperitoneal dexamethasone in preventing lung inflammation (Cui et al. 2010).

Regardless of the type of nucleic acid, atropine exhibits strong antiviral effects against enveloped viruses like herpes simplex virus (HSV) and influenza (Biziagos et al. 1990; Ozçelik et al. 2011; Yamazaki and Tagaya 1980). It interferes with viral attachment to the host cell and impacts both innate and adaptive immune cells (Razani-Boroujerdi et al. 2008). Belladonna 200c, a commonly used homoeopathic medicine made from *Atropa belladonna*, has demonstrated anti-Japanese encephalitis virus (JEV) activity (Chakraborty et al. 2018; Kumar et al. 2020). In a clinical study on animals, it was found that Belladonna 200c enhances the mRNA expression of CCR-5, TLR-3, TLR-7, IFN-α, IFN-β, ISG-15, and IFIT1, while reducing the expression of TLR-4 (Bandyopadhyay et al. 2016; Chakraborty et al. 2019). Another study conducted as a pre-treatment against JEV demonstrated that both pure atropine and atropine sulfate enhance innate immunity, decreasing the disease's severity (Chakraborty et al. 2019). The combination of traditional knowledge and modern evidence has the potential to offer solutions to complex issues such as the SARS-CoV2 pandemic. Ayurveda and homoeopathic medicine have effectively utilized substances like atropine through careful processing and controlled dosages. Therefore, during this crisis, atropine could be utilized to enhance the body's natural defenses and restrict viral replication. As we face the possibility of encountering new mutated viruses in the future, and with ongoing challenges related to vaccine effectiveness worldwide, exploring the potential of atropine in drug development efforts could prove beneficial. While the properties of atropine show promise as a prophylactic or therapeutic option against COVID-19, further research is needed to confirm its efficacy (Devkar et al. 2021).

5.10 Side Effects of Atropine

Atropine has various side effects that depend on the dosage. It is used as an antidote for organophosphate poisoning, a pre-anesthetic agent, and to treat bradycardia. At low doses (0.5–1 mg), atropine primarily affects the central and peripheral autonomic nervous systems. It reduces saliva production, causing dry mouth (xerostomia); increases heart rate (mild tachycardia) by blocking the vagus nerve; and can cause pupil dilation (mydriasis) and blurred vision due to impaired visual accommodation. At moderate doses (1–2 mg), side effects become more pronounced as tachycardia may become more significant, potentially leading to palpitations, inhibiting bladder contraction, causing difficulty in urination (Urinary Retention), and also, anticholinergic effects can slow gastrointestinal motility, leading to constipation. At higher doses (2–5 mg), atropine can cause more severe side effects in the form of heart rates exceeding 120 beats per minute; patients experience confusion, hallucinations, or delirium, particularly in the elderly; increased body temperature and flushed skin may occur due to decreased sweating. In cases of overdose (over 5 mg), atropine toxicity manifests with severe symptoms, including symptoms such as extreme agitation, hallucinations, hyperthermia, and seizures (anticholinergic syndrome), and respiratory depression or failure; as well as life-threatening arrhythmias may develop due to excessive vagal blockade (cardiac arrhythmias) (Sweta and Lakshmi 2015; Ghosh 2018; Kohnen-Johannsen and Kayser 2019; Dyer 2020).

5.11 Scopolamine

Pharmacodynamics: Scopolamine is a medication that can be given intravenously, orally, through the eyes, or applied topically. It is used for various purposes, such as preventing motion sickness and postoperative nausea. It acts by competitively blocking 5-hydroxytryptamine (HT3) receptors, with an IC50 of 2.09 μM; scopolamine is commonly used to induce cognitive dysfunction in animal studies (Lochner and Thompson 2016; Shim et al. 2022).

Mechanism of action: Besides its role as a nonselective mAChR antagonist, scopolamine also inhibits G-protein coupled postganglionic muscarinic receptors. As a result, it acts on both the central nervous system (CNS) and the peripheral nervous system (PNS) (Zhang et al. 2017).

Pharmacokinetics: The duration of scopolamine's effects varies depending on the method of administration. For instance, the half-life is approximately 63.7 ± 1.3 min when taken orally. When administered intravenously, the half-life is around 68.7 ± 1.0 min. Intramuscular administration results in a half-life of approximately 69.1 ± 8.0 min, while subcutaneous administration leads to a half-life of 213 min (Renner et al. 2005; Shim et al. 2022). Similarly, the pharmacokinetics of scopolamine are significantly influenced by the route of administration. The intravenous dose of 0.5 mg demonstrated more favorable outcomes, with a maximum concentration (Cmax) of 5.00 ± 0.43 ng/mL, a time to reach maximum concentration (Tmax) of 5.0 min and an area under the curve (AUC) of 369.4 ± 2.2 ng min/mL. In comparison, the oral dose had a Cmax of 0.54 ± 0.1 ng/mL, a Tmax of 23.5 ± 8.2 min, and an AUC of 50.8 ± 1.76 ng min/mL. However, rapid absorption was observed after intranasal administration, with a Cmax of 1.68 ± 0.23 ng/mL, a Tmax of 2.2 ± 3 min, and an AUC of 167 ± 20 ng min/mL (Renner et al. 2005; Shim et al. 2022).

The FDA has approved the scopolamine transdermal therapy system patch (TTS patch) to address the adverse effects of the drug dependent on the dosage and to achieve therapeutic plasma concentrations for an extended period. The TTS patch ensures that protective levels of scopolamine are reached within 6–8 h, with optimal efficacy lasting up to 72 h, resulting in a constant high plasma concentration (56–245 pg/mL). On the other hand, oral or intravenous doses provide quick effectiveness (within 0.5 h) but have a shorter duration of action (6 h). When both transdermal and oral administration are combined, a peak concentration of approximately 0.37 ng/mL is observed after 1 h (Nachum et al. 2001). Although the specific enzymes responsible for metabolizing scopolamine are unknown, it is believed to undergo oxidative methylation via cytochrome P450 (CYP450), followed by Phase II conjugation (glucuronidation and sulfation), resulting in approximately 2.6% of scopolamine being eliminated in the (Renner et al. 2005; Shim et al. 2022). The metabolism of scopolamine is highly species-specific, with rats producing metabolites such as p-hydroxy-, m-hydroxy-, and p-hydroxy-m-methoxy-scopolamine. In contrast, rabbits and guinea pigs primarily produce tropic acid as a major metabolite. Mice, however, do not produce tropic acid as a metabolite (Wada et al. 1991; Shim et al. 2022).

5.11.1 Therapeutic Uses of Scopolamine

Scopolamine is commonly prescribed to alleviate nausea and excessive drooling in patients with Alzheimer's and Parkinson's diseases (AD and PD) (Dreyfuss et al. 1991). The transdermal application of scopolamine can effectively treat motion sickness, postoperative nausea, and vomiting (Attias et al. 1987; Spinks and Wasiak 2011). Moreover, due to its antimuscarinic properties, it also provides antiemetic and sedative effects (Shim et al. 2022). It exhibited an LD_{50} (Lethal dose for 50% of deaths) of 1880 mg/kg and 1270 mg/kg when administered orally to mice and rats, respectively. However, when administered subcutaneously, the LD_{50} was significantly lower at 1650 mg/kg for mice and 296 mg/kg for rats (Cayman 2022).

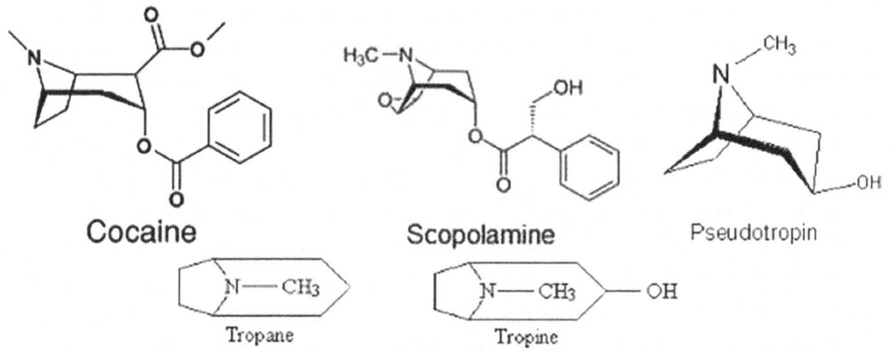

Fig. 5.3 Some of the derivatives of tropane alkaloids are tropane, tropine, tropanone, scopolamine, cocaine, pseudotropine. (After Sweta and Lakshmi 2015)

5.12 Members of Tropane Alkaloids and their Silimars

The structure of tropane alkaloids consists of a bicyclic ring system known as N-methyl-8-a-zabicyclo (Griffin and Lin 2000) octane. In addition to this, solanaceae plants also contain bicyclic aminoalcohols such as di- and tri-hydroxylated tropines, 6,7-epoxide, and corresponding de-N-methylated compounds (N-nortropane derivatives) (Fig. 5.3). Tropane alkaloids are found as esters with carboxylic acids, and tropic acid attends as a diverse marker for alkaloid-containing plants in relationships of chemotaxonomy. Plants which are related to species of solanaceae family have derivatives of tropane alkaloids including tropane, tropine, tropanone, scopolamine, cocaine, pseudotropine and others (Fig. 5.3) (Sweta and Lakshmi 2015).

5.13 Physiological Effects of Tropane Alkaloids

Tropane alkaloids, such as cocaine, scopolamine, and calystegines, have a shared tropane moiety. However, despite this similarity, these compounds have distinct physiological effects on humans. Cocaine exerts its influence on the synaptic cleft by inhibiting the reuptake of dopamine, noradrenaline, and serotonin, while scopolamine acts as a competitive antagonist of muscarinic receptors. Ingesting either of these substances can result in hallucinations, psychoactive effects, or even death (Rothman et al. 2001; Langmead et al. 2008; Kohnen-Johannsen and Kayser 2019). On the other hand, calystegines do not penetrate the central nervous system (CNS) due to their hydrophilic nature, and therefore, they do not produce any psychoactive effects in humans (Dräger 1994; Kohnen-Johannsen and Kayser 2019).

The stereochemical relationship between atropine and hyoscyamine has been studied by Lossen et al. (1864), Wolffenstein (1923), and Kohnen-Johannsen and Kayser (2019). They elucidated the reaction mechanism of the alkaline hydrolysis of hyoscyamine and identified tropic acid and tropine as the cleavage products of both hyoscyamine and atropine. These findings concluded that atropine is the racemate of hyoscyamine (Wolffenstein 1923; Kohnen-Johannsen and Kayser 2019). Additionally, Ladenburg (1879) discovered that the reverse reaction of hydrolysis can be achieved by boiling the reactants in hydrochloric acid. He also established a commonly used method of esterifying tropine with various organic acids (Ladenburg 1879; Kohnen-Johannsen and Kayser 2019).

5.14 Calystegines

To date, there are no available drug products derived from calystegines. Calystegines are hydroxylated nortropane alkaloids derived from the tropane alkaloid biosynthetic pathway. They are strong glycosidase inhibitors and occur in vegetables such as potatoes, tomatoes, and cabbage. Calystegine accumulation in root cultures was described to increase with carbohydrate availability (Richter et al. 2007). The recent study findings revealed that calystegines significantly protect human adipose-derived stromal stem cells (ASCs) against apoptosis, inflammation, and oxidative stress in hyperglycemic conditions. By maintaining the AKT/PI3K/mTOR pathway, calystegines effectively reduce endoplasmic reticulum (ER) stress, ensuring the survival and proper metabolic function of ASCs in a hyperglycemic environment. These results suggest that calystegines hold promise as a potential alternative for managing metabolic conditions such as diabetes mellitus, obesity, and metabolic syndrome associated with hyperglycemia. Calystegines exhibit significant promising capabilities in safeguarding HuASCs from hyperglycemia-induced cellular impairment, with findings indicating that these effects are linked to the enhancement of the PI3K/AKT/mTOR pathway (Kowalczuk et al. 2022). Polyhydroxylated NTAs such as calystegine lack psychoactive properties as they are unable to cross the blood-brain barrier due to their hydrophilic nature. Furthermore, they demonstrate minimal pharmacological activity. Consequently, ancient medicines have not utilized this category of NTAs. A recent suggestion by researchers indicates that these substances hinder mammalian and plant glucosidases. However, they have not been applied pharmacologically and have received minimal research focus (Jocković et al. 2013; Kohnen-Johannsen and Kayser 2019). The initial polyhydroxylated NTAs structures were documented in 1990 (Dräger 2004; Kohnen-Johannsen and Kayser 2019).

5.15 Cocaine

The use of coca leaves dates back to around 3000 BC in Ecuador, where the Inca civilization cultivated and consumed them. Indigenous tribes chewed coca leaves with lime to extract alkaloids for spiritual rituals and to enhance strength and endurance, especially at high altitudes. Initially reserved for tribal leaders, the practice spread with the arrival of Spanish soldiers. Cocaine was isolated in 1855 by F. Gaedcke, who published his findings in *Archiv der Pharmacie*, identifying the substance as erythroxyline (Kohnen-Johannsen and Kayser 2019).

Lossen determined cocaine's chemical formula in 1862 and analyzed atropine. Later, in 1898, Willstätter accomplished the first chemical synthesis of cocaine and revealed its chemical structure. The stereochemistry of cocaine was then elucidated around 50 years later (Hardegger and Ott 1954).

In 1885, S. Freud released his publication "Über Coca" where he endorsed cocaine as a nearly miraculous remedy, possessing local anesthetic qualities, and being the ideal treatment for postpartum depression and morphine dependency an unsafe use for a substance known for its highly addictive nature (Kohnen-Johannsen and Kayser 2019).

5.15.1 Pharmacology of Cocaine

Cocaine pharmacodynamic: Following the completion of Phase 3 trials and Phase 1 pharmacokinetic study, Numbrino (cocaine hydrochloride) nasal spray was approved by the FDA for use as a topical solution for the nasal cavity mucus membrane. Prior to this approval, cocaine was utilized as a local anesthetic agent, but it was eventually substituted with safer alternatives. Upon administration of cocaine (intravenous/pulmonary), rapid attainment of high peak plasma levels occurs (Paly et al. 1982), and the substance is eliminated from the body within 4–6 h (Jatlow 1988).

Pharmacokinetics: Cocaine pharmacokinetics were studied in humans via intravenous (32 mg at 1 mL/min) and intranasal (64 and 96 mg inhaled with a 5-cm straw) routes. A one-compartment model revealed that intravenous administration followed first-order elimination, while intranasal administration showed first-order absorption and elimination. The half-life for intravenous cocaine was approximately 41.4 ± 8.2 min (mean ± SEM), and the area under the curve (AUC) significantly differed between the two methods (Shim et al. 2022).

Volunteers in a related study were given oral doses of 100 mg and 200 mg, as well as an intravenous dose of 40 mg. The analysis of cocaine metabolites was performed using gas chromatography-mass spectrometry (GC-MS) (Coe et al. 2018). The major metabolites identified were benzoylecgonine and ecgonine methyl ester. Minor metabolites included m- and p-hydroxybenzoylecgonine, m- and p-hydroxycocaine, and norcocaine (Jufer et al. 2000; Shim et al. 2022).

The pharmacokinetic parameters were evaluated using non-compartmental analysis and a two-factor model. The bioavailability of cocaine was determined to be 0.32 ± 0.04 and 0.45 ± 0.06 after administering oral doses of 100 mg and 200 mg, respectively. The volume of distribution (Vd) and clearance (CL) were found to be the lowest for the intravenous dose (Vd: 1.3 L/kg; CL: 32.7 mL/min.kg) and highest for the 100-mg oral dose (Vd: 4.2 L/kg; CL: 116.2 mL/min.kg) compared to the 200-mg oral dose (Vd: 2.9 L/kg; CL: 87.5 mL/min.kg). The oral administration exhibited a distinct metabolic profile, with higher concentrations of major and minor metabolites than the intravenous dose (Coe et al. 2018; Shim et al. 2022).

Cocaine has a long history of use, but its therapeutic applications in neurobiology and pharmacology remain largely unexplored due to legal restrictions (Henningfield and Griffiths 1991). Known for its local anesthetic effects and ability to stimulate the central nervous system, cocaine's impacts depend on the method of administration—oral, nasal, chewing, or injection (Henningfield and Griffiths 1991).

The feelings of euphoria and increased mental stimulation associated with cocaine use are primarily caused by elevated dopamine levels, as cocaine inhibits the reabsorption of dopamine, noradrenaline, and serotonin in the limbic system (Rothman et al. 2001). These effects can also lead to changes in electroencephalogram (EEG) readings, which may show heightened or diminished brain activity (Berger 1931).

The topical application of cocaine achieves its local anesthetic properties by blocking ion channels in neural membranes. Once absorbed by the mucosa, cocaine quickly induces paralysis in the peripheral ends of sensory nerves. While it was once commonly used in dentistry as a local anesthetic, safer alternatives have since taken place. Nonetheless, cocaine has paved the way for the development of various local anesthetics and painkillers.

Procaine, the first major cocaine derivative, was mainly used in dentistry but has declined in usage due to the availability of more potent local anesthetics. Its instability arises from susceptibility to ester hydrolysis. Tetracaine, a more advanced derivative, is utilized in minor facial surgeries and ophthalmology; it is nearly ten times more potent than procaine but also has increased toxicity (Dewick 2002).

Lidocaine, developed in 1943 by Swedish chemists Nils Löfgren and Bengt Lundqvist (Sinatra et al. 2010), is a synthetic amino amide derivative of cocaine. It offers increased stability against hydrolysis in aqueous solutions (Dewick 2002) and is used as a local anesthetic. Additionally, lidocaine serves as an Ib-type antiarrhythmic medication, benefiting cardiovascular health.

5.15.2 *Cocaine Side Effects and Toxicity*

Cocaine use leads to lasting structural changes in vital organs, including the brain, heart, lungs, liver, and kidneys. Its mechanisms of injury are complex, primarily involving vasoconstriction in coronary arteries, which increases the risk of atherosclerosis and myocardial infarction. Cocaine also disrupts cardiac function by blocking potassium channels, enhancing calcium channel activity, and inhibiting sodium flow during depolarization. Chronic use is associated with serious heart conditions such as myocarditis, ventricular hypertrophy, dilated cardiomyopathy, and heart failure. Inhalation of crack cocaine is linked to respiratory problems, as the drug affects pulmonary function and respiratory tract alterations. Additionally, cocaine abuse can lead to renal failure, and children born to mothers with a history of cocaine use have a higher incidence of congenital cardiovascular and neurological issues (Riezzo et al. 2012).

5.16 Trospium Chloride

Trospium chloride acts as an antispasmodic agent, aiding in the relaxation of smooth muscles and is commonly prescribed for the treatment of overactive bladder (Rudy et al. 2006). Another tropane derivative, tropisetron, works as a serotonin receptor antagonist. Tropisetron is utilized as an antiemetic and as an analgesic for fibromyalgia (Sorbe et al. 1994; Shim et al. 2022).

N-butylation of scopolamine yields N-butylscopolamine, which is on the WHO "List of Essential Medicines" (WHO 2013). It is an antispasmodic used to treat abdominal cramping, colic pain, and bladder spasms (Tytgat 2007). The FDA has approved its use in animals as Buscopan® (Boehringer Ingelheim's animal-health US Inc.) (Shim et al. 2022).

Tiotropium is a FDA-approved inhalation spray, Spiriva®. It is a bronchodilator used in the management of chronic obstructive pulmonary disease (COPD). The drug binds well with M1–M3 receptors, more potent than ipratropium, and is considered safe (Barnes 2000). The first derivative of cocaine was procaine which was used in dentistry as anesthetic and pain killer.

Tetracaine (STERI-UNIT®, FDA approved), a potent procaine derivative, is commonly employed for local anesthesia in minor surgeries. When combined with oxymetazoline, tetracaine serves as a superior topical anesthetic for nasal procedures compared to cocaine. Additionally, it is utilized as a local ophthalmic anesthetic. Another derivative, lidocaine (Xylocaine, FDA approved), is applied topically to alleviate itching and pain caused by minor burns, eczema, and insect bites. Lidocaine acts as a sodium channel blocker, reducing muscle contraction, vasodilation, hypotension, and irregular heartbeat (Dewick 2002; Noorily et al. 1995).

5.17 Tropane Alkaloids Side Effects

According to Armanini et al. (2002), inhibiting the parasympathetic nervous system by anticholinergic drug substances can lead to various side effects, including a decrease in heart rate, dryness of the mouth, and reduced perspiration. At higher therapeutic oral doses, these drugs can also cause an increase in heart rate, inhibition of respiratory tract secretory activity, bronchodilation, and mydriasis. Additionally, the inhibition of sweating can result in a subsequent increase in body temperature. Disorders such as depression, epilepsy, Parkinson's, and Alzheimer's disease have been linked to a dysfunction of the muscarinic cholinergic system, leading to continued interest in muscarinic system antagonists like TAs as potential CNS drug candidates (Armanini et al. 2002).

Physostigmine, a pyrolo-indole alkaloid, is the remedy for scopolamine and hyoscyamine intoxications. It is derived from the Calabar bean (*Physostigma venenosum*, Fabaceae) and is a parasympathomimetic drug by temporarily inhibiting the cholinesterase (Robenshtok et al. 2002).

Scopolamine is known for its mydriatic, spasmolytic, and local anesthetic effects, but it also has hallucinogenic and potentially lethal side effects. The primary method of scopolamine is through transdermal application, a technology introduced as transdermal therapeutic systems (TTS) in 1981. Scopoderm TTS® is a specific scopolamine delivery system used to treat motion sickness. During World War II, scopolamine was utilized for treating shell shock, psychoactive effects, and motion sickness. However, the numerous peripheral and central nervous system side effects of scopolamine have been a significant drawback. To address these concerns, derivatives of scopolamine have been developed, leading to its classification as a drug lead substance (Kohnen-Johannsen and Kayser 2019).

Scopolamine, hyoscyamine, and atropine all share similar mechanisms of action and effects. The pharmacological activity of these substances is selective based on the different stereoisomers and their affinity for muscarinic receptors. This leads to variations in potency between the S-(−)- and R-(+)-isomers of hyoscyamine, with the S-(−)- isomer being estimated to be 30–300 times more potent than the R-(+)-isomer. However, the S-(−)-isomer of hyoscyamine is not stable and rapidly racemizes into atropine, which is a 1:1 mixture of the two forms. Atropine, being more stable over time, is preferred for medicinal applications over hyoscyamine. Both atropine and scopolamine exhibit characteristic dose-dependent effects on the cardiovascular system, which have clinical significance in resuscitation (Kohnen-Johannsen and Kayser 2019).

Anisodamine, derived from the *Anisodus tanguticus* plant native to Tibet, is known for its lower toxicity compared to atropine and scopolamine. It has been utilized in traditional Asian medicine for its effectiveness in treating septic shock by enhancing blood flow in microcirculation and addressing circulatory and gastric issues similar to those of atropine and scopolamine (Kohnen-Johannsen and Kayser 2019).

Homatropine, cyclopentolate, and tropicamide are three different drugs used in ophthalmology for various purposes. Homatropine, derived from the mandelic acid ester of tropine, is known for its ability to produce a faster and less paralytic effect compared to atropine. This advantage led to its introduction as a new mydriatic by Merck Darmstadt in 1883, making it one of the earliest synthetic drugs. Cyclopentolate, on the other hand, is primarily used for pediatric eye examinations. Lastly, tropicamide has been approved for use in ophthalmology since 2005 (Kohnen-Johannsen and Kayser 2019).

Trospium chloride is a benzylic acid esterified quaternary ammonium 3α-nortropane derivative. This artificial anticholinergic cannot penetrate the blood-brain barrier and effectively eases the bladder's smooth muscle tension. Its primary purpose is to address urgency and reflex incontinence (Rudy et al. 2006). Tropisetron is classified as a serotonin receptor antagonist, despite having a tropane skeleton. It is commonly used in antiemetic therapy to alleviate nausea and vomiting caused by chemotherapy. Additionally, it is utilized as an analgesic in the treatment of fibromyalgia (Sorbe et al. 1994).

N-butylscopolamine has undergone N-butylation to reduce its impact on the central nervous system. As a result, it is unable to cross the blood-brain barrier. This modified form of scopolamine is commonly employed in the treatment of abdominal pain caused by cramping, renal colic, and bladder spasms (Mutschler et al. 2013).

Tiotropium bromide, ipratropium bromide, and oxitropium bromide have been developed as inhalation treatments to reduce the adverse and intoxicating effects of smoking *Datura* leaves for the treatment of asthmatic symptoms. It works by blocking M3 receptors on smooth muscle cells in the bronchi, resulting in bronchodilation. This approach aims to provide a safer and more controlled method of managing asthma symptoms. Tiotropium bromide medication has been on the market since 2005 and is primarily utilized for treating chronic obstructive pulmonary disease (COPD). On the other hand, ipratropium bromide has been available since 1975 and is commonly used for COPD treatment (in conjunction with salbutamol, a β2-adrenergic receptor agonist) as well as for asthma. Oxitropium bromide, although less recognized and utilized compared to the aforementioned medications, also functions as an anticholinergic bronchodilator for managing asthma and COPD (Barnes 2000; Kohnen-Johannsen and Kayser 2019).

Benzatropine, an effective medication, acts as a central nervous system agent by selectively inhibiting the M1 muscarinic acetylcholine receptor. Its chemical structure combines the tropine skeleton from atropine and the benzhydryl skeleton from diphenhydramine. It enhances dopaminergic activity by partially blocking cholinergic activity in the basal ganglia and preventing dopamine reuptake. As a result, it has proven beneficial in managing the initial phases of Parkinson's disease (Schlagmann and Remien 1986).

Scopolamine has potential applications in treating central nervous system (CNS) diseases. It, along with other muscarinic receptor antagonists, influences cognitive processes, sensory functions, and stress responses. Research into scopolamine's effects as a central-acting muscarinic inhibitor has produced mixed results. Some studies indicate a rapid and significant impact (Yu et al. 2019), while others find no advantage over placebo (Park et al. 2018). These inconsistencies highlight the need for further research to confirm scopolamine's efficacy in treating CNS disorders.

5.18 Tropane Alkaloids Toxicity

The presence of tropane alkaloids (TAs) in plants from the Solanaceae family, such as *Atropa, Datura*, and *Hyoscyamus* species, can cause concern. These plants are often considered weeds and can inadvertently contaminate food and feed during the harvesting or processing stages. It has been observed that herbal teas have been found to have high levels of TAs, with *A. belladonna* leaves and berries being the main source of contamination (Chan 2017). Unprocessed cereals (like buckwheat, maize, millet, sorghum, wheat) and the cereal-based foods prepared from them are often adulterated (~7% to 27%) with TAs (atropine, scopolamine) (Chan 2017). In one of the cases, it was confirmed that there was an extremely high level of atropine (>30 mg/g) (Bryson et al. 1978). Surprisingly, baby food samples were also found to contain TAs (11.5 μg/kg of atropine; 2.8 μg/kg of scopolamine), which exceeded the EU regulatory limit of 1 μg/kg for each alkaloid (Marín-Sáez et al. 2019). The most common contaminants in food are the seeds of *Datura stramonium*, followed by *Hyoscyamus niger* seeds and berries of *A. belladonna* (Adams et al. 1982; Benford et al. 2013). These contaminants are responsible for numerous cases of food poisoning worldwide (González-Gómez et al. 2022; Haughey et al. 2021; Mulder et al. 2016).

Toxicity caused by TAs in livestock and other grazing animals is remarkable because the plants containing TAs are generally not edible, leading animals to avoid consuming them. However, if animals consume feed contaminated with TAs, such as millet or corn, it can result in toxicity (Benford et al. 2013). Poisoning from TAs is commonly observed in children who consume attractive and colorful berries from plants in the Solanaceae family, such as *A. belladonna* and *H. niger* (O'Shaughnessy 2011).

The symptoms of TA toxicity typically manifest within 2 h after oral intake (Scavone et al. 1998). The characteristic signs of TAs poisoning can be recognized using the mnemonic "red as a beet" (flushing), "dry as a bone" (dry mouth), "blind as a bat" (mydriasis), "mad as a hatter" (mental confusion), "hot as a hare" (fever), and "full as a flask" (urinary retention) (Broderick et al. 2023). The absence of sweating distinguishes anticholinergic toxicity from sympathomimetic toxicity (Holstege and Borek 2012). The effects of TAs on the CNS tend to last much longer (over 8 h) than the cardiovascular system (Ali-melkkila et al. 1993; Shim et al. 2022).

Poisoning from tetrahydrozoline (TA) requires careful management, including various approaches such as inducing gastric emptying, administering activated charcoal (0.5–1 g/kg in children or 25–100 g in adults) to absorb the toxic substance, and utilizing benzodiazepines to control agitation (Furbee 2009). In cases where both the peripheral nervous system (PNS) and CNS are affected by anticholinergic poisoning, the use of physostigmine (an acetylcholinesterase inhibitor) is recommended (Bronstein et al. 2009; Goldfrank et al. 2007).

5.19 Impact of Anticholinergic Burden

Prolonged use of medications with anticholinergic properties (MACs) can lead to cognitive decline, linked to cholinergic changes and increased amyloid-beta (Aβ) peptide deposition in brain regions like the amygdala, cortex, and hippocampus. A study showed that Parkinson's patients treated with MACs long term had more neurofibrillary tangles than those with short-term or no treatment. However, these findings were not replicated in another study. Research on the reversibility of cognitive effects after stopping MACs is mixed. Some studies indicate cognitive improvement post-withdrawal, while others do not report changes. Additionally, discontinuation may worsen conditions in certain psychotic or bipolar disorders. Anticholinergic burden (AB) negatively impacts the mental health of the elderly, contributing to cognitive impairment, delirium, and an increased risk of falls. A recent study by Reinold et al. (2021) involving around 16 million individuals aimed to accurately assess these cholinergic effects.

5.20 Medicinal Use of Tropane Alkaloids

Solanaceous alkaloids are anticholinergic drugs that affect the central and peripheral nervous systems. Acetylcholine antagonism occurs at or beyond postganglionic nerve endings (Waseem et al. 2023). Accidental toxicity can happen since these alkaloids are found in all parts of the plant. Traditionally recognized for their mood-elevating properties, the primary modern medical use of these alkaloids differs (Christen 2000; Sweta and Lakshmi 2015).

These alkaloids cause mydriasis and cycloplegia, with atropine being preferred for ophthalmic use due to its stability and low central nervous system impact in small doses. They are beneficial for treating bladder spasms, irritable bowel disease, colic, cystitis, pancreatitis, and peptic ulcers (Grynkiewicz and Gadzikowska 2008; Sweta and Lakshmi 2015).

Their effects include decreased respiratory secretions, bronchodilation, heart rate changes (bradycardia at low doses and tachycardia at high doses), and reduced gastric secretions. Atropine is also an antidote for organophosphorus poisoning, while scopolamine hydrobromide (0.5–1.0 mg) is used for motion sickness (Sweta and Lakshmi 2015).

5.21 Tropane Alkaloids and Their Role as Drug Lead Substances

Hyoscyamine and scopolamine are commonly utilized as anticholinergic medications. They exert their effects on the central and peripheral nervous systems by acting as competitive, nonselective antagonists of the muscarinic acetylcholine receptor (mAChR). This prevents the binding of the natural neurotransmitter acetylcholine. In humans, there are two known types of acetylcholine receptors: muscarinic and nicotinic, which derive their names from their respective agonists, muscarine and nicotine (Fig. 5.4). Muscarine is a toxin found in the *Amanita muscaria* mushroom

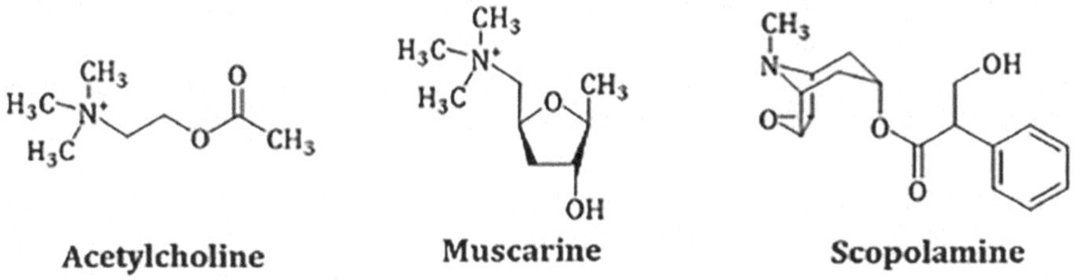

Fig. 5.4 Comparison of the chemical structures of acetylcholine, muscarine and scopolamine. Scopolamine is protonated in the body due to the physiological pH and is present as a quaternary ammonium salt. (After Kohnen-Johannsen and Kayser 2019)

and acts on the mAChR at synapses similarly to acetylcholine, except that it is not metabolized by acetylcholinesterase.

The G-protein coupled receptors (GPCRs) family includes the mAChRs, which consist of five subtypes (M1–M5). The M1, M3, and M5 subtypes are associated with stimulating Gq receptors and produce cytosolic calcium transients through the phospholipase C signaling pathway. On the contrary, the M2 and M4 subtypes interact with the Gi protein and inhibit the adenylyl cyclase (Langmead et al. 2008; Kohnen-Johannsen and Kayser 2019).

M1 receptors are present in the central nervous system and ganglia, playing a role in memory and learning functions. In contrast, M2 receptors are primarily located in the heart and are less abundant compared to M1 receptors. M3 receptors are specifically involved in the contraction of smooth muscles. As for M4 receptors, they have been identified in the forebrain, hippocampus, and striatum, suggesting their involvement in pain processes (Wess et al. 2003). The exact physiological function of M5 receptors remains unclear currently. Nevertheless, it is hypothesized that these receptors play a role in regulating brain microcirculation by facilitating vasoconstriction, vasodilatation, and the activation of nitric oxide synthase (Watts et al. 2008; Kohnen-Johannsen and Kayser 2019).

5.22 Tropane Alkaloids Biopharmaceutics Potentialities

The tropane ring system biosynthesis is homologous in organisms producing these three TA classes. TA biosynthesis initiates with ornithine or arginine amino acids and their intermediate putrescine, progressing toward the universal N-methyl-D1-pyrrolinium cation precursor of all TAs. This serves as the pivotal stage for the biosynthesis of cocaine, hyoscyamine/scopolamine, calystegine, and nicotine (Gadzikowska and Grynkiewicz 2002; Ziegler and Facchini 2008; Kohnen-Johannsen and Kayser 2019).

At the initial stages of TA biosynthesis the various categories of TAs: cocaine, scopolamine/hyoscyamine, and the calystegines all have a shared common precursor biosynthetic pathway that started with the amino acids L-ornithine (Orn) and L-arginine (Arg) Kohnen-Johannsen and Kayser (2019).

The spatial distribution and arrangement of scopolamine biosynthesis in plants is varied and intricate. Within Solanaceae plants, scopolamine biosynthesis occurs in the roots, with the alkaloids subsequently being transferred to the aboveground portions for storage.

The compartmentalization observed in biosynthesis within plants could pose challenges for future endeavors to produce substances in single-celled microbial systems. It is possible that eukaryotic host cells like yeasts or microalgae, which demonstrate organelle compartmentalization and have a track record of successful metabolic engineering of intricate metabolites, may serve as suitable host organisms for biosynthesis (Lauersen 2019; Kohnen-Johannsen and Kayser 2019).

5.23 Conclusion

Tropane alkaloids (TAs) are a class of naturally occurring organic compounds characterized by a bicyclic structure known as tropane found in plants belonging to the Solanaceae family, e.g., *Atropa belladonna*, *Hyoscyamus niger*, and *Datura stramonium*. TAs include hyoscyamine, scopolamine, and cocaine. TAs act as antagonists at muscarinic acetylcholine receptors in the central and peripheral nervous systems. By blocking the action of acetylcholine, they induce a range of physiological responses, including decreased glandular secretions, muscle relaxation, and alterations in heart rate. TA derivatives are used in medicine and have several clinically signific: Scopolamine is used for its antiemetic properties. Scopolamine is effective in preventing motion sickness and postoperative nausea. Hyoscyamine is utilized in treating gastrointestinal disorders, such as irritable bowel syndrome, due to its antispasmodic effects. It helps reduce motility and alleviates cramping. Cocaine was employed as a local anesthetic and vasoconstrictor, particularly in ophthalmology and dentistry. Its use has declined due to its potential for addiction and abuse, but it remains significant in specific medical contexts. Despite their medicinal benefits, tropane alkaloids can be highly toxic, particularly when ingested in large amounts. Severe cases of toxicity can lead to respiratory failure and even death, particularly with plants that contain high concentrations of these alkaloids, such as *Datura stramonium*. Common side effects associated with the therapeutic use of tropane alkaloids include dry mouth, constipation, urinary retention, and CNS depression. Side effects are primarily due to the anticholinergic properties of the compounds, highlighting the importance of careful dosage and monitoring in clinical settings.

References

Aboelsoud NH (2010) Herbal medicine in ancient Egypt. J Med Plants Res 4(2):82–86. https://doi.org/10.5897/JMPR09.013

Adams RG, Verma P, Jackson AJ, Miller RL (1982) Plasma pharmacokinetics of intravenously administered atropine in normal human subjects. J Clin Pharmacol 22(10):477–481. https://doi.org/10.1002/j.1552-4604.1982.tb02638.x

Ahmed MN, Azam K, Nur M (2024) Traditional knowledge and formulations of medicinal plants used by the traditional medical practitioners of Bangladesh to treat schizophrenia like psychosis. Schizophr Res Treatment 2014(1):679810. https://doi.org/10.1155/2014/679810

Alamgir ANM (2018) Therapeutic use of medicinal plants and their extracts: volume 2: phytochemistry and bioactive compounds, Progress in drug research, vol 74. Springer, Cham

Alfaro M, Alfaro I, Angel C (2020) Identification of potential inhibitors of SARS-CoV-2 papain-like protease from tropane alkaloids from *Schizanthus porrigens*: A molecular docking study. Chem Phys Lett 761:138068. https://doi.org/10.1016/j.cplett.2020.138068

Ali-melkkila T, Kanto J, Iisalo E (1993) Pharmacokinetics and related pharmacodynamics of anticholinergic drugs. Acta Anaesthesiol Scand 37(7):633–642. https://doi.org/10.1111/j.1399-6576.1993.tb03780.x

Armanini D, Fiore C, Mattarello MJ, Bielenberg J, Palermo M (2002) History of the endocrine effects of licorice. Exp Clin Endocrinol Diabetes 110(6):257–261. https://doi.org/10.1055/s-2002-34587

Attias J, Gordon C, Ribak J, Binah O, Rolnick A (1987) Efficacy of transdermal scopolamine against seasickness: a 3-day study at sea. Aviat Space Environ Med 58(1):60–62

Bandyopadhyay B, Banerjee A, Das S, Sengupa M, Chakravarty R, Ray K (2016) Role of ultra-diluted belladonna extract in the immune-mediated healing of JE virus infection in mice. World J Pharm Res 5:1615–1627

Barnes PJ (2000) The pharmacological properties of tiotropium. Chest 117:63S–66S. https://doi.org/10.1378/chest.117.2_suppl.63S

Benford D, Ceccatelli S, Cottrill B et al (2013) Scientific opinion on tropane alkaloids in food and feed. EFSA J 11(10):1–113. https://doi.org/10.2903/J.EFSA.2013.3386

Berger H (1931) Archiv Für Psychiatrie Und Nervenkrankheiten. Clin Neurophysiol 28:95–132

Biziagos E, Crance JM, Passagot J, Deloince R (1990) Inhibitory effects of atropine, protamine, and their combination on hepatitis A virus replication in PLC/PRF/5 cells. Antimicrob Agents Chemother 34(6):1112–1117. https://doi.org/10.1128/AAC.34.6.1112

Broderick ED, Metheny H, Crosby B (2023) Anticholinergic toxicity. In: StatPearls. StatPearls Publishing, Treasure Island. https://www.ncbi.nlm.nih.gov/books/NBK534798/

Bronstein AC, Spyker DA, Cantilena LR, Green JL, Rumack BH, Heard SE (2009) 2007 annual report of the American Association of Poison Control Centers' National Poison Data System (NPDS): 25th annual report. Clin Toxicol 47(10):911–941

Brown J, Taylor P (2003) Agonists and antagonists of mucarinic receptors. In: Pharmacological basis of therapeutic, 10th edn. McGraw-Hill, New York, pp 162–181

Bryson PD, Watanabe AS, Rumack BH, Murphy RC (1978) Burdock root tea poisoning. Case report involving a commercial preparation. JAMA 239(20):2157–2157. https://doi.org/10.1001/jama.1978.03280470069028

Carrera MRA, Meijler MM, Janda KD (2004) Cocaine pharmacology and current pharmacotherapies for its abuse. Bioorg Med Chem 12:5019–5030

Cayman (2022) Safety Data Sheetn acc. to OSHA HCS. Available from: https://cdn.caymanchem.com/cdn/msds/14108m.pdf

Chaachouay N, Zidane L (2023) *Atractylis gummifera* (Stemless Atractylis). In: Husen A (ed) Exploring poisonous plants. Boca Raton, CRC Press

Chaachouay N, Zidane L (2024) Plant-derived natural products: A source for drug discovery and development. Drugs Drug Candidates 3(1):184–207. https://doi.org/10.3390/ddc3010011

Chaachouay N, Azeroual A, Zidane L (2023) Ethnobotany, ethnopharmacology, and traditional uses of medicinal and aromatic plants. In: Adnan M, Patel M, Snoussi M (eds) Ethnobotany and ethnopharmacology of medicinal and aromatic plants. CRC Press, Boca Raton. https://doi.org/10.1201/b22842

Chakraborty U, Katoch S, Sinha M (2018) Changes in viral load in different organs of Japanese encephalitis virus-infected chick embryo under the influence of Belladonna 200C. Indian J Res Homoeopathy 12:75–80. https://doi.org/10.4103/ijrh.ijrh_17_18

Chakraborty U, Sinha M, Bhattacharjee A, Nayak D, Khurana A, Manchanda RK, Sarkar D, Das S (2019) Suppression of viral load by Belladonna 200c through modulation of TLR and type-I IFN signaling pathways. Int J Biol Med Res 10:6635–6640

Chan TYK (2017) Worldwide occurrence and investigations of contamination of herbal medicines by tropane alkaloids. Toxins 9(9):284. https://doi.org/10.3390/toxins9090284

Christen P (2000) Tropane alkaloids: old drugs used in modern medicine. In: Rahman A (ed) Studies in natural products chemistry, bioactive natural products, Part C, vol 22. Elsevier Science and Technology, Amsterdam, pp 717–749

Coe MA, Jufer Phipps RA, Cone EJ, Walsh SL (2018) Bioavailability and pharmacokinetics of oral cocaine in humans. J Anal Toxicol 42(5):285–292. https://doi.org/10.1093/jat/bky007

Cui Y, Devillier P, Kuang X, Wang H, Zhu L, Xu Z, Xia Z, Zemoura L, Advenier C, Chen H (2010) Tiotropium reduction of lung inflammation in a model of chronic gastro-oesophageal reflux. Eur Respir J 35(6):1370–1376. https://doi.org/10.1183/09031936.00139909

Devkar SJ, Patange VD, Nema V (2021) Atropine: a new perspective on prophylactic and therapeutic management of COVID-19. Radiol Diagn Imaging 4:1–3. https://doi.org/10.15761/RDI.1000171

Dewick PM (2002) Medicinal natural products: A biosynthetic approach. Wiley, Hoboken

Diaz JL (2010) Sacred plants and visionary consciousness. Phenomenol Cogn Sci 9(2):159–170. https://doi.org/10.1007/s11097-010-9157-z

Dräger B (1994) Identification and quantification of calystegines, polyhydroxyl nortropane alkaloids. Phytochem Anal 6(1):31–37. https://doi.org/10.1002/pca.2800060105

Dräger B (2004) Chemistry and biology of calystegines. Nat Prod Rep 21(2):211–223. https://doi.org/10.1039/B300289F

Dreyfuss P, Vogel D, Walsh N (1991) The use of transdermal scopolamine to control drooling. A case report. Am J Phys Med Rehabil 70(4):220–222. https://doi.org/10.1097/00002060-199108000-00012

Dyer J (2020) Tropane alkaloids: pharmacology and therapeutics. Pharmacol Ther 215:107641

Ebert U, Oertel R, Kirch W (2000) Influence of grapefruit juice on scopolamine pharmacokinetics and pharmacodynamics in healthy male and female subjects. Int J Clin Pharmacol Ther 38(11):523–531. https://doi.org/10.5414/cpp38523

EFSA (2011) EFSA Panel on Contaminants in the Food Chain (CONTAM). Scientific opinion on pyrrolizidine alkaloids in food and feed. EFSA J 9(11):2406

Fuchs C, Schwabe M (1990) Die rektale Prämedikation mit Ketamin-Dehydrobenzperidol-Atropin im Kindesalter [Rectal premedication using ketamine-dehydrobenzperidol-atropine in childhood]. Anaesthesiol Reanim 15(5):322–326

Furbee B (2009) Neurotoxic plants. In: Dobbs MR (ed) Clinical neurotoxicology: syndromes, substances, environments. Elsevier, Philadelphia, pp 523–542

Gadzikowska M, Grynkiewicz G (2002) Tropane alkaloids in pharmaceutical and phytochemical analysis. Acta Pol Pharm 59(2):149–160

Ghosh S (2018) Clinical pharmacology of atropine and its derivatives. J Clin Pharmacol 54(5):467–473

Gilani SA, Cobbin LB (1987) Interaction of himbacine with carbachol at muscarinic receptors of heart and smooth muscle. Arch Int Pharm Ther 290(1):46–53. https://doi.org/10.1111/cts.13753

Goldfrank L, Flowmenbaum N, Nelson L (2007) Goldfrank's toxicologic emergencies, 8th edn. McGraw-Hill Medical, New York

Gomez A, Ingelman-Sundberg M (2009) Pharmacoepigenetics: its role in interindividual differences in drug response. Clin Pharmacol Ther 85(4):426–430. https://doi.org/10.1038/clpt.2009.2

Gomez-Mancilla B, Boucher R, Bedard PJ (1991) Effect of clonidine and atropine on rest tremor in the MPTP monkey model of parkinsonism. Clin Neuropharmacol 14(4):359–366. https://doi.org/10.1097/00002826-199108000-00008

González-Gómez L, Morante-Zarcero S, Pérez-Quintanilla D, Sierra I (2022) Occurrence and chemistry of tropane alkaloids in foods, with a focus on sample analysis methods: a review on recent trends and technological advances. Foods (Basel, Switzerland) 11(3):407. https://doi.org/10.3390/foods11030407

Griffin WJ, Lin GD (2000) Chemotaxonomy and geographical distribution of tropane alkaloids. Phytochemistry 53:623–637. https://doi.org/10.1016/S0031-9422(99)00475-6

Grynkiewicz G, Gadzikowska M (2008) Tropane alkaloids as medicinally useful natural products and their synthetic derivatives as new drugs. Pharmacol Rep 60(4):439–463

Gyermek L (2002) Structure-activity relationships among derivatives of dicarboxylic acid esters of tropine. Pharmacol Ther 96(1):1–21. https://doi.org/10.1016/s0163-7258(02)00296-6

Hardegger VE, Ott H (1954) Konfiguration des Cocains und Derivate der Ecgoninsäure. Helv Chim Acta 38(1):312–320. https://doi.org/10.1002/hlca.19550380137

Harrison PK, Tattersall JE, Gosden E (2006) The presence of atropinesterase activity in animal plasma. Naunyn Schmiedeberg's Arch Pharmacol 373(3):230–236. https://doi.org/10.1007/s00210-006-0054-5

Hassan HMA (2015) A short history of the use of plants as medicines from ancient times. Chimia 69:622–623. https://doi.org/10.2533/chimia.2015.622

Haughey SA, Chevallier O, McVey C, Elliott CT (2021) Laboratory investigations into the cause of multiple serious and fatal food poisoning incidents in Uganda during 2019. Food Control 121:107648. https://doi.org/10.1016/j.foodcont.2020.107648

Heinrich M, Mah J, Amirkia V (2021) Alkaloids used as medicines: Structural phytochemistry meets biodiversity – an update and forward look. Molecules (Basel, Switzerland) 26(7):1836. https://doi.org/10.3390/molecules26071836

Henningfield JE, Griffiths RR (1991) Therapeutic uses of cocaine. Trends Pharmacol Sci 12(1):1–3

Hinderling PH, Gundert-Remy U, Schmidlin O (1985) Integrated pharmacokinetics and pharmacodynamics of atropine in healthy humans. I: pharmacokinetics. J Pharm Sci 74(7):703–710. https://doi.org/10.1002/jps.2600740702

Holstege CP, Borek HA (2012) Toxidromes. Crit Care Clin 28(4):479–498

Hulse EJ, Davies JO, Simpson AJ, Sciuto AM, Eddleston M (2014) Respiratory complications of organophosphorus nerve agent and insecticide poisoning. Implications for respiratory and critical care. Am J Respir Crit Care Med 190(12):1342–1354. https://doi.org/10.1164/rccm.201406-1150CI

Jatlow P (1988) Cocaine: analysis, pharmacokinetics, and metabolic disposition. Yale J Biol Med 61(2):105–113. https://www.ncbi.nlm.nih.gov/pmc/articles/PMC2590277/

Jocković N, Fischer W, Brandsch M, Brandt W, Dräger B (2013) Inhibition of human intestinal α-glucosidases by calystegines. J Agric Food Chem 61(23):5550–5557. https://doi.org/10.1021/jf4010737

John H, Rychlik M, Thiermann H, Schmidt C (2018) Simultaneous quantification of atropine and scopolamine in infusions of herbal tea and Solanaceae plant material by matrix-assisted laser desorption/ionization time-of-flight (tandem) mass spectrometry. Rapid Commun Mass Spectrom 32(22):1911–1921. https://doi.org/10.1002/rcm.8264

Jufer RA, Wstadik A, Walsh SL, Levine BS, Cone EJ (2000) Elimination of cocaine and metabolites in plasma, saliva, and urine following repeated oral administration to human volunteers. J Anal Toxicol 24(7):467–477. https://doi.org/10.1093/jat/24.7.467

Kaddurah-Daouk R, Weinshilboum RM (2014) Pharmacometabolomics: implications for clinical pharmacology and systems pharmacology. Clin Pharmacol Ther 95(2):154–167. https://doi.org/10.1038/clpt.2013.217

Kaddurah-Daouk R, Kristal BS, Weinshilboum RM (2008) Metabolomics: a global biochemical approach to drug response and disease. Ann Rev Pharmacol Toxicol 48:653–683. https://doi.org/10.1146/annurev.pharmtox.48.113006.094715

Kalser SC (1971) The fate of atropine in man. Ann N Y Acad Sci 179(1):667–683

Kanto J, Kentala E, Kaila T, Pihlajamäki K (1989) Pharmacokinetics of scopolamine during caesarean section: relationship between serum concentration and effect. Acta Anaesthesiol Scand 33:482–486. https://doi.org/10.1111/j.1399-6576.1989.tb02950.x

Kohnen-Johannsen KL, Kayser O (2019) Tropane alkaloids: chemistry, pharmacology, biosynthesis and production. Molecules 24(4). https://doi.org/10.3390/molecules24040796

Kowalczuk A, Bourebaba N, Panchuk J, Marycz K, Bourebaba L (2022) Calystegines improve the metabolic activity of human adipose derived stromal stem cells (ASCs) under hyperglycaemic condition through the reduction of oxidative/ER stress, inflammation, and the promotion of the AKT/PI3K/mTOR pathway. Biomol Ther 12(3):460. https://doi.org/10.3390/biom12030460

Kumar S, Maurya VK, Kabir R, Nayak D, Khurana A, Manchanda RK, Gadugu S, Shanker K, Saxena SK (2020) Antiviral activity of belladonna during japanese encephalitis virus infection via inhibition of microglia activation and inflammation leading to neuronal cell survival. ACS Chem Neurosci 11(21):3683–3696. https://doi.org/10.1021/acschemneuro.0c00603

Ladenburg A (1879) Künstliches Atropin. Ber Dtsch Chem Ges 12(1):941–944

Langmead CJ, Watson J, Reavill C (2008) Muscarinic acetylcholine receptors as CNS drug targets. Pharmacol Ther 117(2):232–243. https://doi.org/10.1016/j.pharmthera.2007.09.009

Lauersen KJ (2019) Eukaryotic microalgae as hosts for light-driven heterologous isoprenoid production. Planta 249(1):155–180. https://doi.org/10.1007/s00425-018-3048-x

Lochner M, Thompson AJ (2016) The muscarinic antagonists scopolamine and atropine are competitive antagonists at 5-HT3 receptors. Neuropharmacology 108:220–228. https://doi.org/10.1016/j.neuropharm.2016.04.027

Lossen W, Lossen W, Lossen W, Lossen W (1864) Ueber das Atropin. Annal Chem Pharm 131(1):43–49. https://doi.org/10.1002/jlac.18641310103

Lounasmaa M, Tamminen T (1993) The tropane alkaloids. Alkaloids 44:1–114. https://doi.org/10.1016/S0099-9598(08)60143-1

Marín-Sáez J, Romero-González R, Frenich AG (2019) Reliable determination of tropane alkaloids in cereal based baby foods coupling on-line spe to mass spectrometry avoiding chromatographic step. Food Chem 275:746–753

McLendon K, Preuss CV (2024) Atropine. [Updated 2023 June 23]. In: StatPearls [Internet]. StatPearls Publishing, Treasure Island. Available from: https://www.ncbi.nlm.nih.gov/books/NBK470551/

McQueen DS, Donaldson K, Bond SM, McNeilly JD, Newman S, Barton NJ, Duffin R (2007) Bilateral vagotomy or atropine pre-treatment reduces experimental diesel-soot induced lung inflammation. Toxicol Appl Pharmacol 219(1):62–71

Merriam-Webster (2023) Definition of pharmacology. Retrieved 28 February 2023 https://www.merriamwebster.com/dictionary/pharmacology#:~:text=phar%C2%B7%E2%80%8Bma%C2%B7%E2%80%8Bcol,pharmacokinetics%2C%20therapeutic%20use%2C%20and%20toxicology

Michelon H, Larabi IA, Lemoine J, Alvarez J-C, Genevée A, Lillo-Lelouet A, Azouvi P, Lefèvre C, Paquereau J (2021) Atropine-induced toxicity after off-label sublingual administration of eyedrop for sialorrhoea treatment in neurological disabled patients. Br J Clin Pharmacol 87(8):3364–3369. https://doi.org/10.1111/bcp.14757

Mirakhur RK (1978) Comparative study of the effects of oral and i.m. atropine and hyoscine in volunteers. Br J Anaesth 50(6):591–598. https://doi.org/10.1093/bja/50.6.591

More SR, Dabhade SS, Ghongane BB (2015) Drug audit of intravenous anaesthetic agents in tertiary care hospital. J Clin Diagn Res 9(11):FC25. https://doi.org/10.7860/JCDR/2015/14159.6815

Mulder PJ, Castellari M, Hortos M, MacDonald S, Crews C, Hajslova J, Stranska M (2016) Occurrence of tropane alkaloids in food. EFSA Support Publ 13(12):1140E. https://doi.org/10.2903/sp.efsa.2016.EN-1140

Mutschler E, Geisslinger G, Kroemer HK et al (2013) Mutschler Arzneimittelwirkungen. WVG, Wissenschaftliche Verlagsgesellschaft, Stuttgart

Nachum Z, Shahal B, Shupak A, Spitzer O, Gonen A, Berian I, Lavon H, Eynan M, Dachir S, Levy A (2001) Scopolamine bioavailability in combined oral and transdermal delivery. J Pharmacol Exp Ther 296(1):121–123

Noorily AD, Susan SH, Otto RA (1995) Cocaine, lidocaine, tetracaine: which is best for topical nasal anesthesia? Anesth Analg 81(4):724–727

O'Shaughnessy KM (2011) Cholinergic and antimuscarinic (anticholinergic) mechanisms and drugs. Clinical. In: Pharmacology, 11th edn, pp 372–381. https://doi.org/10.1016/B978-0-7020-4084-9.00061-6

Olsson GL, Bejersten A, Feychting H, Palmér L, Pettersson BM (1983) Plasma concentrations of atropine after rectal administration. Anaesthesia 38(12):1179–1182

Ozçelik B, Karta LM, Orhan I (2011) Cytotoxicity antiviral and antimicrobial activities of alkaloids, flavonoids, and phenolic acids. Pharm J 49:396–402

Pagare S, Bhatia M, Tripathi N, Pagare S, Bansal YK (2015) Secondary metabolites of plants and their role: overview. Curr Trends Biotechnol Pharm 9:293–304

Paly D, Jatlow P, Van Dyke C, Jeri FR, Byck R (1982) Plasma cocaine concentrations during cocaine paste smoking. Life Sci 30(9):731–738. https://doi.org/10.1016/0024-3205(82)90606-3

Park L, Furey M, Nugent A, Farmer C, Ellis J, Szczepanik J, Lenner M, Zarate C (2018) Negative trial of scopolamine in major depressive disorder does not demonstrate neurophysiological changes seen with the antidepressant response of ketamine. Biol Psychiatry 83:S305–S306

Ravina E (2011) The evolution of drug discovery: from traditional medicines to modern drugs. Wiley, Weinheim. https://books.google.com/books?hl=fr&lr=&id=iDNy0XxGqT8C&oi=fnd&pg=PR5&dq=Traditional+

Razani-Boroujerdi S, Behl M, Hahn FF, Pena-Philippides JC, Hutt J, Sopori ML (2008) Role of muscarinic receptors in the regulation of immune and inflammatory responses. J Neuroimmunol 194:83–88

Reinold J, Braitmaier M, Riedel O, Haug U (2021) Anticholinergic burden: First comprehensive analysis using claims data shows large variation by age and sex. PLoS One 16(6):e0253336. https://doi.org/10.1371/journal.pone.0253336

Reis HT (2001) Relationship experiences and emotional well-being. In: Ryff CD, Sinder B (eds) Emotion, social relationships, and health. Oxford University Press, New York, pp 57–86

Renner U, Oertel R, Kirch W (2005) Pharmacokinetics and pharmacodynamics in clinical use of scopolamine. Ther Drug Monit 27:655–665. https://doi.org/10.1097/01.ftd.0000168293.48226.57

Ricard F, Abe E, Duverneuil-Mayer C, Charlier P, de la Grandmaison G, Alvarez JC (2012) Measurement of atropine and scopolamine in hair by LC-MS/MS after *Datura stramonium* chronic exposure. Forensic Sci Int J 223(1–3):256–260. https://doi.org/10.1016/j.forsciint.2012.09.015

Richter U, Sonnewald U, Dräger B (2007) Calystegines in potatoes with genetically engineered carbohydrate metabolism. J Exp Bot 1:13. https://doi.org/10.1093/jxb/erl295

Riezzo I, Fiore C, Carlo D, Pascale N, Neri M, Turillazzi E, Fineschi V (2012) Side effects of cocaine abuse: multiorgan toxicity and pathological consequences. Curr Med Chem 19(33):5624–5646. https://doi.org/10.2174/092986712803988893

Rizkallah MR, Saad R, Aziz RK (2010) The human microbiome project, personalized medicine and the birth of pharmacomicrobiomics. Curr Pharmacogenomics Pers Med 8(3):182–193. https://doi.org/10.2174/187569210792246326

Robenshtok E, Luria S, Tashma Z, Hourvitz A (2002) Adverse reaction to atropine and the treatment of organophosphate intoxication. Isr Med Assoc J 4:535–539

Rothman RB, Baumann MH, Dersch CM, Romero DV, Rice KC, Carroll FI, Partilla JS (2001) Amphetamine-type central nervous system stimulants release norepinephrine more potently than they release dopamine and serotonin. Synapse (New York, N.Y.) 39(1):32–41. https://doi.org/10.1002/1098-2396(20010101)39:1<32::AID-SYN5>3.0.CO;2-3

Rothman RB, Baumann M, Prisinzano TE, Newman AH (2008) Dopamine transport inhibitors based on GBR12909 and benztropine as potential medications to treat cocaine addiction. Biochem Pharmacol 75(1):2–16

Rudy D, Cline K, Harris R, Goldberg K, Dmochowski R (2006) Multicenter phase III trial studying trospium chloride in patients with overactive bladder. Urology 67:275–280. https://doi.org/10.1016/j.urology.2005.08.017

Salmanroghani H, Mirvakili M, Mirjalili M, Baghbanian M, Salmanroghani R (2020) The efficacy and safety of low dose versus usual dose of hyoscine during endoscopic retrograde cholangiopancreatography: a randomized clinical trial. Clin Pharmacol 12:123–130. https://doi.org/10.2147/CPAA.S263531

Sam S (2019) Importance and effectiveness of herbal medicines. J Pharmacogn Phytochem 8:354–357

Scavone JM, Greenblatt DJ, Harmatz JS, Engelhardt N, Shader RI (1998) Pharmacokinetics and pharmacodynamics of diphenhydramine 25 mg in young and elderly volunteers. J Clin Pharmacol 38(7):603–609. https://doi.org/10.1002/j.1552-4604.1998.tb04466.x

Schlagmann C, Remien J (1986) Klinische schrift zur behandlung der parkinson-krankheit. Wien Klin Wochenschr 64:939–942. https://doi.org/10.1007/BF01728622

Scurlock J (2014) Sourcebook for ancient Mesopotamian medicine. SBL Press, Atlanta. https://doi.org/10.2307/j.ctt1287mwm

Shaoul E, Ayalon A, Tal Y, Lotan T (2012) Transdermal delivery of scopolamine by natural submicron injectors: in-vivo study in pig. PLoS One 7(2):e31922. https://doi.org/10.1371/journal.pone.0031922

Shim KH, Kang MJ, Sharma N, An SSA (2022) Beauty of the beast: anticholinergic tropane alkaloids in therapeutics. Nat Prod Bioprospect 12:33. https://doi.org/10.1007/s13659-022-00357-w

Sinatra RS, Jahr JS, Watkins-Pitchford M (2010) The essence of analgesia and analgesics. Cambridge University Press, Cambridge

Sneader W (2005) Drug discovery: a history. Wiley, New York

Sorbe B, Andersson H, Schmidt M, Söderberg M, Högberg T, Wernstedt L, Janson ET, Ehrström B, Kjaer M, Havsteen H (1994) Tropisetron (Navoban) in the prevention of chemotherapy-induced nausea and vomiting—the Nordic experience. Support Care Cancer 2:393–399. https://doi.org/10.1007/BF00344055

Spinks A, Wasiak J (2011) Scopolamine (hyoscine) for preventing and treating motion sickness. Cochrane Database Syst Rev 657:6

Sumner J (2000) The natural history of medicinal plants. Timber Press, London

Sweta VR, Lakshmi TJ (2015) Pharmacological profile of tropane alkaloids. J Chem Pharm Res 7(5):117–119

Takemura Y, Helms MN, Eaton AF, Self J, Ramosevac S, Jain L, Bao H-F, Eaton DC (2013) Cholinergic regulation of epithelial sodium channels in rat alveolar type 2 epithelial cells. Am J Physiol Lung Cell Mol Physiol 304(6):L428–L437

Thirumurugan D, Cholarajan A, Raja SS, Vijayakumar R (2018) An introductory chapter: secondary metabolites. In: Vijayakumar R, Raja SSS (eds) Secondary metabolites-sources and applications. Intech Open, London, pp 3–21

Tytgat GN (2007) Hyoscine butylbromide. Drugs 67(9):1343–1357. https://doi.org/10.2165/00003495-200767090-00007

Vallance P, Smart TG (2006) The future of pharmacology. Brit J Pharmacol 147(Suppl 1 (S1)):S304–S307. https://doi.org/10.1038/sj.bjp.0706454

Van der Meer MJ, Hundt HK, Müller FO (1986) The metabolism of atropine in man. J Pharm Pharmacol 38(10):781–784. https://doi.org/10.1111/j.2042-7158.1986.tb04494.x

Vaou N, Stavropoulou E, Voidarou C, Tsigalou C, Bezirtzoglou E (2021) Towards advances in medicinal plant antimicrobial activity: a review study on challenges and future perspectives. Microorganisms 9(10):2041. https://doi.org/10.3390/microorganisms9102041

Wada S, Yoshimitsu T, Koga N, Yamada H, Oguri K, Yoshimura H (1991) Metabolism in vivo of the tropane alkaloid, scopolamine, in several mammalian species. Xenobiotica 21(10):1289–1300. https://doi.org/10.3109/00498259109043204

Wang X, Zhou ZJ, Zhang XF et al (2010) Comparison of two different doses of rectal ketamine added to 0.5 mg × kg(−1) midazolam and 0.02 mg × kg(−1) atropine in infants and young children. Anaesth Intensive Care 38(5):900–904

Waseem R, Hussein MIH, Salih TSM, Ahmed SMG (2023) Pharmacology of the autonomic nervous system. IntechOpen. https://doi.org/10.5772/intechopen.113006

Watts SW, Kanagy NL, Lombard JH (2008) Receptor-mediated events in the microcirculation. In: Microcirculation, 2nd edn, pp 285–348. https://doi.org/10.1016/B978-0-12-374530-9.00007-3

Wess J, Duttaroy A, Zhang W, Gomeza J, Cui Y, Miyakawa T, Bymaster FP, McKinzie L, Felder C, Lamping KG, Faraci FM, Deng C, Yamada M (2003) M 1-M 5 Muscarinic receptor knockout mice as novel tools to study the physiological roles of the muscarinic cholinergic system. Recept Channels 9:279–290

WHO (2013) Hyoscine butylbromide-EML. WHO, Geneva

Wink MA (1998) Short history of alkaloids. In: Roberts MF, Wink M (eds) Alkaloids. Springer, Boston, pp 11–44. https://doi.org/10.1007/978-1-4757-2905-4_2

Wolffenstein R (1923) Die Pflanzenalkaloide. J Am Med Assoc 80(15):1094. https://doi.org/10.1001/jama.1923.02640420056037

Yamazaki Z, Tagaya I (1980) Antiviral effects of atropine and caffeine. J Gen Virol 50:42

Yu H, Lv D, Shen M, Zhang Y, Zhou D, Chen Z, Wang C (2019) BDNF mediates the protective effects of scopolamine in reserpine-induced depression-like behaviors via up-regulation of 5-HTT and TPH1. Psychiatry Res 271:328–334

Zeng J, Zhao K, Lan X, Liao Z (2016) Analyzing the contents of tropane alkaloids in *Scopolia lurida*, a resource plant species of Tibetan medicines. Sci Technol Tibet 279:60–62

Zhang XC, Farrell N, Haronian T, Hack J (2017) Postoperative anticholinergic poisoning: concealed complications of a commonly used medication. J Emerg Med 53(4):520–523

Zhu YP (1998) Chinese materia medica: chemistry, and applications. Harwood Academic Publishers, Amsterdam

Ziegler J, Facchini PJ (2008) Alkaloid biosynthesis: metabolism and trafficking. Annu Rev Plant Biol 59:735–769

Chapter 6
Tropane Alkaloids in Food Commodities: Occurrence in Food and Feed

Nandini Boregowda, Poojitha Jain, and Parvatam Giridhar

Abstract Tropane alkaloids (TAs) represent a class of secondary metabolites derived from weeds, which have the potential to contaminate vegetables and grains during the harvesting process. Over 200 distinct TAs have been identified throughout various plant parts, such as seeds, fruits, flowers, leaves, and stems. The widespread distribution contributes to frequent cross-contamination due to rapid and mechanical harvesting methods. TAs are secondary metabolites that are found naturally in various plant families. Human exposure to TAs is mainly linked to the dietary intake of certain plants from the Solanaceae family, which often contribute to the cross-contamination of agricultural products during harvesting. Due to the invasive characteristics of TA-producing crops and the distribution of TAs throughout all parts of the plant, there exists a considerable risk of unintentional contamination of edible crops during the harvesting and processing phases. Cereals, pseudo-cereals, legumes, and grains represent the crops most frequently affected by contamination, as demonstrated by many alerts documented on the Rapid Alert System for Food and Feed (RASFF) portal throughout the last 28–30 years. The cross-contamination of various vegetable food products arises from their morphological and botanical similarities to other commonly consumed vegetables, particularly the young leaves of borage and spinach. Their global distribution and invasive

N. Boregowda
Plant Cell Biotechnology Department, CSIR-Central Food Technological Research Institute, Mysuru, Karnataka, India

P. Jain
Plant Cell Biotechnology Department, CSIR-Central Food Technological Research Institute, Mysuru, Karnataka, India

Academy of Scientific and Innovative Research (AcSIR), Ghaziabad, India

P. Giridhar (✉)
Plant Cell Biotechnology Department, CSIR-Central Food Technological Research Institute, Mysuru, Karnataka, India

Academy of Scientific and Innovative Research (AcSIR), Ghaziabad, India

Food Safety and Analytical Quality Control Laboratory, CSIR-Central Food Technological Research Institute, Mysuru, Karnataka, India

T. Pullaiah et al. (eds.), *Tropane Alkaloids*,
https://doi.org/10.1007/978-981-96-5805-3_6

nature further exacerbate this issue. The current chapter presents a complete summary of the current information on the contamination of foods and feeds, as well as the economic concessions available for the production of TAs, ranging from food production to feed production.

Keywords Secondary metabolites · Plant natural toxins · Bioactive properties · Cross-contamination

6.1 Introduction

Tropane alkaloids (TAs) are characterized as compounds with a tropane ring structure (Grynkiewicz and Gadzikowska 2008). Globally distributed Solanaceae plants produce TA. *Datura* plants, originating in Asia and America, grow worldwide except in polar and subpolar climates. *Duboisia* plants are found in Australia and New Caledonia. *Atropa* and *Hyoscyamus* originated in Europe, Asia, and North Africa and spread to the USA, Australia, and Canada (Griffin and Lin 2000). Drugs like cocaine, scopolamine, and calystegines all have a tropane moiety in common with one another. However, these chemicals affect humans physiologically differently. TAs represent a class of secondary metabolites synthesized by a diverse array of plant species, predominantly within the families Solanaceae, Brassicaceae, and Erythroxylaceae (González-Gómez et al. 2022). Additionally, they can be found in other families, including Proteaceae, Euphorbiaceae, Rhizophoraceae, and Convolvulaceae (Gonçalves et al. 2006, 2017; Dräger 2003; Adamse et al. 2014; Adibah and Azzreena 2019). These compounds function as a protective strategy against various organisms and exhibit allelopathic influences on various other species of plants (Croteau et al. 2000; Jank and Rath 2021).

The plant natural toxins known as tropane (TAs), opium (OAs), and pyrrolizidine (PAs), alkaloids have drawn attention recently due to the concerning effects that their consumption may have on the organism, ranging from minor ailments like acute intoxications to major conditions like the onset of chronic diseases or even death (EFSA 2011a, b, 2013a, b, 2018, Wang et al. 2024). Several plant families are commonly farmed for food, including Brassicaceae, Proteaceae, Solanaceae, and Convolvulaceae. There is no evidence of TAs in cultivated Convolvulaceae species like *Ipomoea batatas* (sweet potato) and *Ipomoea aquatica*. However, this plant family is known to produce tropine esters of methoxy-substituted benzoic acids, including convolvine and convolidine. None of the Proteaceae family's commonly consumed species have been found to produce TAs. On the contrary, it is worth noting that the TAs that have been found within the *Solanum* genus do not contain either (−)-hyoscyamine or (−)-scopolamine, as stated by EFSA et al. (2018) and Griffin and Lin (2000). Although no edible Brassicaceae species have been found to possess hyoscyamine or scopolamine, *Brassica oleracea* has been reported to have calystegines at amounts of up to 30 mg kg^{-1} by dry weight (DW) (Griffin and Lin 2000). The Solanaceae family includes over 100 genera and 3000 plant species, with several containing poisonous TAs. Plants commonly ingested include *Capsicum*, *Jaltomata*, *Physalis*, *Solanum*, and *Lycianthes*. Only *Physalis* and *Solanum* have TAs. *Physalis* contains TAs only in its roots and leaves, not edible fruits. The present chapter provides a comprehensive overview of current knowledge on medicinal and pharmaceutical applications of TAs and a comparative analysis of TA production economic concedes from food production to feed.

6.2 Chronology and Chemical Characterisation of Tropane Alkaloids

Historical accounts of states of perception induced by hyoscyamine or scopolamine date back to ancient times. Approximately 3000 years ago, Mandragora extracts were incorporated into beer in Egypt to reduce the alcohol content of these beverages. *Datura* extracts are utilized in China and Russia, while *Hyoscyamus* is incorporated in Europe to augment the stimulating effects of beer. Over 200 compounds in the TA group have a tropane ring system ((1R,5S)-8-methyl-8-azabicyclo) and alkaline-like characteristics (Perharič et al. 2013a, b; Kohnen-Johannsen and Kayser 2019). The tropane ring system is a cyclic structure with fused pyrrolidine and piperidine rings sharing a nitrogen atom and two carbon atoms. The nitrogen atom on carbon-8 is usually methylated; however, demethylated derivatives (nortropane) also exist. Two varieties of TAs can be identified by analyzing the location of the hydroxyl group on carbon 3. Esters of 3-tropanole (tropine) carrying TAs are the most commonly discovered. 3β-tropanole (pseudotropine) has its hydroxyl group in the equatorial position, in contrast to the axial location of the hydroxyl group in the former. The first group includes TAs found in plants of the Solanaceae family, such as hyoscyamine and scopolamine, and the second group includes *Erythroxylum coca* alkaloids, such as cocaine, which are derived from pseudotropine. Polyhydroxylated nortropane alkaloids known as calystegines are in the Solanaceae family and contain calystegines A3, A5, B1, B2, B3, B4, C1, and N1 (Romera-Torres et al. 2018).

Gas chromatography-mass spectrometry (GC-MS) was initially used to reveal the structure of some *Datura* TAs by Berkova and Zayed (2004) and Doncheva et al. (2006). According to EFSA (2008), TAs work biologically by competitively opposing muscarinic cholinergic receptors, which prevents acetylcholine from binding. The anticholinergic activity brings numerous symptoms, including dry mouth, decreased secretions, trouble swallowing and speaking, tachycardia, sleepiness, dizziness, hallucinations, and delirium. Increased dosages of TAs have the potential to cause mortality, circulatory and respiratory failure, depression of the central nervous system, and coma (Kohnen-Johannsen and Kayser 2019; Fatur and Kreft 2020; Khoshnam-Rad et al. 2022).

Certain plants yield vibrant berries or stunning flowers, making them appealing and frequently used for decorative purposes (EFSA 2008). These plants exhibit a significant diversity and are distributed extensively across the globe. Herbal infusions (contaminated or mislabeled) are the most frequently reported route of unintentional human exposure. It is also known that individuals intentionally consume seeds or other plant parts and smoke-dried leaves for recreational purposes. Selective M1 muscarinic acetylcholine receptor antagonists benztropine affect the central nervous system. It combines atropine's tropine and diphenhydramine's benzhydryl skeletons. It partially blocks basal ganglia cholinergic action and blocks dopamine reuptake, increasing availability. This boosts dopaminergic activity and treats early Parkinson's (Schlagmann and Remien 1986). The fresh leaves of *Erythroxylum coca* have a historical precedent as a mild stimulant, traditionally chewed to enhance physical strength over the centuries. In the past couple of decades, there have been numerous reports of intoxication episodes caused by the consumption of contaminated herbal infusions, including nettle (*Urtica*) tea, burdock (*Arctium*) root tea, comfrey (Symphytum) tea, lungwort (*Pulmonaria*) tea, and Paraguay (*Ilex paraguariensis*) tea (Chan 2017).

6.3 Tropane Alkaloids Contaminating Various Foods

6.3.1 TAs Concentration-Dependent Food Risks

Tropane alkaloids exert toxic effects that can range significantly from mild adverse symptoms like dilated pupils, increased heart rate, and hallucinations to severe outcomes, including respiratory failure and death, depending on the dose ingested. The introduction of tropane alkaloids into the human body typically occurs through several pathways, including contamination of food sources, misuse of traditional or alternative medicine, or intentional abuse due to their psychoactive properties (Lugasi 2023). Their intoxicating effects can lead to a dangerous allure, prompting individuals to seek them out despite the associated risks. These alkaloids may inadvertently find their way into food products meant for public consumption. Documented cases of poisoning have been linked to insufficiently cleaned cereals, pulses, or other grain crops, which may harbor these toxic substances. This risk is notably amplified in organic products, as they are often produced without chemical pesticides that could otherwise reduce the presence of such contaminants. Therefore, organic product consumers might be at higher risk if proper agricultural practices and cleaning processes are not meticulously followed. Awareness and education regarding the potential risks of tropane alkaloids are essential for safe consumption and preventing poisoning incidents (Lugasi 2023).

Hyoscyamine is more prevalent in *Datura* and *Atropa*, while scopolamine can be found in higher quantities in *Hyoscyamus* species. In *Datura* species, hyoscyamine levels range from tens to 5900 mg kg^{-1}, but scopolamine levels are lower. *Atropa* species, hyoscyamine levels can surpass 9000 mg kg^{-1}, but scopolamine levels are typically 5–10 times lower. In the *Hyoscyamus* species, hyoscyamine and scopolamine levels rarely surpass 1000 mg kg^{-1}. High amounts of hyoscyamine and scopolamine, up to 30,000 mg kg^{-1} and 7000 mg kg^{-1}, have been recorded (EFSA 2013a, b). This involves identifying food concerns, affected items, geographical distribution, and actions taken (RASFF 2019). The RASFF database was searched for TAs detection alerts (atropine and scopolamine) over the available period. The RASFF database was reviewed for notifications regarding the detection of TAs (atropine and scopolamine) during the available period (Table.6.1).

6.3.2 Prevalence of Tropane Alkaloids in Food and Feed

The literature on the prevalence and levels of TAs in food and feed can be divided into three categories: reports covering the primary food commodities consumed across multiple countries, notifications produced by official authorities in charge of market surveillance, and specific studies spent on the advancement of analytical techniques and their verification.

6.3.2.1 In Cereals and Cereal Products

According to González-Gómez et al. (2022), TAs have been found in many different cereals and pseudocereals, including millet, teff, buckwheat, sorghum, rice, and many more. In order to prevent malnutrition, babies between the ages of 6 and almost 36 months should get vitamins and minerals (European Parliament 2013; Roess et al. 2018). Many cereal species can make cereal-based baby feeds, including millet, barley, corn, oats, quinoa, rice, rye, sorghum, triticale, and wheat (Jeelani et al. 2020). TAs-containing seeds and other plant parts have the potential to infect these grains. Secondary metabolites known as TAs are found naturally in plants belonging to several families, such as Erythroxylaceae, Convolvulaceae, Brassicaceae, and Solanaceae (which includes plants like mandrake, henbane, deadly nightshade, and Jimson weed). Cross-contamination is common as a result of rapid and mechanical harvesting because more than 200 distinct TAs have been identified and are present in every part of a plant, including seeds, fruits, flowers, leaves, and stems (EFSA CONTAM Panel 2013a, b; Kohnen-Johannsen and Kayser 2019). Similarly, despite the widespread belief that PAs are unlikely to be present in these food matrices, several researchers have discovered PAs in infant food samples made of maize and multigrain (García-Juan et al. 2023; Letsyo et al. 2021). TA was identified in 10 of 151 cereal samples in Korea. The detection levels of TA in the ten samples varied from 0.12 to 88.10 μg/kg (Han et al. 2024).

In 2015, a peak of ten notifications was recorded following the 2013 European Food Safety Authority (EFSA) Opinion, which emphasized the necessity of data collection regarding TAs in cereals and oilseeds. This led to the issuance of Recommendation (European Union (EU) 2015/976; EFSA 2013a, b; Commission

Table 6.1 Tropane alkaloids contaminate various foods and their source

S. no.	Name of the food/food product	Name of the TAs contamination	Source	References
1	Bread made from wheat flour	Jimsonweed (*Datura stramonium*)	Wheat flour	Friedman and Levin (1989)
2	Buckwheat	Hyoscyamine and scopolamine	*Datura stramonium* seeds	Perharic (2005)
3	Soybeans	Jimsonweed seeds (*Datura stramonium*)	*Datura* seeds	Dugan et al. (1989)
4	Honeycomb	*Datura innoxia* (suspected)	Wasp honey	Ramirez et al. (1999)
5	Herbal teas (e.g., nettle, comfrey, and burdock)	*Atropa belladonna* berries	Tea contamination	Adamse and van Egmond (2010)
6	Buckwheat flour in baby food	Hyoscyamine and scopolamine	Contaminated buckwheat	Marín-Sáez et al. (2019a, b)
7	Green beans	*Datura stramonium* flower buds	Canned vegetables	Department of health (2010)
8	Sorghum	*Datura stramonium* seeds	Sorghum grain	Onen et al. (2002)
9	Pasta	Hyoscyamine and scopolamine	Buckwheat	Perharič et al. (2013a, b)
10	Millet-based dishes	*Datura stramonium* seeds	Millet grain	Fretz et al. (2007)
11	Cereal products	*Datura stramonium* seeds	General cereal contamination	Mulder et al. (2016)
12	Biscuits	Atropine and scopolamine	Cereal-based products	Marín-Sáez et al. (2019a, b)
13	Herbal teas (e.g., Paraguay tea)	*Hyoscyamus niger* (henbane)	Plant contamination	CDC (1995)
14	Stir-fry mixes	*Datura stramonium* seeds	Mixed vegetables	RASFF (2013)
15	Bread	Atropine and scopolamine	Datura seeds	List et al. (1979)
16	Sunflower seeds	*Datura stramonium* seeds	Bird feed	RASFF (2012)
17	Poppy seeds	*Hyoscyamus niger* (henbane)	Poppy seed contamination	RASFF (2007)
18	Linseed	*Datura stramonium*	Crop seed contamination	EFSA (2008)
19	Corn flour	*Datura stramonium*	Corn flour	Aga and Geyid (1992)
20	Red millet	*Datura stramonium* seeds	Pet food	RASFF (2005)
21	Cereal grains	*Datura* spp. (various seeds)	Contaminated grain samples	Mulder et al. (2016)

(continued)

Table 6.1 (continued)

S. no.	Name of the food/food product	Name of the TAs contamination	Source	References
23	Dried herbal teas	*Atropa belladonna* berries	Herbal tea products	Romera-Torres et al. (2018)
24	Legumes and stir-fry mixes	*Datura* spp. seeds	Processed food samples	RASFF (2013)
25	Infant cereal-based foods	Hyoscyamine and scopolamine	Baby food products	Marín-Sáez et al. (2019a)
26	Processed cereal products	Hyoscyamine and scopolamine	Cereal-based foods	EFSA (2008)
28	Marshmallow root	Hyoscyamine	Herbal supplements	RASFF (2013)
29	Buckwheat-based baby food	Hyoscyamine and scopolamine	Buckwheat contamination	Marín-Sáez et al. (2019b)
30	Herbal infusions (dietary supplements)	Various TAs	Herbal supplement products	Martinello et al. (2017)
31	Wheat-based Super Cereal (SC)	Atropine, scopolamine	Fortified cereal blend	WFP (2020)
32	Super Cereal Plus (SC+)	Atropine, scopolamine	Young children cereal blend	WFP (2020)
33	Cereal grains in animal feed	*Datura* spp. (primarily *D. ferox*)	Animal feed contamination	Directive 2002/32/EC
34	Tea blends (e.g., burdock root tea)	*Hyoscyamus niger* (henbane)	Mixed herbal teas	Bryson et al. (1978)
35	Canned vegetable and bacon stir-fry	*Datura stramonium* seeds	Mixed vegetable product	RASFF (2007)
36	Frozen vegetable and bean mix	*Hyoscyamus niger* (henbane)	Mixed frozen vegetables	RASFF (2013)
37	Processed food for infant nutrition	Hyoscyamine, scopolamine	Cereal-based baby food	Marín-Sáez et al. (2019a, b)
38	Spinach-stuffed puff pastry rolls	*Datura stramonium* and *D. metel*	Spinach pasta	Caprai et al. (2024)
39	Poppy seed products	*Hyoscyamus niger* (henbane)	Seeds in baking products	RASFF (2007)
40	Organic wheat	Scopolamine and atropine	Whole grains	Cirlini et al. (2018)

Recommendation (EU) 2015). In cereals and pastry items, atropine outnumbered scopolamine with an average ratio of 2.9, whereas tea and herbs had a ratio of 3.7. TA contamination primarily affected buckwheat, millet, sorghum, and maize, including popcorn, polenta, and derivative goods. Buckwheat and allied products were tested for TAs like atropine, scopolamine, anisodamine, and homatropine. The analytical method was verified, showing good linearity, low quantification limit, repeatability (RSDr<15%), inter-day precision (RSDR<19%), and recovery (74–113%) (Mateus et al. 2024).

6.3.2.2 In Bakery Products

It is important to observe that the concentrations of PAs identified in bakery products are generally minimal and do not present a significant health risk to consumers. The observed phenomenon could be attributed to the minimal quantities of these ingredients, which are prone to PA contamination in bakery products (Table 6.1). Additionally, it is important to consider that these food items undergo heat treatment, unlike the fresh products analyzed in existing literature for PA determination. On the other hand, as OAs are unique to poppy seeds and products derived from them, they will likely appear in products where poppy seeds are used as toppings more and more, like salads, yoghurts, and baked goods (Casado-Hidalgo et al. 2021; Casado et al. 2023, 2024b). An effective multicomponent analytical method extracts alkaloids from commercial bread items simultaneously. The validated analytical method was applied to 15 samples, demonstrating 100% contained target alkaloids, sometimes exceeding the legal limits. The three alkaloid families were also found in two samples. These findings emphasize the need to monitor dietary alkaloids and include opium alkaloids in present regulations (Vera-Baquero et al. 2024). Gonçalves et al. (2020) consider related matrices to get insight into alternative and advanced analytical methodologies. The scientific opinion on opium alkaloids in poppy seeds is worthy of having an understanding of these metabolite level limits (EFSA et al. 2018).

6.3.2.3 Tea and Herbs

The median concentrations of atropine and scopolamine were elevated in tea and herbs, the primary food types within the category of "cocoa and cocoa preparations, tea, and coffee" compared to cereals and bakery products (Table 6.1). These levels were also greater than those found in baby food, the leading food type in the "dietetic foods, food supplements, and fortified foods." In 2017, organic burdock root (*Arctium lappa*) tea in Germany contained 520 mg kg^{-1} of atropine, which was linked to food poisoning. In 2018, whole cumin seeds in Hungary contained 16.1 and 4.6 mg kg^{-1} of atropine and scopolamine, respectively. The EFSA CONTAM Panel (2013a, b) determined that *Datura* TAs are one of the most significant families of TAs for consumer food safety. Furthermore, large TAs and PAs that may be hazardous to humans have been found in aromatic herbs, spices, teas, and herbal teas (Casado et al. 2023, 2024a; González-Gómez et al. 2022).

6.3.2.4 Infant Foods

Despite few detections in infant food and no scopolamine findings, levels surpassed the Regulation (EU) 2016/239 limit of 1 μg/kg^{-1} by 37-fold. Most of the central European countries where the TAs were found were Germany, the Czech Republic, Slovenia, and Austria, even if some of the products had already been imported from nearby nations. Most of the notifications occurred within the framework of official market regulations, leading to the implicated products' recall or withdrawal.

6.3.2.5 Other Categories of Foods

Interestingly, 8 of the 38 detections happened in organic products, specifically in organic millet and polenta cornmeal. Notifications for the discovery of TA-producing plant parts and seeds in food and feed are also available in the RASFF database. These notifications include the following: 14 notifications of *Datura stramonium* seeds in millet (mostly organic), sunflower seeds for feed, and green beans; 3 notifications of *Hyoscyamus niger* seeds in poppy seeds; 3 notifications of *Atropa belladonna* parts in tea and herbal tea; and 7 notifications of *Solanum nigrum* in food supplements and green beans (RASFF 2019). Many reports overviewed chromatographic methods for finding and measuring pyrrolizidine alkaloids in food, herbal medicines, and plants (Sousa et al. 2022; Klein et al. 2022; and Tábuas et al. 2024).

6.3.2.6 Honey and Borage

Rizzo et al. (2023) investigated the occurrence of 118 PAs in honey; dietary supplements, honey, food supplements, infusions, honey, and teas were tested for 118 pyrrolizidine alkaloids (PAs) by Rizzo et al. After pretreatment, salting-out-assisted liquid-liquid extraction (SALLE) was employed to extract and clean these materials. A water-based extract was salted with magnesium (II) sulphate heptahydrate and sodium sulphate. Lycopsamine N-oxide had the lowest recovery, 61–117%. This preparation process was repeatable. The pyrrolizidine alkaloids found in Borage (*Borago officinalis*) were studied in detail, and a validated liquid chromatography with tandem mass spectrometry (LC-MS-MS) method was developed to quantify it (Sattler et al. 2023).

6.4 Adverse Effects Upon Consumption

The acute food consumption of atropine and scopolamine was evaluated by estimating the daily intake of the Italian population using the TA concentrations found in the spinach sample or presuming that a 500-g spinach pouch contained 1% of the amount of one *Datura stramonium* leaf. Ominously, the exposure levels in both cases were 2–1200 times higher than the acute standard dose of 0.016 μg/kg bw. Compared to adolescents and the elderly, toddlers had the highest acute dietary exposures on average and in the 95th percentile. Caprai et al. (2024) found that to reduce the risk of illness for vulnerable populations, it is necessary to increase monitoring efforts and implement regulatory measures to reduce exposure to TAs from less commonly contaminated food sources. Casado et al. (2024b) insights into an exhaustive examination of the evolution of these food alerts, emphasizing the primary food products that are frequently contaminated with these alkaloids and reviewing the analytical advancements in their determination over the past 3 years (2020–2023).

6.5 Food Processing vs TAs

Tropane alkaloids (TAs) are known for their remarkable thermostability, which means they maintain their chemical structure and potency even when exposed to heat. Studies have shown that TAs degrade more rapidly in solutions with a neutral pH of 7 compared to more acidic environments, such as those with a pH of 4. This characteristic has important implications for food safety, particularly in infant foods containing these compounds. Despite regulatory limits intended to control the presence of TAs in food products, many infant foods have been found in retail environments to exceed these safe levels. This raises significant concerns about the potential health risks of vulnerable populations, such as infants. Furthermore, because TAs can persist even after heat processing methods, such as cooking or sterilization, they represent a serious chemical hazard that requires ongoing monitoring and evaluation in food safety practices (Torrents-Masoliver et al. 2024). This discrepancy may indicate a potential thermal degradation of these substances during the food processing stage (Casado et al. 2023). Consequently, the amount of TAs that possess toxicological significance consumed by humans is restricted when they adhere to conventional dietary practices.

6.6 Possible Route of Contamination of TAs and Outbreaks in the World

On the other hand, exposure might occur through consuming contaminated food that contains plants that have been picked simultaneously. Additionally, livestock animals appear to avoid plants that contain TA because of their odor or taste, indicating that these plants are typically unappealing. According to Griffin and Lin (2000), livestock will only consume these plants if there is a significant lack of nutritious forages available to them or if they are mixed with hay, which causes the plants to lose their flavor, which is repulsive while still preserving their hazardous properties (Table 6.2).

The elderly have lower renal clearance rates, particularly TA clearance, as well as physical problems such as urine retention, paralytic ileus, constipation, and glaucoma, which can exacerbate the anticholinergic impact. In 2013, four Dutch individuals were hospitalized for anticholinergic poisoning symptoms within 2 h after drinking tea contaminated with *A. belladonna* root from marshmallow (*Althaea*

Table 6.2 Some important tropane alkaloids outbreak

S. no.	Type of TAs outbreak	Affected group	Country	Consequences	References
1	Soybean contamination	Animals	USA	Jimsonweed seed toxicity leading to feed recalls	List and Spencer (1976)
2.	Medicated with dried flowers of *Datura metel* instead of dried flowers of *Campsis grandiflora*	Humans	Hong Kong	Anti-cholinergic poisoning, including palpitations and dilated pupils	Department of Health (2012)
3	Organic grain-based products	Humans	France	Visual impairment, ataxia after repeated exposure	Marín-Sáez et al. (2019a, b)
4	Pet food contamination	Animals	EU countries	Adverse reactions in animals due to *Datura* seeds	RASFF (2012)
5	Buckwheat flour	Humans	Hungary	Symptoms of acute poisoning	Adibah and Azzreena (2019)
6	Sorghum grain contamination	Humans	Botswana	Mental confusion, abdominal symptoms, fatal cases	Onen et al. (2002)
7	Green beans and mixed vegetable contamination	Humans	France	Anticholinergic toxicity	Department of Health (2010)
8	Poppy seed contamination with henbane	Humans	EU countries	Neurological poisoning cases	RASFF (2007)
9	Buckwheat flour with *Datura* spp. seeds	Humans	Slovenia	Severe central nervous system effects	Perharič et al. (2013a, b)
11	Bread consumption with Jimsonweed seeds	Humans	Turkey	Anticholinergic poisoning symptoms	Pulewka and Saygin (1949)
12	*Datura* seeds in frozen vegetable-bean-seed mix	Humans	Finland	Poisoning symptoms	RASFF (2013)
13	Millet contamination with *D. stramonium*	Humans	Austria	Nausea, dry mouth, emesis, hallucinations	Fretz et al. (2007)
14	Buckwheat flour in baby food	Humans	France	Visual disturbances, ataxia, urinary retention	Glaizal et al. (2013)
15	Sorghum grain contamination	Humans	Botswana	Mental confusion, abdominal pain, nausea	Onen et al. (2002)
16	Cereal-based products for children	Humans	European countries	Acute foodborne toxicity	Mulder et al. (2016)

(continued)

Table 6.2 (continued)

S. no.	Type of TAs outbreak	Affected group	Country	Consequences	References
17	Stew made from self-picked Jimsonweed	Humans	USA	Food poisoning symptoms	Russel et al. (2010)
18	Buckwheat flour	Humans	Hungary	Acute poisoning due to high TA concentration	Adibah and Azzreena (2019)
19	Red millet contaminated with *D. stramonium*	Animals	Hungary	Toxicity and adverse effects in pets	RASFF (2005)
20	Bread made with contaminated flour	Humans	Ethiopia	Symptoms include dryness, blurred vision, delirium	Aga and Geyid (1992)
21	Herbal tea contamination	Humans	USA	Neurological symptoms	Bryson et al. (1978)
23	Organic buckwheat flour	Humans	France	Symptoms such as xerostomia, visual disorders	Marín-Sáez et al. (2019a, b)
25	Cereal and legume-based products	Humans	European countries	Varied toxicity effects	EFSA (2008)
26.	Spinach contamination	Humans	Italy	Acute poisoning symptoms	Caprai et al. (2024)
27	Cornmeal contaminated with *D. stramonium*	Humans	Ethiopia	Central nervous system symptoms	Naude et al. (2005)
29	Canned green beans with *Datura* buds	Humans	France	Food poisoning with anticholinergic symptoms	Department of Health, France (2010)
31	Sunflower seed feed contamination	Animals	EU countries	Toxicity in birds due to high TA content	RASFF (2012)
32	Infant formula contamination	Humans	EU countries	Adverse reactions in infants; regulations enforced	EU Commission (2016)
33	Mixed canned vegetable and meat products	Humans	France	Neurological symptoms, primarily dizziness and nausea	RASFF (2007)
34	Honey from wasp hives	Humans	Venezuela	Hallucinations, cardiovascular symptoms	Ramirez et al. (1999)
35	Buckwheat pasta	Humans	Slovenia	Dry mouth, flushing, visual disturbances	Perharič et al. (2013a)
36	Red millet in pet food	Animals	France	Adverse effects in pets	RASFF (2005)
37	Spinach contamination	Humans	Germany	Gastrointestinal and neurological symptoms	Sherif et al. (2023)
38	Marshmallow root products	Humans	EU countries	Cases of mild to moderate TA poisoning symptoms	RASFF (2013)

(continued)

Table 6.2 (continued)

S. no.	Type of TAs outbreak	Affected group	Country	Consequences	References
40	Honey from *Datura* spp. areas	Humans	Turkey	Mild to severe central nervous symptoms	Pulewka and Saygin (1949)
41	Contaminated canned green beans	Humans	France	Severe anticholinergic poisoning symptoms	Department of Health (2010)
42	Jimsonweed contamination in buckwheat grains	Humans	Hungary	Visual disturbances, nausea, hallucinations	Perharič et al. (2013a, b)
43	Atropine contamination in animal feed	Animals	Spain	Adverse symptoms in livestock	Lamp et al. (2021)

officinalis) root (Adamse et al. 2014). Romera-Torres et al. (2018) report amounts up to hundreds of mg kg^{-1}. A small amount of plant material can contaminate significant volumes of grains or herbs, potentially causing anticholinergic effects in people. Children and the elderly are more susceptible to TA poisoning. For instance, newborns and young children consume more cereals (measured in grams per kg body weight per day) (Marín-Sáez et al. 2019a, b). Atropine- and scopolamine-contaminated leafy vegetables are usually linked to seeds from *Datura* genus poison plants; toxic leaves from these plants can also contaminate food, though this is not always the case. Indeed, due to their global distribution, invasiveness, and morphological and botanical similarities to other commonly consumed vegetables (particularly young leaves of borage and spinach), leaves of *Datura* plant (including *Datura stramonium*, *Datura innoxia*, and *Datura metel*) may be the cause of TA cross-contamination of various vegetable food products (Cornara et al. 2018; Abia et al. 2021; de Nijs et al. 2023). Moreover, food items that come from organic farming seem to be more vulnerable to TA contamination. This results from specific weed management techniques used in crop fields, which increase the possibility of inadvertent co-harvesting with *Datura* plants if not done correctly (Cirlini et al. 2018). A current instance of TA poisoning was reported in France, where a family went through a series of severe neurological symptoms like agitation and hallucinations after eating spinach that had been tainted with atropine and scopolamine (Table 6.2). The same thing happened in Athens, where seven individuals ended up in the hospital with anticholinergic syndrome after eating boiled blites tainted with *Datura innoxia*. Furthermore, in 2021, more than 100 individuals became intoxicated after eating deep-frozen spinach puree that was tainted with 3446 μg/kg of atropine and 3860 μg/kg of scopolamine, both of which were caused by the unintentional presence of *Datura* plant residues (Plackova et al. 2022; RASFF 2023). Based on a large sample size, data on contamination rates and levels in milk from individual dairy farms are presented with a maximum contamination of 5.6 mg/L. Senecionine-, lycopsamine-, and heliotrine-type 1,2-dehydropyrrolizidine alkaloids/PA and their corresponding *N*-oxides (PA/PANO) were found. Four samples (1.8%) showed atropine levels up to 0.066 mg/L. Daily exposure does not affect toxin levels in milk samples (Klein et al. 2024).

6.7 Regulatory Aspects of Tropane Alkaloids Notifications

Detecting TAs in food and feed is crucial due to the presence of atropine and scopolamine in cereal and tea/herbal infusion samples, which pose a risk to human health, especially for newborns and young children (EFSA 2013a, b). The European Union has recently published relevant laws. Moreover, the European Medicines Agency (EMA) and EFSA have concluded that human exposure to residues of TAs through the consumption of animal food products (milk, meat, or eggs) resulting from the legal application of *A. belladonna* and atropine (a racemic mixture of hyoscyamine enantiomers) as authorized veterinary medicines, or from the ingestion of contaminated feed, is unlikely to pose a risk (Griffin and Lin 2000). Human exposure to TAs can be categorized into three types: unintentional intake, intentional intake, and intentional poisoning (Beyer et al. 2009). The Regulation (EC) 2023/915 defines limit values for atropine and scopolamine in cereal-based diets for infants, including sorghum, millet, buckwheat, or their derived products due to their toxicity. This limit aimed to investigate, as feasible factors in thermal processing to reduce this chemical hazard in cereal-based infant food, the impact of temperature and pH on the stability of TAs.

The EU database, RASFF was established in 1979 to efficiently and quickly share severe and immediate health and safety risks related to food and feed among concerned parties (national food safety authorities, European Commission, EFSA, ESA, partner countries). Because of possible harmful consequences, the European Commission established regulatory limitations for *Datura*-type TAs in processed cereals for babies and toddlers that contain millet, sorghum, buckwheat, or products derived from these grains (European Commission 2016). According to the European Commission (2023), atropine and scopolamine cannot be present more than 1 μg/kg in these goods. Other than atropine and scopolamine, which have been extensively investigated TAs, other TAs may be found in food (Boros et al. 2010; Caligiani et al. 2011; Chen et al. 2017; González-Gómez et al. 2022). Torrents-Masoliver et al. (2022) found that cereal-based diets for infants and children had atropine and scopolamine levels beyond these limits. New European regulation 2023/915, which updates the highest permissible limits of these alkaloid compounds in a wide variety of food products, was published recently in response to the exponential number of food alerts regarding concerning levels of these alkaloids in food (Casado et al. 2024a). The Commission Implementing Regulation (European Union (EU) 2023), also established additional requirements for collecting and analyzing these plant poisons in food, which will be enforced in April 2024. Naturally contaminated millet flour turned out to include atropine, amidoamine, scopolamine, scopine, and nor scopolamine. Fascinatingly, naturally occurring atropine was more thermostable than spiked, and the cereal matrix showed a protective impact on TA degradation. There is a need for appropriate monitoring of TAs in raw materials since this chemical danger may remain in newborn cereal-based food even after intense heat processing (Torrents-Masoliver et al. 2024). Also, it is worth having a glance at the recent guidelines issued by Codex (2024) regarding tropane alkaloids.

6.8 Implications of Tropane Alkaloid on the Economy

Unintentional exposure to TAs can occur when consuming contaminated food or plant components mistaken for edible plants. Ingesting a few henbanes (*Hyoscyamus niger*) or *A. belladonna* berries can induce severe intoxication and even death in small infants. *D. stramonium* seeds have been detected as contaminants in vital crops, including soybean, millet, sunflower, buckwheat, linseed, and their products (EFSA 2013a, b). In order to determine the safe levels of TAs in food, the EFSA set a health-based reference dose (ARfD) of 0.016 μg/kg bw for the combined effects of hyoscyamine and scopolamine. The study results on healthy young men, including the central nervous system symptoms and a reduced heart rate, were used to derive this number (EFSA 2013a, b; Perharič et al. 2013a, b). The European Commission's Recommendation (EU) 2015/976 involves all member states to establish surveillance strategies for TAs in vulnerable food products, and the Regulation (EU) 2023/915 sets maximum levels for atropine and scopolamine in processed and unprocessed cereal-based foods, herbal infusions, and infant and young child cereal-based foods. EFSA's 2018 large-scale acute exposure evaluation prompted these regulations. EFSA analyzed over 40,000 food samples for TAs from 2009 to 2016 (EFSA 2013a, b). Using a probabilistic approach, cereal-based food and tea/herbal infusion were the leading sources of atropine and scopolamine exposure across all age groups and dietary categories (EU 2015, 2021). Even though high consumers of all age groups surpassed the ARfD for atropine and scopolamine, newborns, toddlers, and other children had the most significant average acute intake levels of 1 to 21 ng/kg bw per day (EFSA 2018). The cereal industry rapidly develops gluten-free, whole-grain, and seed-enriched products due to increased awareness of gluten-related illnesses and chronic diseases (Benítez et al. 2018; Estévez and Araya 2016). Bakeries are increasingly using cereal flours other than wheat (*Triticum aestivum*), as well as seeds, aromatic herbs, and spices like clove, aniseed, and curry (Angioloni and Collar 2011; Skendi et al. 2019). Contamination episodes have caused numerous notable poisoning outbreaks. The most recent one occurred in 2019 in Uganda and involved a fortified blended product called "Super Cereal" distributed by the World Food Programme (WFP 2019). Other studies have examined this product (Adamse et al. 2014; Abia et al. 2021; Haughey et al. 2021; de Nijs et al. 2023; EFSA 2022, 2023; González-Gómez et al. 2022).

TAs intakes across 13 EU countries' diverse populations (EFSA 2018). EFSA (2018) reported mean acute exposure values of 0.001 to 0.019 μg/kg bw/day and P95 acute exposure values of 0.0005 to 0.053 μg/kg bw/day, based on age groups. Bread and cereal-based meals accounted for most of the atropine and scopolamine exposure in this study (EFSA 2018). Despite worries about contamination, herbal teas, infusions, and supplements tend to have low TA intake. These items' TA consumption contributed just 2–8% of the ARfD (Cirlini et al. 2019). The Food and Agriculture Organization/World Health Organization (FAO/WHO) used a probabilistic technique to quantify acute dietary exposure to TAs from a general diet in Bangladesh, Bolivia, Laos, the Philippines, Uganda, and Zambia in 2020. The study found that mean acute dietary exposure to TAs was below 0.001 μg/kg bw in all countries except Zambia, where young children and women had exposures of 0.018 and 0.0046 μg/kg bw, respectively. Exposure to P95 varied from 0.0025 to 0.0035 μg/kg bw, with small children in Zambia experiencing 0.0038 μg/kg bw and women 0.01 μg/kg bw (FAO/WHO 2020). The mean and P95 consumption estimates calculated by the FAO are much lower than those in the current study. However, the FAO/WHO could not use incidence data from the stated countries due to the lack of food surveillance and analysis, especially for TA detection in food samples (FAO/WHO 2020). At the 2020 FAO-WHO joint expert meeting, a probabilistic technique was used to quantify dietary exposure to TAs across WFP demographic groups. Rice, maize, and sorghum were the primary sources of dietary exposure in all countries except Zambia, with mean values <1 ng/kg bw per day (FAO/WHO 2020). There are no set limits for TAs in leafy vegetables like spinach and spinach-based infant food, despite numerous studies showing high levels (Adamse et al. 2014; Cornara et al. 2018; Castilla-Fernández et al. 2021; González-Gómez et al. 2022). In the FAO/WHO expert meeting (2020) regarding tropane alkaloids and the varying perspectives of Members, the 15th session of the Codex Committee on Contaminants in Foods (CCCF) recognized the necessity of forming an Electronic Working Group (EWG) to develop a discussion paper on tropane alkaloids. This paper will assess the need and feasibility of potential follow-up actions for consideration at CCCF16. Nevertheless, no member state undertook the mission. CCCF16 (2023) reviewed this item and formed an EWG, led by China and co-chaired by Saudi Arabia, to develop a discussion paper on tropane alkaloids, assessing the necessity and feasibility of potential follow-up actions for CCCF171's consideration.

6.9 Conclusions

TAs are found occurring naturally in various plants belonging to the Solanaceae family. There are a few tropane alkaloids that are notoriously known for their psychoactive properties, as well as their associated usage and cultural connections. These include cocaine and scopolamine. Certain tropane alkaloids, such as these, possess pharmacological qualities and can operate as stimulants or anticholinergics. On the contrary, due to their therapeutic applications, the first group's TAs are the most interesting type of TAs in current research and can yield significant economic benefits. However, excessive TA consumption has been linked to several poisoning cases; as a result, TAs in food are also thought to pose a risk to human health (Zile Wang et al. 2024). Dizziness, dilated pupils, blurred vision, redness of the skin, dry mouth, vomiting, cramping in the muscles, confusion, hypothermia, tachycardia, hallucinations, delirium, and mortality are among the most frequent side effects of toxic doses in humans by TAs. Given that, now more focus is on the samples collection for TA identification and to curtail TA prevalence in food. Nevertheless, most TA-related core identification materials are specific and cannot satisfy the requirements of several TAs being detected simultaneously. This problem could make TA detection more difficult. In the evolution of detection technology, quick on-site screening of TAs in different matrices will probably become the new standard.

References

Abia WA, Montgomery H, Nugent AP, Elliott CT (2021) Tropane alkaloid contamination of agricultural commodities and food products in relation to consumer health: learnings from the 2019 Uganda food aid outbreak. Compr Rev Food Sci Food Saf 20(1):501–525

Adamse P, Van Egmond HP (2010) Tropane alkaloids in food. RIKILT Rep 011:1–24. http://edepot.wur.nl/160741

Adamse P, Van Egmond HP, Noordam MY, Mulder PPJ, De Nijs M (2014) Tropane alkaloids in food: poisoning incidents. Qual Assur Saf Crops Foods 6(1):15–24

Adibah KZM, Azzreena MA (2019) Plant toxins: alkaloids and their toxicities. GSC Biol AFSSA (2008) Agritox. Retrieved from http://www.dive.afssa.fr/agritox/php/fiches.php. Accessed on 2 Sept 2024

Aga A, Geyid A (1992) An outbreak of acute toxicity caused by eating food contaminated with *Datura stramonium*. Ethiop J Health Dev 6(1):Article no. 4

Angioloni A, Collar C (2011) Nutritional and functional added value of oat, Kamut®, spelt, rye and buckwheat versus common wheat in breadmaking. J Sci Food Agric 91(7):1283–1292

Benítez V, Esteban RM, Moniz E, Casado N, Aguilera Y, Mollá E (2018) Breads fortified with wholegrain cereals and seeds as source of antioxidant dietary fibre and other bioactive compounds. J Cereal Sci 82:113–120

Berkova S, Zayed R (2004) Comparison of tropane alkaloid spectra between *Datura innoxia* grown in Egypt and Bulgaria. Z Naturforsch 59C:184

Beyer J, Drummer OH, Maurer HH (2009) Analysis of toxic alkaloids in body samples. Forensic Sci Int 185(1–3):1–9

Boros B, Farkas A, Jakabová S, Bacskay I, Kilár F, Felinger A (2010) LC-MS quantitative determination of atropine and scopolamine in the floral nectar of *Datura* species. Chromatographia 71:43–49

Bryson PD, Watanabe AS, Rumack BH, Murphy RC (1978) Burdock root tea poisoning: case report involving a commercial preparation. JAMA 239:2157

Caligiani A, Palla G, Bonzanini F, Bianchi A, Bruni R (2011) A validated GC–MS method for the detection of tropane alkaloids in buckwheat (*Fagopyron esculentum* L.) fruits, flours and commercial foods. Food Chem 127(1):204–209

Caprai E, Prizio I, Peloso M, Sonfack GM, Bonan S, Benini N, Ghidini S, Varra MO, Zanardi E, Lanz GT, Fedrizzi G (2024) Case reports of tropane alkaloid contamination in spinach from Italy and its potential implications for consumer health. Food Control 160:110334

Casado N, Casado-Hidalgo G, González-Gómez L, Morante-Zarcero S, Sierra I (2023) Insight into the impact of food processing and culinary preparations on the stability and content of plant alkaloids considered as natural food contaminants. Appl Sci 13(3):1704

Casado N, Gañán J, Morante-Zarcero S, Sierra I (2024a) Recent food alerts and analytical advances related to the contamination of tropane and pyrrolizidine alkaloids in food. Front Chem Biol 3:1360027

Casado N, Gañán J, Morante-Zarcero S, Sierra I (2024b) Occurrence of pyrrolizidines and other alkaloids of plant origin in foods. In: Smithers IGW (ed) Encyclopedia of food safety, 2nd edn. Academic Press, Oxford, pp 518–528. https://doi.org/10.1016/b978-0-12-822521-9.00184-2

Casado-Hidalgo G, Morante-Zarcero S, Pérez-Quintanilla D, Sierra I (2021) Opium alkaloids in food products: Current and future perspectives. Trends Food Sci Technol 108:92–102

Castilla-Fernández D, Moreno-González D, García-Reyes JF, Ballesteros E, Molina-Díaz A (2021) Determination of atropine and scopolamine in spinach-based products contaminated with genus *Datura* by UHPLC–MS/MS. Food Chem 347:129020

CDC (1995) Anticholinergic poisoning associated with an herbal tea—New York City (1994). Morb Mortal Wkly Rep 44(11):193–195. Retrieved from http://wwcdcw.cdc.gov/mmwr/preview/mmwrhtml/00036554.htm

Chan TYK (2017) Worldwide occurrence and investigations of contamination of herbal medicines by tropane alkaloids. Toxins (Basel) 9(9):284. https://doi.org/10.3390/toxins9090284

Chen H, Marín-Sáez J, Romero-González R, Frenich AG (2017) Simultaneous determination of atropine and scopolamine in buckwheat and related products using modified QuEChERS and liquid chromatography tandem mass spectrometry. Food Chem 218:173–180

Cirlini M, Demuth TM, Biancardi A, Rychlik M, Dall'Asta C, Bruni R (2018) Are tropane alkaloids present in organic foods? Detection of scopolamine and atropine in organic buckwheat (*Fagopyron esculentum* L.) products by UHPLC–MS/MS. Food Chem 239:141–147

Cirlini M, Cappucci V, Galaverna G, Dall'Asta C, Bruni R (2019) A sensitive UHPLC-ESI-MS/MS method for the determination of tropane alkaloids in herbal teas and extracts. Food Control 105:285–291

Codex (2024) Codex Alimentarius Commission. Discussion paper on tropane alkaloids. Joint FAO/WHO Food Standards Programme Codex Committee on Contaminants in Foods. CX/CF 24/17/11. Panama City

Cornara L, Smeriglio A, Frigerio J, Labra M, Di Gristina E, Denaro M, Mora E, Trombetta D (2018) The problem of misidentification between edible and poisonous wild plants: Reports from the Mediterranean area. Food Chem Toxicol 119:112–121

Croteau R, Kutchan TM, Lewis NG (2000) Natural products (secondary metabolites). Biochem Mol Biol Plants 24:1250–1319

de Nijs M, Crews C, Dorgelo F, MacDonald S, Mulder P (2023) Emerging issues on tropane alkaloid contamination of food in Europe. Toxins 15(2):98

Department of Health and Sports France (2010) Presence of *Datura stramonium* in canned green beans. http://www.sante-sports.gouv.fr/rappel-de-produits-presence-de-daturastramonium-dans-des-conserves-de-haricots-verts-des-marques-u-et-notre-jardin.html. Accessed 27 July 2010

Department of Health (2012) Anticholinergic poisoning. Poisoning Watch 5(1). Available at: http://www.chp.gov.hk/files/pdf/poisoning_watch_vol5_eng_20121231.pdf.

Doncheva T, Berkov S, Philipov S (2006) Comparative study of the alkaloids in tribe Datureae and their chemosystematic significance. Biochem Syst and Ecol 34(6):478–488

Dräger B (2003) Chemistry and biology of calystegines. Nat Prod Rep 21(2):211–223

Dugan GM, Gumbmann MR, Friedman M (1989) Toxicological evaluation of jimson weed (*Datura stramonium*) seed. Food Chem Toxicol 27(8):501–510

EFSA (2008) Scientific opinion of the panel on contaminants in the food chain on a request from the European commission on tropane alkaloids (from *Datura* sp.) as undesirable substances in animal feed. EFSA J 6(8):1–55. https://doi.org/10.2903/j.efsa.2008.691

EFSA Panel on Contaminants in the Food Chain, (CONTAM) (2011a) Scientific opinion on pyrrolizidine alkaloids in food and feed. EFSA J 9(11):2406

EFSA Panel on Contaminants in the Food Chain, (CONTAM) (2011b) Scientific opinion on the risks for public health related to the presence of opium alkaloids in poppy seeds. EFSA J 9(11):2405. https://doi.org/10.2903/j.efsa.2011.2405

EFSA Panel on Contaminants in the Food Chain (CONTAM) (2013a) Scientific opinion of the panel on contaminants in the food chain on tropane alkaloids in food and feed. EFSA J 11(10):3386

EFSA Panel on Contaminants in the Food Chain (CONTAM) (2013b) Scientific opinion on tropane alkaloids in food and feed. EFSA J 11(10). https://doi.org/10.2903/J.EFSA.2013.3386

EFSA Panel on Contaminants in the Food Chain (CONTAM) (2018) Scientific report on human acute exposure assessment to tropane alkaloids. EFSA J 16(2):1–29. https://doi.org/10.2903/j.efsa.2018.5160

EFSA Panel on Contaminants in the Food Chain (CONTAM) (2022) Scientific report on the assessment of the conclusions of the joint FAO/WHO expert meeting on tropane alkaloids. EFSA J 20(4):1–17. https://doi.org/10.2903/j.efsa.2022.7229

EFSA Panel on Contaminants in the Food Chain (CONTAM), Knutsen HK, Alexander J, Barregård L, Bignami M, Brüschweiler B, Ceccatelli S et al (2018) Update of the scientific opinion on opium alkaloids in poppy seeds. EFSA J 16(5):e05243. https://doi.org/10.2903/j.efsa.2018.5243

Estévez V, Araya M (2016) La dieta sin gluten y los alimentos libres de gluten. Rev Chil Nutr 43(4):428–433

EU (2015) Commission Recommendation (EU) 2015/976 of 19 June 2015 on the monitoring of the presence of tropane alkaloids in food. Off J Eur Union 11:97–98

EU (2021) Commission Regulation (EU) 2021/1408 of 27 August 2021 amending Regulation (EC) No 1881/2006 as regards maximum levels of tropane alkaloids in certain foodstuffs. Off J Eur Union L304(1):1–4

European Commission (2016) Commission Regulation (EU) 2016/239 of 19 February 2016 amending Regulation (EC) No 1881/2006 as regards maximum levels of tropane alkaloids in certain cereal-based foods for infants and young children (Text with EEA relevance) [Online]. Off J Eur Union. http://data.europa.eu/eli/reg/2016/239/oj. Accessed on 7 Nov 2023

European Commission (2023). Commission Regulation (EU) 2023/915 of 25 April 2023 on maximum levels for certain contaminants in food and repealing Regulation (EC) No 1881/2006 (Text with EEA relevance). [Online]. Official Journal of the European Union http://data.europa.eu/eli/reg/2023/915/oj. Accessed on 07.11.2023

European Parliament and Council (2013) Regulation (EU) No 609/2013 of the European Parliament and of the Council of 12 June 2013 on food intended for infants and young children, food for special medical purposes, and total diet replacement for weight control and repealing Council Directive 92/52/EEC, Commission Directives 96/8/EC, 1999/21/EC, 2006/125/EC and 2006/141/EC, Directive 2009/39/EC of the European Parliament and of the Council and Commission Regulations (EC) No 41/2009 and (EC) No 953/2009. Off J Eur Union 181:35–56

Fatur K, Kreft S (2020) Common anticholinergic solanaceous plants of temperate Europe – a review of intoxications from the literature (1966–2018). Toxicon 177:52–88

Food and Agriculture Organization & World Health Organization (FAO & WHO) (2020) Joint FAO/WHO expert meeting on tropane alkaloids—30 March—3 April 2020, vol 11. Food Safety and Quality Series, Rome. https://doi.org/10.4060/cb1857en

Fretz R, Schmid D, Brueller W, Girsch L, Pichler AM, Riediger K, Safer M, Allerberger F (2007) Food poisoning due to jimson weed mimicking *Bacillus cereus* food intoxication in Austria. Int J Infect Dis 11:557–558

Friedman MLC, Levin CE (1989) Composition of jimson weed (*Datura stramonium*) seeds. J Agric Food Chem 37(4):998–1005

García-Juan A, León N, Armenta S, Pardo O (2023) Development and validation of an analytical method for the simultaneous determination of 12 ergot, 2 tropane, and 28 pyrrolizidine alkaloids in cereal-based food by LC-MS/MS. Food Res Int 174:113614

Glaizal M, Schmitt C, Tichadou L, Sapori JM, Hayek-Lanthois M, de Haro L (2013) Intoxications induites par de la farine de sarrasin Bio contaminée par *Datura* sp.: rôle d'un centre antipoison et de toxicovigilance. Presse Med 42:1412–1420

Gonçalves C, Dernovics M, Moreno-Gonzalez D, Bouten K, Garcia-Reyes JF, Stroka J (2006) Occurrence and determination of tropane alkaloids in food and feed. In: Encyclopedia of analytical chemistry: applications, theory and instrumentation. Wiley, Chichester, pp 1–32

Gonçalves CO, Cubero-Leon E, Miscke C, Bratinova S, Stroka J (2017) Report on the 2016 proficiency test of the European union reference laboratory for mycotoxins for the network of national reference laboratories: determination of tropane alkaloids in tea and herbal infusions. JRC Technical Report, EUR 28662 EN https://data.europa.eu/doi/10.2760/52884

Gonçalves C, Cubero-Leon E, Stroka J (2020) Determination of tropane alkaloids in cereals, tea and herbal infusions: Exploiting proficiency testing data as a basis to derive interlaboratory performance characteristics of an improved LC-MS/MS method. Food Chem 30;331:127260. https://doi.org/10.1016/j.foodchem.2020.127260

González-Gómez L, Morante-Zarcero S, Pereira JA, Câmara JS, Sierra I (2022) Improved analytical approach for determination of tropane alkaloids in leafy vegetables based on μ-QuEChERS combined with HPLC-MS/MS. Toxins 14(10):650

Griffin WJ, Lin GD (2000) Chemotaxonomy and geographical distribution of tropane alkaloids. Phytochemistry 53(6):623–637

Grynkiewicz G, Gadzikowska M (2008) Tropane alkaloids as medicinally useful natural products and their synthetic derivatives as new drugs. Pharmacol Rep 60(4):439

Han S, Jang S, Oh S, Lee J, Lee HJ, Koo YE, Kim BH (2024) Occurrence and health risk assessment of tropane alkaloids in cereal foods consumed in Korea. Food Chem Toxicol 186:114589

Haughey SA, Chevallier OP, McVey C, Elliott CT (2021) Laboratory investigations into the cause of multiple serious and fatal food poisoning incidents in Uganda during 2019. Food Control 121:107648

Jank B, Rath J (2021) Emerging tropane alkaloid contaminations under climate change. Trends Plant Sci 26(11):1101–1103

Jeelani P, Ghai A, Saikia N, Kathed M, Mitra A, Krishnan A, Sharma A, Chidambaram R (2020) Baby foods based on cereals. In: Gutiérrez TJ (ed) Food science, technology and nutrition for babies and children. Springer, Cham, pp 59–97

Khoshnam-Rad N, Heydari M, Mohammadi K, Mashayekhi M, Sahraei Z, Gholami K (2022) Datura poisoning in a family: case series and literature review. Clin Case Reports 10(7):e6091

Klein LM, Gabler AM, Rychlik M, Gottschalk C, Kaltner F (2022) A sensitive LC–MS/MS method for isomer separation and quantitative determination of 51 pyrrolizidine alkaloids and two tropane alkaloids in cow's milk. Anal Bioanal Chem 414(28):8107–8124

Klein LM, Lamp J, Schopf C, Gabler AM, Kaltner F, Guldimann C, Rychlik M, Schwake-Anduschus C, Knappstein K, Gottschalk C (2024) Pyrrolizidine alkaloids and tropane alkaloids in milk samples from individual dairy farms of the German federal states of Bavaria and Schleswig-Holstein. Food Addit Contam Part A 41(6):629–647. https://doi.org/10.1080/19440049.2024.2336054

Kohnen-Johannsen KL, Kayser O (2019) Tropane alkaloids: chemistry, pharmacology, biosynthesis and production. Molecules 24(4):796

Lamp J, Knappstein K, Walte H-G, Krause T, Steinberg P, Schwake-Anduschus C (2021) Transfer of tropane alkaloids (atropine and scopolamine) into the milk of subclinically exposed dairy cows. Food Control 126:108056

Letsyo E, Adams ZS, Dzikunoo J, Asante-Donyinah D (2021) Uptake and accumulation of pyrrolizidine alkaloids in the tissues of maize (*Zea mays* L.) plants from the soil of a 4-year-old *Chromolaena odorata* dominated fallow farmland. Chemosphere 270:128669

List GR, Spencer GF (1976) Fate of jimson weed seed alkaloids in soybean processing. J Am Oil Chem Soc 53:535–536

List GR, Spencer GF, Hunt WH (1979) Toxic weed seed contaminants in soybean processing. J Am Oil Chem Soc 56(8):706–710

Lugasi A (2023) Tropane alkaloids in food—real and perceived risks. Élelmiszervizsgálati Közlemények 69(1):4317

Marín-Sáez J, Romero-González R, Frenich AG (2019a) Reliable determination of tropane alkaloids in cereal based baby foods coupling on-line spe to mass spectrometry avoiding chromatographic step. Food Chem 275:746–753

Marín-Sáez J, Romero-González R, Garrido Frenich A (2019b) Degradation of tropane alkaloids in baked bread samples contaminated with Solanaceae seeds. Food Res Int 122:585–592

Martinello M, Borin A, Stella R, Bovo D, Biancotto G, Gallina A, Mutinelli F (2017) Development and validation of a QuEChERS method coupled to liquid chromatography and high resolution mass spectrometry to determine pyrrolizidine and tropane alkaloids in honey. Food Chem 234:295–302

Mateus ARS, Crisafulli C, Cruz Barros S, Pena A, Sanches Silva A (2024) Development and validation of an analytical method based on QuEChERS followed by UHPLC–ToF-MS for the determination of tropane alkaloids in buckwheat (*Fagopyrum esculentum* L.) and buckwheat products. Food Addit Contam Part A 41(6):648–663

Mulder PP, Nijs M, Castellari M, Hortos M, MacDonald S, Crews C, Hajslova J, Stranska M (2016) Occurrence of tropane alkaloids in food. External scientific report. EFSA Support Publ 2016-EN-1140:200

Naude TW, Gerber R, Smith RJ, Botha CJ (2005) *Datura* contamination of hay as the suspected cause of an extensive outbreak of impaction colic in horses. J S Afr Vet Assoc 76:107–112

Onen CL, Othol D, Mbwana SK, Manuel IL (2002) *Datura stramonium* mass poisoning in Botswana. S Afr Med J 92(3):213–214

Perharic L (2005) Mass tropane alkaloid poisoning due to buckwheat flour contamination. Clin Toxicol 43:413

Perharič L, Juvan KA, Stanovnik L (2013a) Acute effects of a low-dose atropine/scopolamine mixture as a food contaminant in human volunteers. J Appl Toxicol 33(9):980–990

Perharič L, Koželj G, Družina B, Stanovnik L (2013b) Risk assessment of buckwheat flour contaminated by thorn-apple (*Datura stramonium* L.) alkaloids: a case study from Slovenia. Food Addit Contam Part A 30(2):321–330

Plackova S, Caganova B, Otrubova O, Puchon E, Batora I (2022) Anticholinergic poisoning associated with frozen spinach. Clin Toxicol 60(1):30

Pulewka P, Saygin M (1949) Die Aufklärung einer ungewöhnlichen durch *Datura stramonium* in Brotmehl hervorgerufenen Massenvergiftung. Klin Wochenschr 27:672–674

Ramirez M, Rivera E, Ereu C (1999) Fifteen cases of atropine poisoning after honey ingestion. Vet Hum Toxicol 41:19–20

RASFF (2005) The Rapid Alert System for Food and Feed: annual report 2005. Publications Office. https://op.europa.eu/en/publication-detail/-/publication/cb15fc98-7fe9-4ed3-a35c-126267242a36/language-en. Accessed on 3 Oct 2024

RASFF (2007) The Rapid Alert System for Food and Feed: annual report. Publications Office. https://data.europa.eu/doi/10.2772/63727. Accessed on 3 Oct 2024

RASFF (2012) The Rapid Alert System for Food and Feed: annual report. Publications Office. https://data.europa.eu/doi/10.2772/48887. Accessed on 3 Oct 2024

RASFF (2013) The Rapid Alert System for Food and Feed: annual report. Publications Office. https://data.europa.eu/doi/10.2772/56447

RASFF (2019) European Commission: Directorate-General for Health and Food Safety. Annual report 2018. Publications Office. https://data.europa.eu/doi/10.2875/914558

RASFF (2023) RASFF portal—food and feed safety alerts. Available online: https://webgate.ec.europa.eu/rasff-window/portal/? Accessed 22 Mar 2023

Rizzo S, Celano R, Piccinelli AL, Russo M, Rastrelli L (2023) Target screening method for the quantitative determination of 118 pyrrolizidine alkaloids in food supplements, herbal infusions, honey and teas by liquid chromatography coupled to quadrupole orbitrap mass spectrometry. Food Chem 423:136306

Roess AA, Jacquier EF, Catellier DJ, Carvalho R, Lutes AC, Anater AS, Dietz WH (2018) Food consumption patterns of infants and toddlers: findings from the Feeding Infants and Toddlers Study (FITS) 2016. J Nutr 148:1525S–1535S

Romera-Torres A, Romero-González R, Vidal JLM, Frenich AG (2018) Analytical methods, occurrence and trends of tropane alkaloids and calystegines: an update. J Chromatogr A 1564:1–15

Russel J, Edwards C, Jordan C, Luckman E, Chu A, Blythe D, Krick J (2010) Jimsonweed poisoning associated with a homemade stew—Maryland, 2008. Morb Mortal Wkly Rep 59(4):102–104

Sattler M, Müller V, Bunzel D, Kulling SE, Soukup ST (2023) Pyrrolizidine alkaloids in borage (*Borago officinalis*): comprehensive profiling and development of a validated LC-MS/MS method for quantification. Talanta 258:124425

Schlagmann C, Remien J (1986) Klinische Schrift Zur Behandlung der Parkinson-Krankheit. Klin Wochenschr 64:939–942

Sherif ZA, Deverapalli M, Challa SR, Martirosyan Z, Whitesell P, Pizuorno AM, Naqvi Z, Tulloch IK, Oskrochi G, Brim H, Ashktorab H (2023) Potential long-term neurological and gastrointestinal effects of COVID-19: A review of adult cohorts. World J Methodol 13(4):323–336

Skendi A, Irakli M, Chatzopoulou P, Papageorgiou M (2019) Aromatic plants of Lamiaceae family in a traditional bread recipe: effects on quality and phytochemical content. J Food Biochem 43(11):e13020

Sousa A, Cláudia R, Sandra L (2022) Pyrrolizidine Alkaloids: the underestimated natural toxicants. RevSALUS - Revista Científica da Rede Académica das Ciências da Saúde da Lusofonia 4:10.51126/revsalus.v4iSup.291

Tábuas B, Cruz Barros S, Diogo C, Cavaleiro C, Sanches Silva A (2024) Pyrrolizidine Alkaloids in foods, herbal drugs, and food supplements: chemistry, metabolism, toxicological significance, analytical methods, occurrence, and challenges for future. Toxins (Basel) 16(2):79. https://doi.org/10.3390/toxins16020079

Torrents-Masoliver B, Sandjong D, Jofré A, Ribas-Agustí A, Muñoz I, Felipe X, Castellari M, Meurillon M, Den Besten HM, Engel E, Bover-Cid S (2022) Hazard control through processing and preservation technologies for enhancing the food safety management of infant food chains. Glob Pediatr 2:100014

Torrents-Masoliver B, Terriente-Palacios C, Bover-Cid S, Jofré A, Castellari M, Ribas-Agustí A (2024) Effect of pH and temperature on tropane alkaloids within a processing strategy to provide safe infant cereal-based food. Food Res Int 188:114439

Vera-Baquero FL, Casado N, Morante-Zarcero S, Sierra I (2024) Improving the food safety of bakery products by simultaneously monitoring the occurrence of pyrrolizidine, tropane and opium alkaloids. Food Chem 460:140769

Wang Z, Chen S, Zheng P, Ren Z, Zhang H, Zhang J, Shao B, Wu C, Jiang H (2024) Emerging tropane alkaloids: Global development and potential health threats. Food Qual Saf 8:fyad043

WFP (2019) WFP to dispose of compromised food stocks. World Food Programme. https://www.wfp.org/news/wfp-dispose-compromised-food-stocks. Accessed 5 Oct 2023

WFP (2020) Joint FAO/WHO expert meeting on tropane alkaloids. Executive summary, 30 March–3 April 2020, pp 1–17

Chapter 7
Edible Plant Sources Containing Low and Non-toxic Levels of Tropane and Nortropane Alkaloids

Nandini Boregowda, Kiran Suresh Mawale, and Parvatam Giridhar

Abstract Tropane alkaloids (TAs) are a unique category of alkaloids with a tropane ring in their chemical structure. They are often considered important lead compounds in pharmacology. While many people are familiar with the well-known plant sources of tropane alkaloids, there are also lesser-known plants that contain TAs, some of which may be edible or used in traditional medicine through oral consumption. This chapter primarily focuses on these edible or lesser-known plants recognized as sources of tropane alkaloids. Some nortropanes exhibited inhibitory effects on glucosidases and galactosidases at their low levels. When present at non-toxic levels, tropane alkaloids in these plants could provide beneficial health effects and contribute substantially to health protection. However, further studies in this regard are needed.

Keywords Cochlearine · Calystegines · Goji berry · *Morus alba* · Nortropanes

N. Boregowda
Plant Cell Biotechnology Department, CSIR-Central Food Technological Research Institute, Mysuru, Karnataka, India

K. S. Mawale
Plant Cell Biotechnology Department, CSIR-Central Food Technological Research Institute, Mysuru, Karnataka, India

Academy of Scientific and Innovative Research (AcSIR), Ghaziabad, India

P. Giridhar (✉)
Plant Cell Biotechnology Department, CSIR-Central Food Technological Research Institute, Mysuru, Karnataka, India

Academy of Scientific and Innovative Research (AcSIR), Ghaziabad, India

Food Safety and Analytical Quality Control Laboratory, CSIR-Central Food Technological Research Institute, Mysuru, Karnataka, India

T. Pullaiah et al. (eds.), *Tropane Alkaloids*,
https://doi.org/10.1007/978-981-96-5805-3_7

7.1 Introduction

Plants play a crucial role in synthesizing organic molecules that support vital functions. They produce secondary metabolites through shared metabolic pathways, facilitating various plant functions. Nutraceuticals are considered distinct food products that emphasize the health benefits of the nutrients they contain, potentially aiding in the prevention or treatment of illnesses (Bommakanti et al. 2023; Taroncher et al. 2021). Nutraceutical products enhance conventional dietary and lifestyle practices to maximize people's overall health (Das et al. 2020; Hurkul et al. 2024; Pandey et al. 2024). There is considerable interest in investigating the biological effects of plant-based compounds in common foods and the plant-derived chemicals used as pharmaceutical components. Seeds, roots, and leaves are rich in various nutrients and bioactive compounds that offer health benefits. Alkaloids, a complex category of naturally occurring chemical compounds, contain basic nitrogen atoms (Bhambhani et al. 2021). Heterocyclic ring types, such as indole, purine, quinoline, isoquinoline, tropane, and imidazole, are particularly noteworthy among these categories and exhibit biological activities like antiproliferative, antibacterial, and antioxidant effects. (Amirkia and Heinrich 2014; Dey et al. 2020; Rupani and Chavez 2018). Plants in the Erythroxylaceae (coca) and Solanaceae (including mandrake, henbane, datura, deadly nightshade, potato, and tomato) families naturally contain tropane alkaloids, which are well-researched and documented. The key tropane alkaloids include atropine, hyoscyamine, and scopolamine. While these are the most commonly recognized plants that produce tropane alkaloids, there are others as well (Robins et al. 1997; Griffin and Lin 2000) that contain TAs or their derivatives in low quantity and are often considered non-toxic, hence, could still be a good vehicle for nutraceuticals.

One key criterion for considering a bioactive compound as a nutraceutical is its dose-dependent response. It would benefit researchers in this field to gather information about the capacity of edible or lesser-known plants to produce tropane alkaloids (TAs) and their related ethnobotanical or food applications. Tropane alkaloids are a class of chemical compounds that undergo biochemical processes to produce hydroxylated derivatives known as calystegines produced in Convolvulaceae, Solanaceae, and Moraceae (Asano et al. 1997; Watson et al. 1999). These calystegines primarily accumulate in the leaves, roots, and developing tubers of the potato plant (*Solanum tuberosum* L.), and in the leaves of the eggplant (*S. melongena*), and sweet potato (*Ipomoea batatas*), which has raised concerns regarding the safety of these vegetables in the human diet. One of the key metabolites in this pathway is tropinone. Although it is found in relatively low concentrations throughout the potato plant, tropinone plays a crucial role in plant metabolism. It is transported from the roots to the aerial parts of the plant, including the stems, leaves, and stolon, which connect the roots and tubers. The enzyme tropinone reductase (TR) facilitates this metabolic process and is active across all parts of the potato plant (James and James 1954). This enzyme has two isoforms: tropinone reductase I (TR I) and tropinone reductase II (TR II). Research indicates that TR II's activity surpasses TR

I's in most plant tissues, except tuber tissues, where the activity levels differ (Hashimoto and Yamad 2003).

7.2 Some Lesser-Known Plants Contain Tropanes

Some of the food plants that contain TA are tomatoes (*Solanum lycopersicum*), wolfberries (*Lycium barbarum*), Cape gooseberries (*Physalis peruviana*), potatoes (*Solanum tuberosum*), Kava (*Piper methysticum*), and, to a minor extent, mulberry fruit (*Morus alba*).

*7.2.1 Bitter-Grass (*Tabernanthe iboga *Baill)*

Iboga is consumed in large doses by initiates of this spiritual practice and in smaller doses during rituals and tribal dances performed at night. The primary alkaloid in *Tabernanthe iboga* is called ibogaine. When taken in high doses, ibogaine can be toxic and may cause serious health issues, especially when combined with opioids or prescription medications. The United States Drug Enforcement Administration (DEA) classifies ibogaine as a controlled substance. The root of the shrub *Tabernanthe iboga*, belonging to the Apocynaceae family and native to central West Africa, is the source of this alkaloid. The root bark contains up to 6% (w/w) of indole alkaloids, with ibogaine being the most prevalent. Smaller amounts of other alkaloids, such as tabernanthine, ibogaine, and ibogaine, are also present (Alper et al. 2008). Compared to noribogaine, which has a much longer half-life and thus results in substantially greater blood concentrations, ibogaine has a comparatively shorter pharmacokinetic half-life of about 5–8 h (Cheze et al. 2008).

7.2.2 Cape Gooseberry

Physalis is a Solanaceae family genus that includes annual and perennial plants. These plants produce globular fruits encased in a bladder-like husk that turns papery as it matures (Mericli 2011). Of the more than 70 species in this genus, only a few have significant economic value. One of the most popular species is the Cape gooseberry (*Physalis peruviana* L.), known for its exceptional fruit. This fruit is commonly used in pies, sauces, and preserves, particularly in mild-temperate regions. Cape gooseberry's roots and aerial parts contain compounds such as tigloidine and other secotropane alkaloids (Kasali et al. 2022). In Colombia, a decoction made from the leaves is a diuretic and have anti-asthmatic properties. The fruit can be enjoyed in various forms, including canned, made into jams, sauces, pies, puddings, chutneys, ice cream, fresh salads, and drinks. Secotropane alkaloids are reported from the leaves and roots of *P. peruviana* (Sahia and Ray 1980). They are A3, B1, B2,C1,3β-acetoxy-tropane,(±)-Physoperuvine,(+)-N,N-dimethylphysoperuvinium, Phygrine, N-methylpyrrolidinylhygrine isomers, Cuscohygrine.

7.2.3 Cochlearia officinalis

Cochlearia officinalis (Brassicaceae) plant is native to North and West Europe, commonly known as common scurvygrass or spoonwort, is a highly nutritious and edible plant. Its leaves are excellent when eaten raw and can effectively enhance the flavor of salads or raw vegetables. With a pungent, spicy, and acrid taste that closely resembles horseradish, it adds a distinct kick to dishes. It earned its common name for its vital role in combating scurvy, a deficiency disease resulting from a lack of fresh vegetables. Sailors often relied on it to restore their health after long voyages. Cochlearine, recognized as the 3-hydroxybenzoate ester of tropine, was the first tropane alkaloid discovered in *C. officinalis* (Liebisch et al. 1973). In studies, both *C. officinalis* and *C. arctica* exhibited cochlearine concentrations of approximately 0.05% in their roots and 0.02% in their aerial parts. Furthermore, cochlearine has been identified in extracts from all other *Cochlearia* species examined, consistently showing low concentrations akin to those observed in *C. officinalis* (Brock et al. 2006). This consistent presence of cochlearine across various species highlights the potential for further research into its ecological roles and applications within the Brassicaceae family.

7.2.4 *Gogi Berry or Wolfberry*

Berry foods are often considered healthy due to their sweet taste (Tang et al. 2012; Skenderidis et al. 2018; Bora et al. 2019; Mocan et al. 2018). In the regions where they are cultivated, goji berries (*Lycium barbarum* L.) are primarily consumed fresh (Protti et al. 2017). However, around the world, they are mostly enjoyed in their dried form (Magalhães et al. 2022) or incorporated into various products such as juices, herbal teas, yoghurt, granola, powders, and tablets (Pai et al. 2013; Pires et al. 2018). This popularity is largely due to their antioxidant properties, which can enhance mental and neurological functioning and improve digestive health (Chinese Pharmacopoeia Commission 2015; Bora et al. 2019). *L. barbarum* contains alkaloids found in the seeds' exocarp and endosperm. Recent research has identified about 20 tropane alkaloids in the fruit, all at non-toxic levels. *Lycium chinense* contains polyhydroxytropanes, a type of nortropane alkaloid characterized by a methyl group on the nitrogen atom. Fourteen nortropane alkaloids have been extracted from the root bark of *L. chinense* (Asano et al. 1997). The roots of *Lycium chinense* (Solanaceae) are reported to contain 14 calystegines. Calystegines A7 and B5, A3, A5, A6, B1, B2, B3, B4, C1, C2 and N1, were isolated and determined as 1alpha,

2beta, 4alpha-trihydroxy-nortropane and 1α, 2α,4α, 7α-tetrahydroxy-nortropane, respectively. Also, two polyhydroxytropanes bearing a methyl group on the nitrogen atom, such as N-methylcalystegines B2 and C1, were reported in *L. chinense* (Asano et al. 1997).

7.2.5 Mulberry Fruit

In *Morus alba* fruits, five new nortropane alkaloids (1–5) viz., 2α,3β-dihydroxynortropane (1), 2β,3β-dihydroxynortropane (2), 2α,3β,6*exo*-trihydroxynortropane (3), 2α,3β,4α-trihydroxynortropane (4), 3β,6*exo*-dihydroxynortropane (5), along with nor-ψ-tropine were reported (Kusano et al. 2002). Knowing how such plants are used in traditional food preparations is important.

7.2.6 Potato

Nortropanes are a rare class of alkaloids characterized by the absence of a methylated nitrogen atom. They are bicyclic heterocycles formed from tropane by removing a methylene group. The structure of the nortropane ring system is believed to result from the fusion of a five-membered pyrrolidine ring and a six-membered piperidine ring. Potato tubers (*Solanum tuberosum* L.) accumulate significant amounts of calystegines upon the onset of sprouting (Dräger et al. 1995; Keiner and Dräger 2000; Dräger 2004; Petersson et al. 2013). The rhizosphere of potato plants may contain steroidal glycoalkaloids found in the roots and sprouts of potatoes (Friedman and Dao 1992; Asano et al. 1996; Friedman et al. 2003). The cysteine-deprived *Solanum tuberosum* tropinone reductase (StTRII) RNAi plants allow for natural potato cysteine function studies (Kuster et al. 2017). Analytical studies show two prominent potato trisaccharide glycoalkaloids, α-solanine and α-chaconine, and their hydrolysis products are characterized by varying carbohydrate groups. Also, the water-soluble nortropane alkaloids cysteine A3 and B2 are found in potatoes, inhibiting digestive enzymes in potatoes. These nortropanes survive cooking and are stable enough to be found in processed potato products (Watson et al. 1999).

7.2.7 Tomato

Considering the significant consumption of tomatoes and tomato-based products, a study was conducted to examine the presence of 7 nortropane alkaloids (calystegines A3, A5, B1, B2, B3, B4, and C1) within this matrix. Twelve distinct samples were analyzed: four crushed tomatoes, four fried tomatoes, and four jam tomatoes. Calystegines A3 and B2 were present in all analyzed samples. Calystegine B3 was detected in all fried tomatoes, two crushed tomatoes, and one jam tomato, with concentrations ranging from 0.4 mg/kg for calystegine B2 to 19.0 mg/kg for calystegine A3. Additionally, calystegine A5 was found in all samples; however, only three samples had concentrations exceeding the limit of quantification. Furthermore, calystegine B1 was identified in one jam tomato at a concentration of 1.9 mg/kg (Romera-Torres et al. 2018).

7.3 Conclusion

Tropane alkaloids and nortropanes are a type of secondary metabolite that mainly originates in the Solanaceae family of plants. Even though they are well-known anticholinergic medications and possess potent anticholinergic activity, there have been reports that they are also toxin-producing plants. Although considerable revelations are known for the dose-dependent response of the tropanes, their analogues, and nortropanes in various plants, their pretty low levels in such plants considered them non-toxic. As per the reports, calystegines A3, B1, B2, and C1 were identified in all the evident edible fruits and vegetables. This group of compounds was consistently detected across various produce, including sweet peppers, chili peppers, and various types of potatoes. Additionally, eggplants and tomatoes showed similar results, as did *Physalis* fruits, known for their unique flavour and nutritional benefits. The reports also included sweet potatoes and mulberries, plants belonging to the Moraceae family, further underscoring the widespread occurrence of these calystegines in commonly consumed crops. Calystegines B1 and C1 are strong competitive beta-glucosidase inhibitors in bovine, human, and rat tissues. Calystegine B2 effectively inhibits alpha-galactosidase activity in all liver samples. All four nortropanes also inhibit human beta-xylosidase. Calystegines A3 and B2 selectively inhibit rat liver beta-glucosidase. These findings raise concerns about potential toxicity in humans consuming large quantities of plants containing these compounds (Asano et al. 1997). These findings highlight the significance of these compounds in the diet, suggesting that they are prevalent in many fruits and vegetables that people commonly eat. However, further studies on these tropanes and nortropanes in their respective matrices must be investigated.

References

Alper KR, Lotsof HS, Kaplan CD (2008) The ibogaine medical subculture. J Ethnopharmacol 115(1):9–24

Amirkia V, Heinrich M (2014) Alkaloids as drug leads–a predictive structural and biodiversity-based analysis. Phytochem Lett 10:xlviii–liii. https://doi.org/10.1016/j.phytol.2014.06.015

Asano M, Goto N, Isshiki K (1996) Simple and rapid analysis of potato glycoalkaloids. Nippon Shokuhin Kogaku Kaishi 43:593–597

Asano N, Kato A, Miyauchi M, Kizu H, Tomimori T, Matsui K, Nash RJ, Molyneux RJ (1997) Specific α-galactosidase inhibitors, N-methylcalystegines structure/activity relationships of calystegines from *Lycium chinense*. Eur J Biochem 248(2):296–303

Bhambhani S, Kondhare KR, Giri AP (2021) Diversity in chemical structures and biological properties of plant alkaloids. Molecules 26(11):3374

Bommakanti V, Puthenparambil Ajikumar A, Sivi CM, Prakash G, Mundanat AS, Ahmad F, Haque S, Prieto MA, Rana SS (2023) An overview of herbal nutraceuticals, their extraction, formulation, therapeutic effects and potential toxicity. Separations 10(3):177

Bora P, Ragaee S, Abdel-Aal ESM (2019) Effect of incorporation of goji berry by-product on biochemical, physical and sensory properties of selected bakery products. LWT 112:108225

Brock A, Herzfeld T, Paschke R, Koch M, Dräger B (2006) Brassicaceae contain nortropane alkaloids. Phytochemistry 67(18):2050–2057

Chèze M, Lenoan A, Deveaux M, Pépin G (2008) Determination of ibogaine and noribogaine in biological fluids and hair by LC–MS/MS after *Tabernanthe iboga* abuse: Iboga alkaloids distribution in a drowning death case. Forensic Sci Int 176(1):58–66

Chinese Pharmacopoeia Commission (2015) Pharmacopoeia of the People's Republic of China part I. Chemical Industry Press, Beijing

Das AK, Islam MN, Faruk MO, Ashaduzzaman M, Dungani R (2020) Review on tannins: extraction processes, applications and possibilities. S Afr J Bot 135:58–70

Dey P, Kundu A, Kumar A, Gupta M, Lee BM, Bhakta T, Dash S, Kim HS (2020) Analysis of alkaloids (indole alkaloids, isoquinoline alkaloids, tropane alkaloids). In: Silva AS, Nabavi SF, Nabavi SM (eds) Recent advances in natural products analysis. Elsevier, pp 505–567

Dräger B (2004) Chemistry and biology of calystegines. Nat Prod Rep 21(2):211–223

Dräger B, van Almsick A, Mrachatz G (1995) Distribution of calystegines in several Solanaceae. Planta Med 61(06):577–579

Friedman M, Dao L (1992) Distribution of glycoalkaloids in potato plants and commercial potato products. J Agric Food Chem 40(3):419–423

Friedman M, Roitman JN, Kozukue N (2003) Glycoalkaloid and calystegine contents of eight potato cultivars. J Agric Food Chem 51(10):2964–2973

Griffin WJ, Lin GD (2000) Chemotaxonomy and geographical distribution of tropane alkaloids. Phytochemistry 53(6):623–637

Hashimoto T, Yamada Y (2003) New genes in alkaloid metabolism and transport. Curr Opin Biotechnol 14(2):163–168

Hurkul MM, Cetinkaya A, Yayla S, Ozkan SA (2024) Advanced sample preparation and chromatographic techniques for analyzing plant-based bioactive chemicals in nutraceuticals. J Chromatogr Open 100131

James WO, James GM (1954) The use of solanaceous grafts in the study of alkaloid formation. In: Proceedings of the International Botanical Congress, Stockholm, 1950, vol. 7, pp 774–775

Kasali FM, Tuyiringire N, Peter EL, Ahovegbe LY, Ali MS, Tusiimire J, Engeu Ogwang P, Ntokamunda Kadima J, Agaba AG (2022) Chemical constituents and evidence-based pharmacological properties of *Physalis peruviana* L.: an overview. J Herbmed Pharmacol 11(1):35–47

Keiner R, Dräger B (2000) Calystegine distribution in potato (*Solanum tuberosum*) tubers and plants. Plant Sci 150(2):171–179

Kusano G, Orihara S, Tsukamoto D, Shibano M, Coskun M, Guvenc A, Sibel Erdurak S (2002) Five new nortropane alkaloids and six new amino acids from the fruit of *Morus alba* LINNE growing in Turkey. Chem Pharmaceut Bull 50(2):185–192

Kuster N, Rosahl S, Dräger B (2017) Potato plants with genetically engineered tropane alkaloid precursors. Planta 245:355–365

Liebisch HW, Bernarsch H, Schuette HR (1973) Biosynthesis of tropane alkaloids. 12. Biosynthesis of cochlearine. Z Chemie 13:372–373

Magalhães V, Silva AR, Silva B, Zhang X, Dias AC (2022) Comparative studies on the anti-neuroinflammatory and antioxidant activities of black and red goji berries. J Funct Foods 92:105038

Mericli F (2011) Bilimsel acidan altin cilek meyveleri (*Physalis peruviana*). Fitomed 20:18–22

Mocan A, Moldovan C, Zengin G, Bender O, Locatelli M, Simirgiotis M, Atalay A, Vodnar DC, Rohn S, Crişan G (2018) UHPLC-QTOF-MS analysis of bioactive constituents from two Romanian Goji (*Lycium barbarum* L.) berries cultivars and their antioxidant, enzyme inhibitory, and real-time cytotoxicological evaluation. Food Chem Toxicol 115:414–424

Pai PG, Habeeba PU, Ullal S, Shoeb PA, Pradeepti MS, Ramya K (2013) Evaluation of hypolipidemic effects of *Lycium barbarum* (goji berry) in a murine model. J Nat Rem 13(1):4–8

Pandey R, Pandey VS, Pandey VN (2024) Nutraceutical metabolites, value addition and industrial products for developing entrepreneurship through edible fleshy fungi (chapter 14). In: Shukla AC (ed) Entrepreneurship with microorganisms. Academic Press, pp 293–328

Petersson EV, Arif U, Schulzova V, Krtková V, Hajšlová J, Meijer J, Andersson HC, Jonsson L, Sitbon F (2013) Glycoalkaloid and calystegine levels in table potato cultivars subjected to wounding, light, and heat treatments. J Agric Food Chem 61(24):5893–5902. https://doi.org/10.1021/jf400318p

Pires TC, Dias MI, Barros L, Calhelha RC, Alves MJ, Santos-Buelga C, Ferreira IC (2018) Phenolic compounds profile, nutritional compounds and bioactive properties of *Lycium barbarum* L.: a comparative study with stems and fruits. Ind Crops Prod 122:574–581

Protti M, Gualandi I, Mandrioli R, Zappoli S, Tonelli D, Mercolini L (2017) Analytical profiling of selected antioxidants and total antioxidant capacity of goji (*Lycium* spp.) berries. J Pharm Biomed Anal 143:252–260

Robins RJ, Abraham TW, Parr AJ, Eagles J, Walton NJ (1997) The biosynthesis of tropane alkaloids in *Datura stramonium*: the identity of the intermediates between N-methylpyrrolinium salt and tropinone. J Am Chem Soc 119(45):10929–10934

Romera-Torres A, Romero-González R, Vidal JLM, Frenich AG (2018) Analysis of calystegines in tomato-based products by liquid chromatography–Orbitrap mass spectrometry. J Chromatogr A 1576:51–57

Rupani R, Chavez A (2018) Medicinal plants with traditional use: ethnobotany in the Indian subcontinent. Clin Dermatol 36(3):306–309

Sahia M, Ray AB (1980) Secotropane alkaloids of *Physalis peruviana* C. J Org Chem 45(16):3265–3268

Skenderidis P, Kerasioti E, Karkanta E, Stagos D, Kouretas D, Petrotos K, Hadjichristodoulou C, Tsakalof A (2018) Assessment of the antioxidant and antimutagenic activity of extracts from goji berry of Greek cultivation. Toxicol Rep 5:251–257

Tang WM, Chan E, Kwok CY, Lee YK, Wu JH, Wan CW, Chan RY, Yu PH, Chan SW (2012) A review of the anticancer and immunomodulatory effects of *Lycium barbarum* fruit. Inflammopharmacol 20:307–314

Taroncher M, Vila-Donat P, Tolosa J, Ruiz MJ, Rodriguez-Carrasco Y (2021) Biological activity and toxicity of plant nutraceuticals: an overview. Curr Opin Food Sci 42:113–118

Watson AA, Davies DR, Asano N, Winchester B, Kato A, Molyneux RJ, Stegelmeier BL, Nash RJ (1999) Calystegine alkaloids in the potato and other food plants. ACS Symp Ser 745:129–139. https://doi.org/10.1021/bk-2000-0745.ch009

Chapter 8
Tropane Alkaloids: Biosimilar and Biopharmaceutics

Aziza Mahrous Amer and Mohamed M. Amer

Abstract Tropane alkaloids (TAs) are naturally occurring compounds that play significant roles in pharmacology and medicinal chemistry. These alkaloids are derived from various plants, particularly those in the Solanaceae family, such as belladonna and henbane. Their unique chemical structure, characterized by a bicyclic tropane ring, contributes to their diverse pharmacological effects, including anticholinergic and central nervous system activities. Biosimilars-biopharmaceuticals' development has emerged as a significant trend in the pharmaceutical industry. While most biosimilars focus on protein-based drugs, the principles can extend to alkaloids, particularly in establishing equivalence in efficacy and safety. The regulatory frameworks for biosimilars encourage the development of alternative formulations of existing drugs, potentially offering more affordable options for patients. Biosimilars have great potential to improve patient care by reducing costs and making access to biological treatments easier. They will handle questions from patients and other healthcare providers about biosimilars and ensure they are effectively integrated into medical practice. The biosynthesis of TAs (TAs biosimilars) involves a series of enzymatic reactions that convert simple amino acids into complex alkaloid structures. The pathway typically begins with the amino acid ornithine, which is transformed into putrescine. Enzymatic modifications lead to the formation of tropin and, eventually, tropane derivatives. Biosynthetic TAs have a wide range of pharmacological applications. Biosynthesized atropine is commonly used clinically as an antimuscarinic agent to treat bradycardia, reduce salivation during surgery, and counteract organophosphate poisoning. Biosynthesized scopolamine, another prominent tropane alkaloid biosimilar, is primarily used for its antiemetic properties, particularly in motion sickness, antidepressant, and post-operative nausea. The effectiveness of these compounds stems from their ability to inhibit the action of acetylcholine at muscarinic receptors. The field of biopharmaceutics is rapidly evolving, focusing on improving the efficacy and safety of drug delivery systems. TA biosimilars present unique challenges and opportunities in this

A. M. Amer (✉) · M. M. Amer
Faculty of Veterinary Medicine, Cairo University, Giza, Egypt
e-mail: aziza_amer@cu.edu.eg; profdramer@cu.edu.eg

T. Pullaiah et al. (eds.), *Tropane Alkaloids*,
https://doi.org/10.1007/978-981-96-5805-3_8

context. Their lipophilic nature allows for easy penetration through biological membranes, which can also lead to variable absorption rates and bioavailability.

Keywords Tropane alkaloids · Pharmacological applications · Biopharmaceutics · Biosimilars

Abbreviations

CNS	Central nervous system
EFSA	European Food Safety Authority
EMA	The European Medicine Agency
FDA	The United States Food and Drug Administration
HRCs	Hairy root cultures
TA	Tropane alkaloid
WHO	The World Health Organization

8.1 Introduction

Biologicals related to biology and living organisms include antibodies, interleukins, and vaccines. In medicine, this term refers to substances derived from living entities or their products. These biologicals can be used for various purposes, such as preventing, diagnosing, treating, or easing the symptoms of illnesses. Unlike small-molecule drugs created through chemical processes, biological drugs are made by living cells and require complex manufacturing methods (Morrow and Felcone 2004).

Biosimilars are biological medicines derived from living systems, whether naturally occurring or produced through genetic engineering. They range from simple hormones like insulin to complex molecules such as monoclonal antibodies (Heinemann et al. 2022; Mascarenhas-Melo et al. 2024). In the U.S.A. and Europe, recombinant products are key candidates for biosimilars. The success of biosimilars relies on strong scientific evidence, sound regulatory decisions, and public trust in their safety and effectiveness (Tariman 2018; Cordeiro et al. 2024). Advances in biology have improved treatments for serious diseases like cancer (Arga and Sinha 2021). Biosimilars enhance healthcare access, improve patient outcomes, and lower costs (McCamish and Woollett 2012; Tariman 2018).

8.2 Tropane Alkaloids Biosimilars and Biopharmaceutics Potentialities

Three groups of TAs can be distinguished: the calystegines group, which are polyhydroxylated nortropane alkaloids (NTAs) present in Convolvulaceae, Solanaceae, Moraceae, Erythroxylaceae, and Brassicaceae; TAs found in Solanaceae plants, such as hyoscyamine and scopolamine; and coca alkaloids, such as cocaine from *Erythoxylum coca* (Kohnen-Johannsen and Kayser 2019).

Creating the tropane ring system is similar in different organisms that produce the three classes of TAs. This biosynthesis begins with the amino acid's ornithine or arginine and their intermediate putrescine. From there, it moves to the common precursor, N-methyl-D1-pyrrolinium cation, which is essential to produce cocaine, hyoscyamine/scopolamine, calystegine, and nicotine (Grynkiewicz and Gadzikowska 2008; Ziegler and Facchini 2008; Kohnen-Johannsen and Kayser 2019). The initial stages of TAs biosynthesis involve a common pathway for different TAs, including cocaine, scopolamine/hyoscyamine, and calystegines. This process starts with the amino acids L-ornithine (Orn) and L-arginine (Arg) as precursors (Kohnen-Johannsen and Kayser 2019). Scopolamine is produced in Solanaceae family plant roots and then transported to the above-ground parts for storage.

Biosimilars hold significant promises for enhancing patient care by lowering costs and increasing access to biologic treatments. They will be responsible for addressing inquiries from patients and fellow healthcare professionals about biosimilars and overseeing their effective incorporation into clinical practice (Tinsley et al. 2018).

8.3 Members of Tropane Alkaloids and Tropane Alkaloids Biosimilar

8.3.1 Structure of Tropane Alkaloids and Biosimilars

TAs have a bicyclic ring structure called N-methyl-8-a-zabicyclo-octane (Griffin and Lin 2000). These compounds are found in Solanaceae plants, which also contain bicyclic amino alcohols, including di- and tri-hydroxylated tropines, 6,7-epoxide, and de-N-methylated compounds known as N-nortropane derivatives (Fig. 8.1). Tropane alkaloids occur as esters with carboxylic acids, with tropic acid serving as a significant marker for alkaloid-rich plants in chemotaxonomy, particularly among species within the Solanaceae family. Derivatives of tropane alkaloids include tropane, tropine, tropanone, scopolamine, cocaine, pseudotropine, and others (Sweta and Lakshmi 2015).

8.3.2 Sources of Tropane Alkaloids and Biosimilars

TAs are a specific class of alkaloids, defined as molecules that contain a tropane ring system (Fig. 8.2) for the tropane skeleton structure and the three major TA groups (Grynkiewicz and Gadzikowska 2008; Kohnen-Johannsen and Kayser 2019). The tropane ring system biosynthesis is homologous in organisms that produce these three TA classes. Beginning with the amino acids ornithine or arginine and their

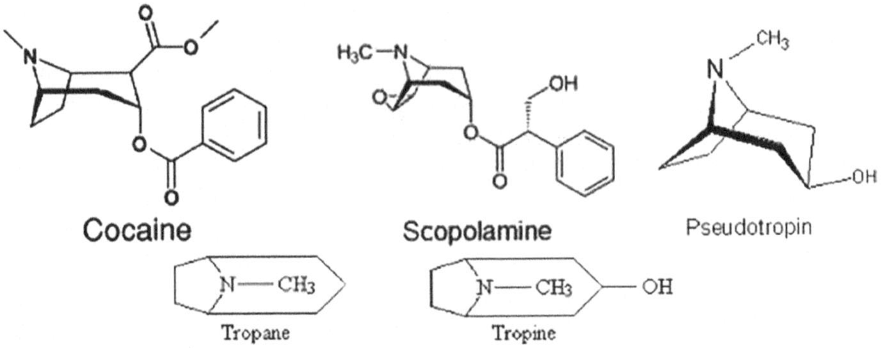

Fig. 8.1 Some members of the derivatives of tropane alkaloids: Tropane, Tropanone, Scopolamine, Cocaine, Pseudotropine. (After Sweta and Lakshmi 2015)

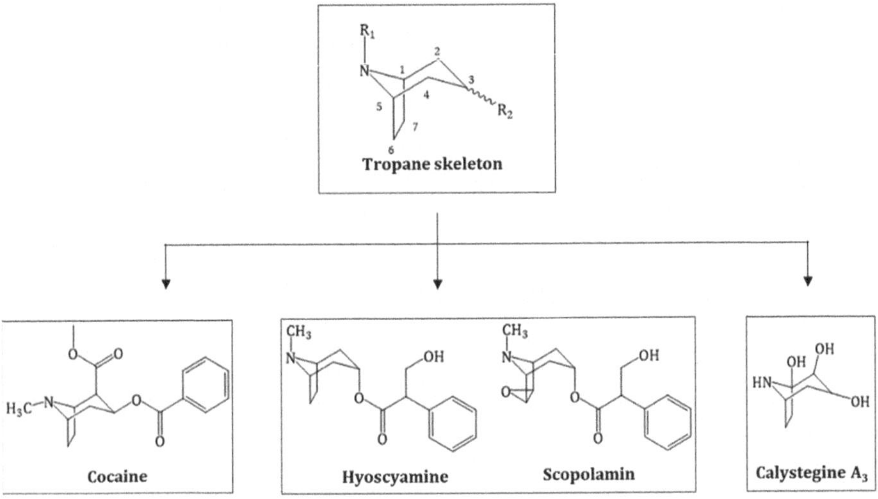

Fig. 8.2 Structure of the tropane skeleton and the three major groups of TAs derived from this skeleton. (After Kohnen-Johannsen and Kayser 2019)

intermediary putrescine, TAs biosynthesis proceeds to the common precursor of all TAs, N-methyl-D1-pyrrolinium cations. This is the branch point for the biosynthesis of nicotine, cocaine, calystegine, and hyoscyamine/scopolamine (Ziegler and Facchini 2008). Even though all TAs have a tropane ring, which gives them a high degree of structural similarity, these compounds' pharmacological actions vary greatly. The blood-brain barrier can be crossed by cocaine and hyoscyamine/scopolamine, which can cause dose-dependent hallucinations and psychoactive effects. Calystegines are hydrophilic, unable to cross this barrier, and do not produce these effects (Kohnen-Johannsen and Kayser 2019).

With a few exceptions, it is unlawful to grow coca plants, extract cocaine, make other drugs that contain cocaine, and trade them. Cocaine is the second most often used illegal substance in the world (Schultze-Kraft 2016). Much research hasn't been done on calystegines, a recently identified class of TAs with little pharmacological, medical, or economic value.

Scopolamine is of significant economic interest due to its various pharmacological applications. The World Health Organization (WHO) has recognized scopolamine as one of its Essential Medicines. In Queensland, Australia, *Duboisia* plants are cultivated on large plantations to extract hyoscyamine and scopolamine (Grynkiewicz and Gadzikowska 2008). However, the pharmaceutical industry faces challenges in consistently producing high quantities of scopolamine due to climate change and the new biotic and abiotic factors that arise from it. Alternative production methods have also been explored (Kohnen-Johannsen and Kayser 2019).

Datura stramonium: It is utilized for various therapeutic purposes, including the treatment of asthma, motion sickness, and Parkinson's disease (Akbar 2020; Thapa et al. 2022). The pharmacological properties of *D. stramonium* are primarily attributed to its alkaloids, scopolamine and atropine, which function as anticholinergic agents like ATs by inhibiting muscarinic receptors within the central nervous system (CNS) (Devi et al. 2011; Sharma et al. 2021; Rasi et al. 2024). These alkaloids are predominantly synthesized in the roots and transported to the above-ground parts, including the leaves and flowers (Cinelli and Jones 2021).

Scopolamine features a unique epoxide bridge in its tropane ring, distinguishing it from atropine (Huang et al. 2021; Wang et al. 2023). Atropine, a racemic mixture of (−)-hyoscyamine and (+)-hyoscyamine, has pharmacological efficacy primarily attributed to the (−)-hyoscyamine isomer (Kohnen-Johannsen and Kayser 2019). Numerous TAs are effective anticholinergic medications and muscarinic receptor antagonists. A few of them have psychedelic properties. Different medical indications demand the therapeutic use of atropine, hyoscyamine, and scopolamine (Melton et al. 2019).

8.4 Biological and Pharmacological Activity of Tropane Alkaloid and/or Biosimilar

TAs and their biosimilars have several biological and pharmacological activities that promote their clinical use in many health problems and diseased conditions (Table 8.1).

Table 8.1 Biological and pharmacological activity of tropane alkaloid and/or biosimilar

Activity	Mode of action	References
Effect on bronchial asthma	Inhaling atropine can improve lung mucociliary function in people.	Sur et al. (1990), Groth et al. (1991) and Dey et al. (2020)
Activity against organophosphorus compounds	When used with a cholinesterase reactivator, atropine counteracts the toxic effects of organophosphates. C-scopolamine can be utilized through positron emission tomography to image muscarinic cholinergic receptors in living subjects.	Ligtenstein and Moes (1991), Tsung-Ming (1991), Solana et al. (1991) and Dey et al. (2020)
Effects against hyperglycemia and parkinsonism	Atropine reduces high blood sugar caused by neostigmine. It affects the vagus nerve in diabetic people. Atropine also reduces tremors. Hyoscine blocks the activity of certain receptors in the brain, with increased dopamine levels that are useful for treating Parkinson's disease.	Julu and Hondo (1992), Houghton and Howes (2008) and Dey et al. (2020)
Activity on ANS	Atropine works by blocking the effects of acetylcholine binding all muscarinic receptors found in smooth muscles, ganglia, and certain neurons.	Fodor and Dharanipragada (1994) and Dey et al. (2020)
Activities on CVS	Atropine prevents the action of muscarinic receptors in heart muscles. It is also used to treat heart problems.	Fodor and Dharanipragada (1994) and Dey et al. (2020)
Effect on nausea, drooling, and Alzheimer disease	Scopolamine is used to help treat nausea and drooling in people with Alzheimer's and Parkinson's diseases. Reduce nausea and vomiting after surgery as applied through skin.	Horimoto et al. (1991), Dreyfuss et al. (1991), Grau et al. (1991) and Dey et al. (2020)
Activity for treating motion sickness	Transdermal scopolamine helps reduce motion sickness. When placed behind the ear 5 hs before a boat trip, this medication effectively prevents seasickness.	Attias et al. (1987) and Dey et al. (2020)
Antispasmodic and antiallergic activity	Hyoscyamine and atropine help relieve spasms and reduce allergic reactions by blocking the muscarinic acetylcholine receptor.	Kinghorn (2001) and Dey et al. (2020)
Hyperactivity: Effect on slowing	Scopolamine causes hyperactivity in mice, involving both the cholinergic and dopaminergic systems. THA blocks acetylcholinesterase and, due to scopolamine, affects the neocortical brain. Scopolamine, oxotremorine, and physostigmine significantly impact the production of acetylcholine in the brain.	Riekkinen Jr et al. (1991), Valjakka et al. (1991) and Dey et al. (2020)
Central activity: Sedative, antiemetic, and amnestic effects	Scopolamine blocks brain muscarinic receptors by stopping acetylcholine from attaching to them. This stops signals from effectively traveling through the body. Therefore, scopolamine can help with problems like nausea.	Renner et al. (2005) and Dey et al. (2020)
Miscellaneous	Atropine affects the pelvic pouch and anal sphincter work. When given centrally it influences the radial arm maze rats perform	Hallgren et al. (1991), Sala et al. (1991) and Dey et al. (2020)

8.5 Anticholinergic Property of Tropane Alkaloids Biosimilars

Because tropane alkaloids biosimilars disrupt the neurotransmitter molecule acetylcholine at synapses in the central and/or peripheral nervous systems, they exhibit anticholinergic characteristics (Kohnen-Johannsen and Kayser 2019). The three TAs atropine, hyoscyamine, and scopolamine are found in the genera *Atropa* (deadly nightshade plants), *Hyoscyamus* (henbane plants), *Mandragora* (mandrake plants), *Latua* (sorcerer's tree), and *Datura* (jimson weed plant). Additionally, all of these acetylcholine-inhibiting TAs are known to be produced by the genus *Brungmansia* (angel trumpet plants) and plant members of the Solanoideae subfamily of the Solanaceae, including the species *Anisodus*. When ingested, TAs and their biosimilars are quickly absorbed from the gut and transported into tissues, where they undergo cytochrome P450-mediated oxidative methylation for metabolism. They are typically eliminated from the body as renal waste rather than accumulating (Adamse and van Egmond 2010; EFSA 2013).

Certain mental health problems, e.g., depression, anxiety, symptoms of epilepsy, Parkinson's, and Alzheimer's are linked to the muscarinic acetylcholine receptor system in the body (Szczurowska et al. 2022). Some plant-based substances like cocaine can boost the activity of the CNS and overall body function (Koob 2013).

Many pharmaceutical companies are involved in the global tropane alkaloid market. The significant market value of plant TAs has spurred their large-scale production around the globe. Traditional farming alone cannot satisfy this demand, leading to the development of synthetic versions of these alkaloids (Azarbad 2022). Even with advancements in synthetic methods, most plant tertiary TAs used by pharmaceutical companies are still grown through traditional farming.

8.6 Tropane Alkaloid Biosimilars Biosynthesis and Commercial Production

8.6.1 Tropane Alkaloids (TAs) and Biosimilars Biosynthesis Pathways

TAs and biosimilars are a fascinating group of substances found in plants, known for having a specific ring structure called an 8-azabicyclo [3.2.1] octane (nortropane) (National Center for Biotechnology Information 2024). Some well-known examples are hyoscyamine, scopolamine, and cocaine, which have been used historically as poisons, hallucinogens, and anesthetics. Since the complete structure of the tropane ring was discovered in 1901, scientists in organic and biochemistry have been curious about how these unique compounds are formed in living organisms and in lab settings. In 2020, researchers were able to fully outline the biosynthetic pathways of hyoscyamine and scopolamine and even produced them from scratch using yeast (Fig. 8.3) (Huang et al. 2021).

TAs and biosimilars are a type of chemical compounds that have a specific structure known as the tropane nucleus (Fig. 8.4). These compounds come from two sources: ornithine and acetoacetate. Pyrrolines serve as the building blocks for these alkaloids. Most of them are esters formed from mono, di, or trihydroxytropane, and they can have many different arrangements of hydroxyl groups. Since the nineteenth century, substances like cocaine, atropine, and scopolamine, along with their derivatives, have been studied extensively due to their significant effects on health (Biastoff and Drager 2007; Gossauer 2003; Hemscheidt 2000; Dey et al. 2020).

8.7 Tropane Alkaloids and Biosimilars Commercial Production

Before the progress in synthetic chemistry and the creation of new drug molecules, plants were the most abundant and renewable source of medicinal compounds through their different metabolic processes (Chaachouay and Zidane 2024). Plant-origin TAs come from various families and have important medical uses. They mainly work as anticholinergics and stimulants, making them suitable for many medical uses (Dey et al. 2020).

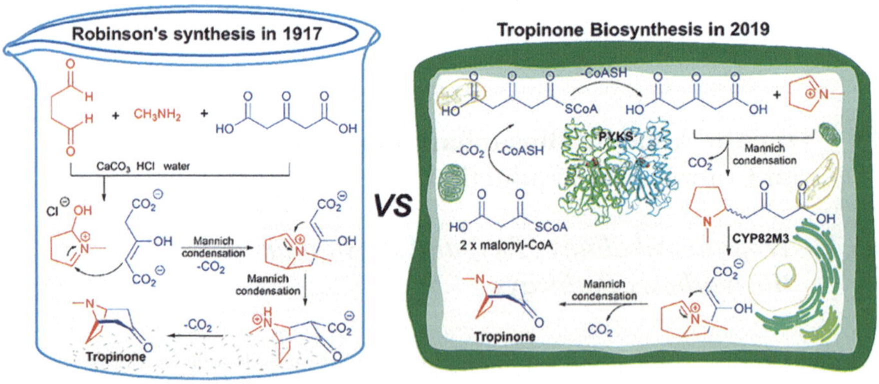

Fig. 8.3 Tropane alkaloids and biosimilars biosynthess. (After Huang et al. 2021)

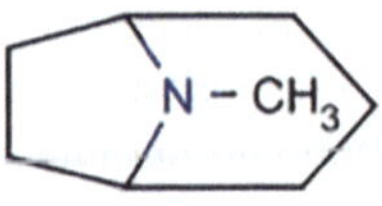

Fig. 8.4 The basic structure of the tropane nucleus. (After Dey et al. 2020)

TAs are defined as alkaloids that have a bicyclic amine with a pyrrolidine, piperidine, and common nitrogen atom (Kohnen-Johannsen and Kayser 2019). Drugs for local and general anesthesia often contain tropane alkaloids, which are in high demand. Traditional farming methods struggle to meet this demand, and careless harvesting harms the environment. To tackle these issues, biotechnological methods are being employed to produce these compounds through lab-based cultures. (Raklami et al. 2022).

TAs and their native plants are now used as a primary ingredients in a variety of contemporary and conventional medications intended to treat physical and mental illnesses after being chemically investigated for their pharmacological effects in the field of pharmaceuticals (Srivastava et al. 2021).

8.7.1 Genetic and Biotechnological Methods for TAs Biosimilar Production

Biotechnology has made it easier to produce TAs, and it also offers a reliable method to grow, preserve, and protect the genetic material of important and at-risk plants that produce TAs as the market demand is still not being fully met and the long time it takes for plants to grow harvestable biomass (Davis and Choisy 2024). The inconsistent farming practices and challenges from climate and environment have led to exploring new ways of producing plant TAs using biotechnology (Lahlali et al. 2024).

8.7.2 Alternative TA and Biosimilars Production System: In Vitro Cultures

Production of therapeutic TAs and biosimilars (Fig. 8.5) primarily occurs in callus, cell suspension cultures, and hairy root cultures (HRCs) after specific infection with *Agrobacterium rhizogenes* (Mehrotra et al. 2015; Hedayati et al. 2020). The first two methods often require phytohormones during growth, while hairy root cultures (HRCs) do not, making HRCs a more cost-effective option for producing tropane alkaloids (TAs) (Srivastava et al. 2018, 2021). Cell suspension cultures are popular due to their scalability. Key TAs produced include hyoscyamine, scopolamine, atropine, and cocaine, all of which have important therapeutic uses. Research is focused on starting and studying callus and cell suspension cultures in TA-producing plants (Yamada and Endo 1984; Herouart et al. 1991; Raoufa et al. 2008).

To explore different ways of producing TAs biosimilars, researchers started shoot cultures of various plants, including *Hyoscyamus albus, Duboisia leichhardtii, D. myoporoides, Datura metel,* and *Scopolia parviflora* (Kang et al. 2004). The initial shoot cultures showed very little hyoscyamine and scopolamine.

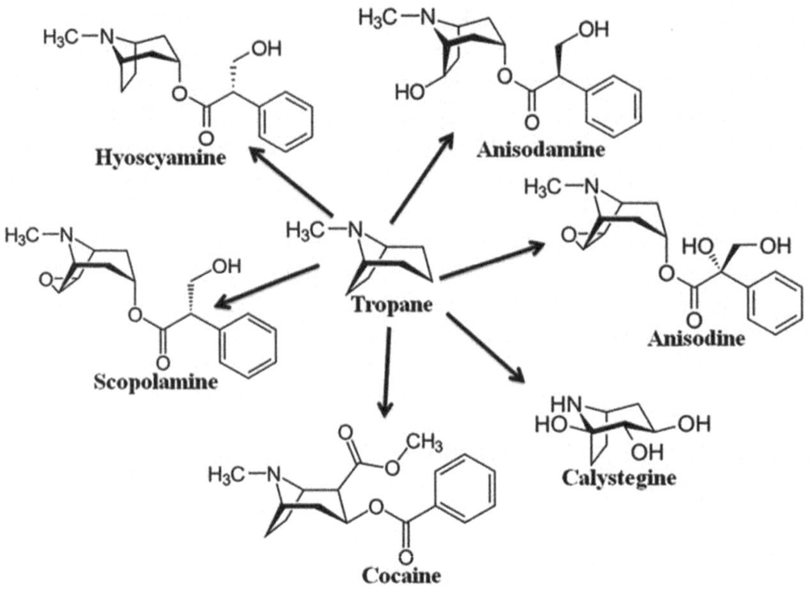

Fig. 8.5 Structural details of major tropane alkaloids and biosimilars. (After Srivastava et al. 2021)

However, in the case of *D. leichhardtii,* when roots developed, and root biomass increased, the amounts of hyoscyamine and scopolamine reached 0.36% and 0.30% of the dry weight of the shoots, respectively. This study highlighted roots' crucial role in producing TAs (Georgiev et al. 2013; Srivastava et al. 2021).

8.7.3 Biotransformation Method for Production of Tropane Alkaloids Biosimilars

In plant laboratory studies, most biotransformation research has focused on cell suspension and hairy root cultures (Giri et al. 2001; Srivastava et al. 2017a). These in vitro systems are not affected by seasonal changes or diseases, making them a sustainable option for using resources. These cultures can perform various biotransformation reactions like hydroxylation, glycosylation, oxidoreduction, and hydrolysis (Giri et al. 2001; Banerjee et al. 2012; Srivastava et al. 2017a). There are three main types of biotransformation based on how cultures are grown and what substances are used: precursor feeding, co-culture, and non-specific/exogenous molecule biotransformation. In plants that produce tropane alkaloids, all these methods have been tested, but non-specific/exogenous biotransformation is particularly common (Srivastava et al. 2017b, 2021).

Precursor feeding has been recognized as an effective way to understand the biosynthetic pathway of tropane alkaloids or to increase the flow of early metabolites. This method uses both nearby precursors, like hyoscyamine for scopolamine production, and distant precursors, such as phenylalanine for scopolamine and tropinone for calystegine production (Srivastava et al. 2021).

8.7.4 Phytoremediation Method for Production of Tropane Alkaloids Biosimilars

Plants can process both organic and inorganic pollutants, which allows them to either remove these harmful substances or convert them into less dangerous forms. Studies have also shown that in vitro cultures of plants that produce tropane alkaloids can be used for cleaning up pollutants (Doran 2009; Srivastava et al. 2021).

8.7.5 Molecular Farming Method for Production of Tropane Alkaloids Biosimilars

Plant molecular farming refers to using whole plants or their cultured cells and tissues in vitro to produce valuable materials or recombinant proteins (biosimilars). This approach provides an economical alternative to traditional production systems that relieve cultivating microbial or mammalian cells in large-scale bioreactors (Schillberg et al. 2013). Molecular farming has also been explored in plants that bear transgenic and biosimilar traits (Banerjee et al. 2002).

Relevant research has shown that tetraploid *H. muticus* plants perform better in producing biomass and scopolamine biosimilar. These autotetraploid plants have a unique alkaloid profile, yielding more scopolamine biosimilar while slightly reducing hyoscyamine biosimilar. This change leads to a higher scopolamine-to-hyoscyamine biosimilars ratio, making them more useful for pharmaceutical purposes. While the total scopolamine biosimilar in these tetraploid plants is still lower than that of species like *Duboisia*, which are known for high scopolamine biosimilar production, they represent a valuable genetic resource. This resource can help optimize growth conditions in lab settings and lead to more detailed research on *Hyoscyamus* species (Dehghan and Ghotbi 2021).

The biosynthesis of tropane alkaloids (TAs) biosimilar is similar across organisms and begins with the amino acids ornithine or arginine, leading to putrescine, and then to the N-methyl-D1-pyrrolinium cation, a key precursor for TAs. This is the point where the biosynthesis of compounds like cocaine, hyoscyamine, scopolamine, calystegine, and nicotine diverge (Ziegler and Facchini 2008).

8.8 Tropane Alkaloids Biosimilar Production: Cocaine Intermediates Synthesis

In the literature, two different pathways towards cocaine biosynthesis have been reported. According to the classical hypothesis, the bridgehead carbon atom C-1 of methylecgonine is derived from an N-methyl-D1-pyrrolinium cation, and that of C-2 originates from acetoacetate (Eich 2008). However, feeding experiments with labelled N-methyl-D1-pyrrolinium cation were inconclusive in planta and could not confirm this theory. As a result, it was proposed that the observed regiochemistry of incorporation of the labelled N-methyl-D1-pyrrolinium cation into cocaine was compatible with the stepwise introduction of C2 units into the ecgonine skeleton, derived from acetate (Leete et al. 1990). Consequently, a new intermediate in cocaine biosynthesis, N-methyl-2-pyrrolidineacetic acid, was suggested. Although this compound was detected in several plants, all attempts at incorporating it into the ester or thioester forms have been unsuccessful. From a chemical perspective, the nucleophilic addition of the first acetyl-CoA moiety reaction is assumed to be a Manniche-like reaction using the enolate anion, and the side-chain extension occurs via Claisen condensation. The (S)-enantiomer cyclizes and forms the bicyclic structure of the cocaine tropane ring skeleton by an intramolecular Mannich reaction (Hoye et al. 2000).

8.9 Tropane Alkaloids Biosimilar Production: Milking the Plant

An advanced stage in the development of hairy root cultures involves the utilization of TAs biosimilars through a process known as "milking the plant." Aeroponically cultivated plants are stimulated in this method, and their roots are "extracted" by immersing them in a physiological extraction medium without causing harm or destruction. The desired secondary metabolites are then isolated from the roots. Following the "milking" process, the plants are returned to their cultivation apparatus to regenerate and produce more secondary metabolites, which can be extracted in subsequent cycles (Bourgaud et al. 2013). Another avenue for TA production involves the generation and cultivation of transgenic plants. For instance, Xia et al. (2016) successfully overexpressed pmt. from *Nicotiana tabacum* (NtPMT) and h6h from *H. niger* (HnH6H) in *A. belladonna*, resulting in high scopolamine levels (2.94–5.13 mg g^{-1} DW) under field conditions.

8.10 Tropane Akaloids Biosimilar Production: Microbial Production and Enzyme Engineering Approaches

Microbial cultures, such as bacteria (e.g., *E. coli*) or yeast (e.g., *Saccharomyces cerevisiae*), are easier to cultivate and are well-studied model organisms with fully developed molecular toolkits (Vanderwaeren et al. 2022). These organisms can be easily scaled in existing fermentation infrastructure, making their cultivation more economically advantageous compared to plant tissue culture (Chandran et al. 2020). As a result, the heterologous production of compounds like scopolamine in these hosts may offer an appealing alternative, especially if the transfer of molecular pathways is feasible (Kohnen-Johannsen and Kayser 2019).

8.11 Other Methods of Tropane Alkaloids Biosimilar Production

To increase the scopolamine level in plants, researchers have developed polyploid plants and determined their yields. An impressive example of this was published by Dehghan et al. (2017). The authors successfully produced stable tetraploid hairy root lines of *Hyoscyamus muticus*. These tetraploid plants exhibited lower biomass production compared to diploids, but they showed higher levels of scopolamine (13.87 mg L^{-1}) and hyoscyamine (107.7 mg L^{-1}), up to 200% more scopolamine than in diploid plants. However, the total yield of scopolamine from these plants was relatively low due to their slow growth rates. The results were only reported for growth conditions in optimized Murashige and Skoog (MS) or Gamborg's B5 media, which are not competitive with conventional field cultivation. Additionally, other species of *Hyoscyamus*, such as *H. senecionis*, have been found to have higher levels of scopolamine than hyoscyamine in their leaves, as indicated by a study (Dehghan et al. 2012). This finding could be advantageous for breeding and optimization purposes. Initial trials have shown promise, and it is worth exploring the potential of polyploid plants from other species, such as *D. myoporoides*, as viable alternatives. Naik et al. (2017) significantly developed the TA-producing endophytes in *D. metel* including *Colletotrichum boninense*, *Phomopsis* sp., *Fusarium solani, Colletotrichum incarnatum, Colletotrichum siamense,* and *Colletotrichum gloeosporioides,* were found to possess the enzymes PMT, TR-I, and H6H.

Recent reports have indicated that these fungi can produce a significant quantity of scopolamine (4.1 mg L^{-1}) and hyoscyamine (3.9 g L^{-1}). The independent cultivation of these fungal species has the potential to serve as a viable natural alternative to using heterologous hosts or traditional agricultural cultivation methods (Naik et al. 2017).

8.12 Use of "Omics" in Plant Science and Pathway Explanation for Tropane Alkaloids Biosimilars Production

Many steps had significant progress in biosynthesis calystegine through metabolic steps involving genomic, transcriptomic, metabolomic, and proteomic data analysis, combined with biochemical and heterologous enzyme characterization (Wienkoop et al. 2008; Kohnen-Johannsen and Kayser 2019; Naik et al. 2023). The biosynthesis of TAs has not been fully elucidated, with important steps such as the reaction of hygrine to tropinone and the synthesis of littorine remaining unexplained. The application of "omics" techniques for candidate gene isolation and detailed process understanding can shed light on these unresolved aspects (Kohnen-Johannsen and Kayser 2019).

8.13 Conclusions

TAs, or tropane alkaloids biosimilars, are a diverse group of plant secondary metabolites, which can be categorized into three main groups: hyoscyamine and scopolamine, cocaine, and calystegines. The demand for TAs is on the rise due to its various therapeutic applications and increasing market demand. Investigations have been carried out in alternative production systems, including engineered plants, climate-independent production systems such as plant cell cultures, and heterologous microbial production.

Currently, alternative methods for producing tertiary alkaloids (TAs) from agricultural plants are not economically competitive with conventional techniques. The primary challenge is the limited understanding of the biosynthetic pathways for various TAs. Many enzymes are not well characterized, their genes are often unidentified, and there's a lack of clarity on rate-limiting steps and metabolic interactions.

The application of "omics" techniques is expected to help clarify these pathways, as has been done for other chemical classes in plants. However, challenges arise when transferring TA biosynthesis to heterologous hosts; the production and storage sites of TAs in plants may differ, necessitating compartmentalized systems like yeast or algae. Additionally, only specific plant cells express the necessary biosynthetic pathways, limiting the utility of undifferentiated plant cell cultures.

Recent studies suggest that endophytic fungi could also produce TA biosimilars, offering a novel source independent of plants. The field of scopolamine production, whether from native plants, endophytes, or microbial hosts, is rapidly evolving, with significant potential for advancements through modern "omics" technologies.

TAs represent an intriguing and valuable area of exploration within pharmacology and biopharmaceutics. The complexity of their biosynthetic pathways, coupled with their notable therapeutic applications, emphasizes the importance of sustained research in this field. As we witness advancements in biotechnology and drug formulation techniques, there is a promising opportunity to develop new biosimilars and enhance existing tropane alkaloid therapies. Such progress could significantly improve patient outcomes and expand the range of therapeutic options available. By deepening our understanding of these compounds, researchers may pave the way for innovative drug delivery methods and treatment strategies that leverage the distinctive properties of tropane alkaloids.

References

Adamse P, van Egmond HP (2010) Tropane alkaloids in food. RIKILT-Institute of Food Safety, Wageningen. https://www.researchgate.net/publication/254834358

Akbar S (2020) *Datura stramonium* L. (Solanaceae). In: Handbook of 200 medicinal plants. Springer, Berlin/Heidelberg, pp 857–868

Arga KY, Sinha R (2021) Recent developments in cancer systems biology: lessons learned and future directions. J Pers Med 11(4):271. https://doi.org/10.3390/jpm11040271

Attias J, Gordon C, Ribak J, Binah O, Rolnick A (1987) Efficacy of transdermal scopolamine against seasickness: a 3-day study at sea. Aviat Space Environ Med 58:60–62

Azarbad H (2022) Conventional vs. organic agriculture–which one promotes better yields and microbial resilience in rapidly changing climates? Front Microbiol 13:903500. https://doi.org/10.3389/fmicb.2022.903500

Banerjee S, Shang TQ, Wilson AM, Moore A, Strand SE, Gordon MP, Doty SL (2002) Expression of functional mammalian P450 2E1 in hairy root cultures. Biotech Bioeng 77(4):462–466. https://doi.org/10.1002/bit.10151

Banerjee S, Singh S, Rahman LU (2012) Biotransformation studies using hairy root cultures – a review. Biotechnol Adv 30(3):461–468. https://doi.org/10.1016/j.biotechadv.2011.08.010

Biastoff S, Drager B (2007) Calystegines. Alkaloids Chem Biol 64:49–102

Bourgaud F, Benoit M, De Picardie U, Je Laine J-M, Gontier E, Fevre J-P (2013) Développement D' un Nouveau Procédé de Production D'actifs Pharmaceutiques à Partir de Plantes Médicinales: La Technologie des Plantes à Traire. Conference Paper: Les Rencontres du Végétal. Angers, France, https://www.researchgate.net/profile/Eric-Gontier/publication/294872849

Chaachouay N, Zidane L (2024) Plant-derived natural products: a source for drug discovery and development. Drugs Drug Candidates 3(1):184–207. https://doi.org/10.3390/ddc3010011

Chandran H, Meena M, Barupal T, Sharma K (2020) Plant tissue culture as a perpetual source for production of industrially important bioactive compounds. Biotech Rep 26:e00450. https://doi.org/10.1016/j.btre.2020.e00450

Cinelli MA, Jones AD (2021) Alkaloids of the genus *Datura*: review of a rich resource for natural product discovery. Molecules (Basel, Switzerland) 26(9):2629. https://doi.org/10.3390/molecules26092629

Cordeiro MA, Vitorino C, Sinogas C, Sousa JJ (2024) A regulatory perspective on biosimilar medicines. Pharmaceutics 16(3):321. https://doi.org/10.3390/pharmaceutics16030321

Davis CC, Choisy P (2024) Medicinal plants meet modern biodiversity science. Curr Biol 34(4):R158–R173. https://doi.org/10.1016/j.cub.2023.12.038

Dehghan I, Ghotbi E (2021) Production of Tropane Alkaloids (TAs) in plants and in vitro cultures of different ploidy levels. In: Srivastava V et al (eds) Tropane alkaloids. Springer, Singapore, pp 105–111

Dehghan EH, Oksman-Caldentey S, Ahmadi K-M, Ahmadi FS (2012) Production of tropane alkaloids in diploid and tetraploid plants and in vitro hairy root cultures of Egyptian henbane (*Hyoscyamus muticus* L.). Plant Cell, Tissue and Organ Culture 110:35–44. https://doi.org/10.1007/s11240-012-0127-8

Dehghan E, Reed DW, Covello PS, Hasanpour Z, Palazon J, Oksman-Caldentey K-M, Ahmadi FS (2017) Genetically engineered hairy root cultures of *Hyoscyamus senecionis* and *H. muticus:* ploidy as a promising parameter in the metabolic engineering of tropane alkaloids. Plant Cell Rep 36(10):1615–1626. https://doi.org/10.1007/s00299-017-2178-0

Devi MR, Bawari M, Paul SB, Sharma GD (2011) Neurotoxic and medicinal properties of *Datura stramonium* L.–review. Assam Univ J Sci Tech 7(1):139–144. https://www.bilaspuruniversity.ac.in/PDF/ResearchPub lications/212.pdf

Dey P, Kundu A, Kumar A, Gupta M, Lee B, Bhakta T, Dash S, Kim HS (2020) Analysis of alkaloids (indole alkaloids, isoquinoline alkaloids, tropane alkaloids). In: Seyed Mohammad Nabavi SM, Saeedi M, Nabavi SF, ATS S (eds) Recent advances in natural products analysis, 1st edn. Elsevier, Amsterdam, pp 505–567

Doran PM (2009) Application of plant tissue cultures in phytoremediation research: incentives and limitations. Biotechnol Bioeng 103(1):60–76. https://doi.org/10.1002/bit.22280

Dreyfuss P, Vogel D, Walsh N (1991) The use of transdermal scopolamine to control drooling. A case report. Am J Phys Med Rehab 70(4):220–222. https://doi.org/10.1097/00002060-199108000-00012

EFSA (2013) Scientific opinion on tropane alkaloids in food and feed. EFSA J 9:1–134

Eich E (2008) Solanaceae and convolvulaceae: secondary metabolites biosynthesis, chemotaxonomy, biological and economic significance (A handbook). Springer, Berlin/Heidelberg

Fodor G, Dharanipragada R (1994) Tropane alkaloids. Nat Prod Rep 11:443–450. https://doi.org/10.1039/NP9941100443

Georgiev V, Marchev A, Berkov S, Pavlov A (2013) Plant in vitro systems as sources of tropane alkaloids. In: Ramawat K, Mérillon JM (eds) Natural products. Springer, Berlin/Heidelberg. https://doi.org/10.1007/978-3-642-22144-6_6

Giri A, Dhingra V, Giri CC, Singh A, Ward OP, Narasu ML (2001) Biotransformations using plant cells, organ cultures and enzyme systems: current trends and future prospects. Biotech Adv 19(3):175–199. https://doi.org/10.1016/s0734-9750(01)00054-4

Gossauer A (2003) Monopyrrolic natural compounds including tetramic acid derivatives. Fortschritte der Chemie organischer Naturstoffe = Progress in the chemistry of organic natural products. Progres dans la chimie des substances organiques naturelles 86:1–188

Grau JW, Illich PA, Chen P-S, Meagher MW (1991) Role of cholinergic systems in pain modulation. I. Impact of scopolamine on environmentally induced hypoalgesia and pain reactivity. Behav Neurosci 105(1):62–81. https://doi.org/10.1037/0735-7044.105.1.62

Griffin WJ, Lin GD (2000) Chemotaxonomy and geographical distribution of tropane alkaloids. Phytochemistry 53(6):623–637. https://doi.org/10.1016/s0031-9422(99)00475-6

Groth M, Langenback E, Foster W (1991) Influence of inhaled atropine on lung mucociliary function in humans. Am Rev Respir Dis 144(5):1042–1047. https://doi.org/10.1164/ajrccm/144.5.1042

Grynkiewicz G, Gadzikowska M (2008) Tropane alkaloids as medicinally useful natural products and their synthetic derivatives as new drugs. Pharmacol Rep 60(4):439–463

Hallgren T, Fasth S, Delbro D, Nordgren S, Oresland T, Hultén L (1991) The effects of atropine or benzilonium on pelvic pouch and anal sphincter functions. Scand J Gastroenterol 26(5):563–571. https://doi.org/10.3109/00365529108998581

Hedayati A, Hosseini B, Palazon J, Maleki R (2020) Improved tropane alkaloid production and changes in gene expression in hairy root cultures of two *Hyoscyamus* species elicited by silicon dioxide nanoparticles. Plant Physiol Biochem 155:416–428. https://doi.org/10.1016/j.plaphy.2020.07.029

Heinemann L, Davies M, Home P, Forst T, Vilsbøll T, Schnell O (2022) Understanding biosimilar insulins – development, manufacturing, and clinical trials. J Diabetes Sci Tech 17(6):1649. https://doi.org/10.1177/19322968221105864

Hemscheidt T (2000) Tropane and related alkaloids, biosynthesis. Springer, pp 175–206

Herouart D, Gontier E, Sangwan RS, Sangwan-norree BS (1991) Analysis of the potential use of androgenic *Datura innoxia* for the development of cell cultures producing high amounts of tropane alkaloids. J Exp Bot 42(8):1073–1076. https://www.jstor.org/stable/23693585

Horimoto Y, Tomie H, Hanzawa K, Nishida Y (1991) Scopolamine patch reduces postoperative emesis in paediatric patients following strabismus surgery. Can J Anaesth 38(4 Pt 1):441–444. https://doi.org/10.1007/BF03007580

Houghton PJ, Howes MJ (2008) Natural products and related compounds of realized and potential use in treating neurodegenerative disease. In: Selected topics in the chemistry of natural products. World Scientific, pp 377–426. https://doi.org/10.1142/9789812790781_0012

Hoye TR, Bjorklund JA, Koltun DO, Renner MK (2000) N-methylputrescine oxidation during cocaine biosynthesis: study of prochiral methylene hydrogen discrimination using the remote isotope method. Org Lett 2(1):3–5. https://doi.org/10.1021/ol990940s

Huang J-P, Wang Y-J, Tian T, Wang L, Yan Y, Huang S-X (2021) Tropane alkaloid biosynthesis: a centennial review. Nat Prod Rep 38(9):1634–1658. https://doi.org/10.1039/d0np00076k

Julu PO, Hondo RG (1992) Effects of atropine on autonomic indices based on electrocardiographic R-R intervals in healthy volunteers. J Neurol Neurosurg Psychiatry 55(1):31–35. https://doi.org/10.1136/jnnp.55.1.31

Kang YM, Min JY, Moon HS, Karigar CS, Prasad DT, Lee CH, Choi MS (2004) Rapid in vitro adventitious shoot propagation of *Scopolia parviflora* through rhizome cultures for enhanced production of tropane alkaloids. Plant Cell Rep 23(3):128–133. https://doi.org/10.1007/s00299-004-0820-0

Kinghorn AD (2001) Pharmacognosy in the 21st century. J Pharm Pharmacol 53(2):135–148. https://doi.org/10.1211/0022357011775334

Kohnen-Johannsen KL, Kayser O (2019) Tropane alkaloids: chemistry, pharmacology, biosynthesis and production. Molecules 24(4):796. https://doi.org/10.3390/molecules24040796

Koob GF (2013) Negative reinforcement in drug addiction: the darkness within. Curr Opin Neurobiol 23(4):559–563. https://doi.org/10.1016/j.conb.2013.03.011

Lahlali R, Taoussi M, Laasli S, Gachara G, Ezzouggari BZ, Aberkani K, Assouguem A, Meddich A, El Jarroudi M, Barka EA (2024) Effects of climate change on plant pathogens and host-pathogen interactions. Crop Environ 3(3):159–170. https://doi.org/10.1016/j.crope.2024.05.003

Leete E, Endo T, Yamada Y (1990) Biosynthesis of nicotine and scopolamine in a root culture of *Duboisia leichhardtii*. Phytochemistry 29:1847–1851. https://doi.org/10.1016/0031-9422(90)85026-C

Ligtenstein DA, Moes GW (1991) The synergism of atropine and the cholinesterase reactivator HI-6 in counteracting lethality by organophosphate intoxication in the rat. Toxicol Appl Pharmacol 107(1):47–53. https://doi.org/10.1016/0041-008x(91)90329-d

Mascarenhas-Melo F, Diaz M, Gonçalves MBS, Vieira P, Bell V, Viana S, Nunes S, Paiva-Santos AC, Veiga F (2024) An overview of biosimilars—development, quality, regulatory issues, and management in healthcare. Pharmaceuticals 17(2):235. https://doi.org/10.3390/ph17020235

McCamish M, Woollett G (2012) The state of the art in the development of biosimilars. Clin Pharmacol Ther 91(3):405–417. https://doi.org/10.1038/clpt.2011.343

Mehrotra S, Srivastava V, Ur Rahman L, Kukreja AK (2015) Hairy root biotechnology indicative timeline to understand missing links and future outlook. Protoplasma 252(5):1189–1201. https://doi.org/10.1007/s00709-015-0761-1

Melton L, Shahidi F, Varelis P (2019) Encyclopedia of food chemistry. Elsevier

Morrow T, Felcone LH (2004) Defining the difference: what makes biologics unique. Biotechnol Healthc 1(4):24. https://pmc.ncbi.nlm.nih.gov/articles/PMC3564302/

Naik T, Vanitha SC, Rajvanshi PK, Rajvanshi PK, Chandrika M, Subban Kamalraj S, Jayabaskaran C (2017) Novel microbial sources of tropane alkaloids: first report of production by endophytic fungi isolated from *Datura metel* L. Curr Microbiol 75(2):206–212. https://doi.org/10.1007/s00284-017-1367-y

Naik B, Kumar V, Rizwanuddin S, Chauhan M, Choudhary M, Gupta AK, Kumar P, Kumar V, Saris PEJ, Rather MA, Bhuyan S, Neog PR, Mishra S, Rustagi S (2023) Genomics, proteomics, and metabolomics approaches to improve abiotic stress tolerance in tomato plant. Int J Mol Sci 24(3):3025. https://doi.org/10.3390/ijms24033025

National Center for Biotechnology Information (2024) PubChem Compound Summary for CID 136097, 8-Azabicyclo(3.2.1)octane. https://pubchem.ncbi.nlm.nih.gov/compound/8-Azabicyclo_3.2.1_octane

Raklami A, Meddich A, Oufdou K, Baslam M (2022) Plants-microorganisms-based bioremediation for heavy metal cleanup: recent developments, phytoremediation techniques, regulation mechanisms, and molecular responses. Int J Mol Sci 23(9):5031. https://doi.org/10.3390/ijms23095031

Raoufa A, El-Wakil H, Gabal A, El-Said A, Khlifa H.D (2008) Production of scopolamine and hyoscyamine in calli and regenerate cultures of *Datura metel* (L.). J Appl Sci Res 4(12): 1858–1866. https://www.aensiweb.com/old/jasr/ jasr/2008/1858-1866.pdf

Rasi A, Sabokdast M, Naghavi MR, Jariani P, Dedičová B (2024) Modulation of tropane alkaloids' biosynthesis and gene expression by methyl jasmonate in *Datura stramonium* L.: a comparative analysis of scopolamine, atropine, and hyoscyamine accumulation. Life (Basel, Switzerland) 14(5): 618. https://doi.org/10.3390/life14050618

Renner UD, Oertel R, Kirch W (2005) Pharmacokinetics and pharmacodynamics in clinical use of scopolamine. Ther Drug Monit 27(5):655–665. https://doi.org/10.1097/01.ftd.0000168293.48226.57

Riekkinen P Jr, Jäkälä P, Sirviö J, Koivisto E, Miettinen R, Riekkinen P (1991) The effects of THA on scopolamine and nucleus basalis lesion-induced EEG slowing. Brain Res Bull 26(4):633–637. https://doi.org/10.1016/0361-9230(91)90107-u

Sala M, Braida D, Calcaterra P, Leone MP, Arianna Comotti F, Gianola S, Gori E (1991) Effect of centrally administered atropine and pirenzepine on radial arm maze performance in the rat. Eur J Pharmacol 194(1):45–49. https://doi.org/10.1016/0014-2999(91)90122-7

Schillberg S, Raven N, Fischer R, Twyman RM, Schiermeyer A (2013) Molecular farming of pharmaceutical proteins using plant suspension cell and tissue cultures. Curr Pharm Des 19(31):5531–5542. https://doi.org/10.2174/1381612811319310008

Schultze-Kraft M (2016) Evolution of estimated coca cultivation and cocaine production in South America (Bolivia, Colombia and Peru) and of the actors, modalities and routes of cocaine trafficking to Europe; Background paper commissioned by the EMCDDA for the 2016 EU Drug Marke. Ed. Lisbon: OEDT / EMCDDA, European Monitoring Centre for Drugs and Drug Addiction: Lisbon, Portugal, pp 1–15. https://bdoc.ofdt.fr/index.php?lvl=notice_display&id=77085

Sharma M, Dhaliwal I, Rana K, Delta AK, Kaushik P (2021) Phytochemistry, pharmacology, and toxicology of *Datura* species – a review. Antioxidants (Basel, Switzerland) 10(8):1291. https://doi.org/10.3390/antiox10081291

Solana RP, Gennings C, Carter WH, Anderson D, Lennox WJ, Carchman RA, Harris LW (1991) Efficacy comparison of two cholinolytics, scopolamine and azaprophen, when used in conjunction with physostigmine and pyridostigmine for protection against organophosphate exposure. J Am Coll Toxicol 10(2):215–222. https://doi.org/10.3109/10915819109078631

Srivastava V, Mehrotra S, Verma PK (2017a) Biotechnological interventions for production of therapeutic secondary metabolites using hairy root cultures of medicinal plants. In: Current developments in biotechnology and bioengineering; crop modification, nutrition, and food production. Elsevier, Amsterdam, pp 259–282

Srivastava V, Mehrotra S, Mishra S (2017b) Biotransformation through hairy roots: perspectives, outcomes, and major challenges. In: Jha S (ed) Transgenesis and secondary metabolism. Springer, Cham, pp 1–24

Srivastava V, Mehrotra S, Mishra S (2018) Hairy roots: an effective tool of plant biotechnology. Springer, Cham

Srivastava V, Mishra S, Lhamo S, Chowdhary AA, Mehrotra S (2021) Involvement of various biotechnological contrivances for tropane alkaloid biosynthesis and applications of tropane alkaloid-bearing in vitro cultures. In: Srivastava V, Mehrotra S, Mishra S (eds) Tropane alkaloids. Springer, Singapore. https://doi.org/10.1007/978-981-33-4535-5_4

Sur S, Mohiuddin A, Vichyanond P, Nelson HS (1990) A random double-blind trial of the combination of nebulized atropine methylnitrate and albuterol in nocturnal asthma. Ann Allergy 65(5):384–388

Sweta VR, Lakshmi T (2015) Pharmacological profile of tropane alkaloids. J Chem Pharm Res 7(5):117–119. https://www.jocpr.com/articles /pharmacological-profile-of-tropane-alkaloids.pdf

Szczurowska E, Szánti-Pintér E, Chetverikov N, Randáková A, Kudová E, Jakubík J (2022) Modulation of muscarinic signalling in the central nervous system by steroid hormones and neurosteroids. Int J Mol Sci 24(1):507. https://doi.org/10.3390/ijms24010507

Tariman JD (2018) Biosimilars: exploring the history, science, and progress. Clin J Oncol Nurs 22(5):5–12. https://doi.org/10.1188/18.CJON.S1

Thapa A, Ali Y, Madan S, Verma P, Verma P, Gaurav N (2022) An assessment of in vitro propagation and medicinal properties of *Datura stramonium* (Dhatura). Sci Temper 13:293–301. https://doi.org/10.58414/SCIENTIFICTEMPER.2022.13.2.43

Tinsley SM, Grande C, Olson K, Plato L, Jacobs I (2018) Potential of biosimilars to increase access to biologics: considerations for advanced practice providers in oncology. J Adv Pract Oncol 9(7):699. https://pmc.ncbi.nlm.nih.gov/articles/PMC6570521/

Tsung-Ming S (1991) Cholinergic actions of diazepam and atropine sulfate in soman poisoning. Brain Res Bull 26(4):565–573. https://doi.org/10.1016/0361-9230(91)90097-4

Valjakka A, Lukkarinen K, Koivisto E, Riekkinen P, Miettinen R, Airaksinen MM, Lammintausta R, Riekkinen P (1991) Modulation of EEG rhythmicity and spike activity in the rat hippocampus by systemically administered tetrahydroaminoacridine, scopolamine and atipamezole. Brain Res Bull 26(5):739–745. https://doi.org/10.1016/0361-9230(91)90169-k

Vanderwaeren L, Dok R, Voordeckers K, Nuyts S, Verstrepen KJ (2022) *Saccharomyces cerevisiae* as a model system for eukaryotic cell biology, from cell cycle control to DNA damage response. Int J Mol Sci 23(19):11665. https://doi.org/10.3390/ijms231911665

Wang Z, Chen S, Zheng P, Ren Z, Zhang H, Zhang J, Shao B, Wu C, Jiang H (2023) Emerging tropane alkaloids: global development and potential health threats. Food Qual Saf 8. https://doi.org/10.1093/fqsafe/fyad043

Wienkoop S, Morgenthal K, Wolschin F, Scholz M, Selbig J, Weckwerth W (2008) Integration of metabolomic and proteomic phenotypes: analysis of data covariance dissects starch and RFO metabolism from low and high temperature compensation response in *Arabidopsis thaliana*. Mol Cell Proteomics 7(9):1725. https://doi.org/10.1074/mcp.M700273-MCP200

Xia K, Liu X, Zhang Q, Qiang W, Guo J, Lan X, Chen M, Liao Z (2016) Promoting scopolamine biosynthesis in transgenic *Atropa belladonna* plants with pmt and h6h overexpression under field conditions. Plant Physiol Biochem 106:46–53. https://doi.org/10.1016/j.plaphy.2016.04.034

Yamada Y, Endo T (1984) Tropane alkaloid production in cultured cells of *Duboisia leichhardtii*. Plant Cell Rep 3:186–188. https://doi.org/10.1007/BF00270196

Ziegler J, Facchini PJ (2008) Alkaloid biosynthesis: metabolism and trafficking. Annu Rev Plant Biol 59:735–769. https://doi.org/10.1146/annurev.arplant.59.032607.092730

Chapter 9
Tropane Alkaloids In Vitro Production, Current Status, and Perspectives

Kiran S. Mawale, Arnab S. Mahapatra, Haripriya Pakala, Parvatam Giridhar, Ashwani Sharma, and Nagashree N. Rao

Abstract This chapter emphasizes the challenges of extracting tropane alkaloids from their natural sources and the need for alternative methods, such as in vitro synthesis, to meet the growing demand. It covers the advancements in in vitro synthesis and other production techniques like cell cultures, organ cultures, and genetic engineering. Further, this chapter evaluates the advantages and disadvantages of each approach and provides insight into the current state of tropane alkaloid synthesis in vitro. It also discusses the factors affecting tropane alkaloid synthesis, including precursor availability, culture conditions, and biotechnological interventions. Additionally, it covers optimization strategies such as medium composition, elicitation techniques, and bioreactor systems to enhance alkaloid yields and productivity.

Keywords Tropane alkaloid · In vitro cultures · Drugs · Secondary metabolites · Metabolic engineering

---

K. S. Mawale · A. S. Mahapatra · H. Pakala
Plant Cell Biotechnology Department, CSIR-Central Food Technological Research Institute, Mysuru, Karnataka, India

Academy of Scientific and Innovative Research (AcSIR), Ghaziabad, India

P. Giridhar (✉)
Plant Cell Biotechnology Department, CSIR-Central Food Technological Research Institute, Mysuru, Karnataka, India

Academy of Scientific and Innovative Research (AcSIR), Ghaziabad, India

Food Safety and Analytical Quality Control Laboratory, CSIR-Central Food Technological Research Institute, Mysur, Karnataka, India

A. Sharma · N. N. Rao
Department of Biotechnology, R.V. College of Engineering, Bengaluru, Karnataka, India

T. Pullaiah et al. (eds.), *Tropane Alkaloids*,
https://doi.org/10.1007/978-981-96-5805-3_9

9.1 Introduction

Tropane, one of the world's oldest plant medicines, has numerous ethnopharmacological and medicinal applications, including analgesia, hallucinogens, and poisons. These ornithine-derived compounds are found in many plant families, such as Solanaceae, Erythroxylaceae, Convolvulaceae, Brassicaceae, and Euphorbiaceae. Tropane alkaloids are essential for treating neuromuscular diseases (Grynkiewicz and Gadzikowska 2008; World Health Organization 2019). Due to legalities, research focuses on biosynthesis pathway elucidation, but large-scale commercial production research has not been conducted. One of the important tropanes, scopolamine cultivation and manufacture, is of significant commercial interest due to its numerous medicinal uses. Global demand for this compound is indeed expanding. Furthermore, scopolamine is one of the World Health Organization's (WHO) essential medicines (World Health Organisation 2019). Hyoscyamine and scopolamine are produced from *Duboisia* plants growing on massive plantations in Queensland, Australia (Endo and Yamada 1985; Grynkiewicz and Gadzikowska 2008). Climate change and the resulting new biotic and abiotic factors make it harder for the pharmaceutical industry to manufacture massive amounts of scopolamine continuously (Srinivasan and Smolke 2019; Shi et al. 2022). In plants such as *Anisodus tanguticus*, the variation in tropane alkaloid content in plants from different geographical regions was shown by employing multivariate statistical analysis (Chen et al. 2022a, b). Similarly, the role of N, P, and K fertilizers on tropane alkaloids was shown (Chen et al. 2023).

Tropane alkaloids are classified into three primary classes based on their distinctive bicyclic tropane ring system: (1) cocaine, (2) hyoscyamine and scopolamine, and (3) calystegines (Fig. 9.1).

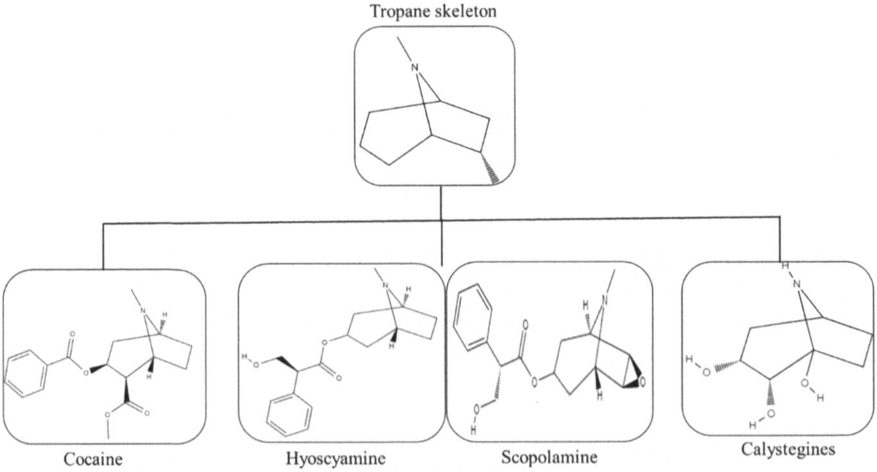

Fig. 9.1 Tropane skeleton structure and key classes of derived TAs

The primary structural feature of most Tropane Alkaloids (TAs) is a general tropane skeletal structure with pyrrolidine and piperidine rings sharing nitrogen and two carbons. These compounds are known as 8-methyl-8-azabicyclo [3.2.1.] octane through IUPAC naming system (National Center for Biotechnology Information 2023). Tropane alkaloids (TAs) are classified as unique alkaloids, defined as all molecules possessing a tropane ring system. The classifications can be structurally highlighted as either esters of 3α-tropanole (tropine) or, aside from that, 3β-tropanole (pseudo-tropine). As previously stated and illustrated in the figure, these three types are primarily from Solanaceae, which are hyoscyamine and scopolamine, from *Erythroxylum coca*, coca alkaloids such as cocaine, and polyhydroxylated nortropane alkaloids from the Convolvulaceae family, which are the most recent to be discovered in all three groups. Biosynthesis begins with the amino acid ornithine or arginine and their intermediary putrescine, which leads to the N-methyl-D1-pyrrolinium cation precursor. The divergence of distinct tropane alkaloids generation for cocaine, hyoscyamine/scopolamine, and calystegines occurs from this branch point. The same pathway produces another well-known compound, nicotine (Fig. 9.2).

9.2 Current Market Demand for Tropane Alkaloids

Current market reports from web sources have projected an encouraging market for all the TAs, mostly in medicinal use such as Hyoscyamine Sulfate injection, where the CAGR projection is 3.5% up to 2028 from an existing market of 1 billion USD in 2017 (Anonymous from Dataintelo 2021). For Cocaine, CAGR is projected at 6.5% till 2030, with a revenue forecast of 10 billion USD from 6 billion in 2022 (Anonymous from Prescient and Strategic Intelligence 2023). Scopolamine market size forecast is USD 1001.1 billion by 2037 from USD 511.7 billion in 2024 (https://www.researchnester.com/reports/scopolamine-market/3312 accessed on 16 April 2025). For Belladonna, from USD 500 million in 2018, the CAGR is projected at 6.85% till 2031 (Anonymous from Growth Plus reports 2023). For Benztropine Mesylate, Anisole, and Atropine, the projected CAGR are 5.4% (till 2028), 4% (till 2032, revenue forecast of USD 5.18 billion), and 1.8% (till 2028) consecutively from a current market value of USD 1 billion, USD 3.64 billion, and USD 580 million (Anonymous from Dataintelo 2022; Anonymous from Polaris Market Research 2022; Anonymous from Adroit Market Research 2022).

9.3 Tissue Culture Approaches in TA Biosynthesis

Tropane alkaloids (TAs) are anticholinergic and stimulating chemicals used in medications for various conditions, including central nervous system problems, postoperative nausea, pain, and Parkinson's disease (Shi et al. 2022). They are crucial

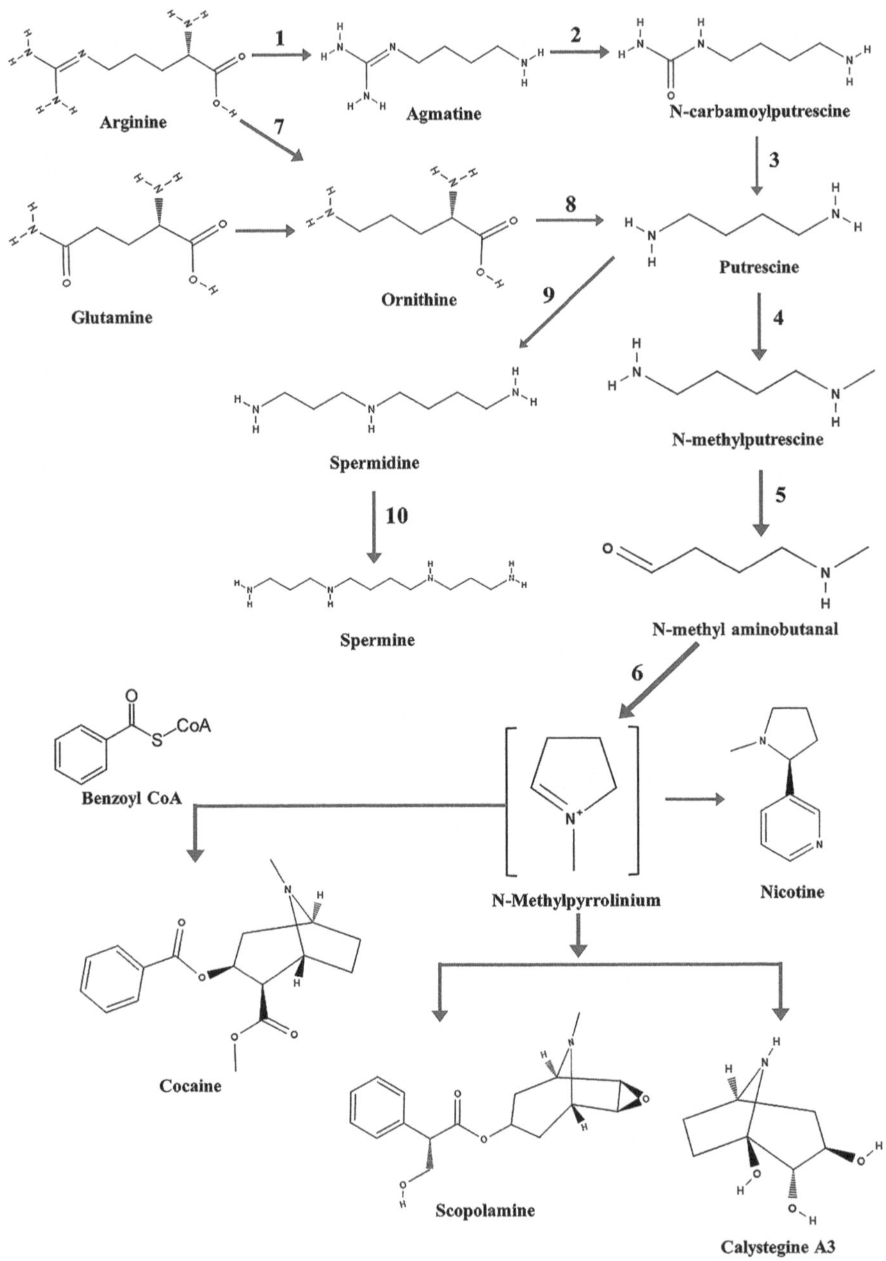

Fig. 9.2 Early stages of Tropane Alkaloid biosynthesis; *1* arginine decarboxylase, *2* agmatine deiminase, *3* N-carbamoylputrescine amidase, *4* putrescine N-methyltransferase, *5* N-methylputrescine oxidase, *6* spontaneous cyclization, *7* arginase, *8* ornithine decarboxylase, *9* spermidine synthase, *10* spermine synthase

components in localized and general anaesthetic medicines. With increasing commercial demand, plant-based bioactive chemicals are needed in pharmaceutical manufacturing lines. Traditional cultivation leads to an unmet supply-to-demand ratio, and modern biotechnological approaches aim to modulate production. Synthesis is mostly underrated and takes over 3 years (Zhou and Wu 2006), leading to the popularity of alternative methods. Modern biotechnological approaches are employed to modulate the production of tropane alkaloids (Savita et al. 2021). Angiosperm plants are susceptible to *Agrobacterium*-mediated transformation and hairy root development. Solanaceae family genera, such as *Atropa, Anisodus, Brugmansia, Datura, Duboisia, Hyoscyamus,* and *Scopolia*, have hairy roots when infected with *Agrobacterium* strains. Hairy root cultures boost secondary metabolite synthesis, such as tropane alkaloids (Table 9.1). Tropane alkaloid production in *Hyoscyamus* root cultures was explored along with optimization of culture conditions (Hashimoto et al. 1986; Hashimoto and Yamada 1987). Kitamura (1988) studied in vitro regeneration and the production of tropane and pyridine alkaloids in *Duboisia* spp. Similarly, alkaloid composition and atropine esterase activity in callus and differentiated tissues of *Duboisia myoporoides* was reported (Kitamura et al. 1985). Genetic transformation with *Agrobacterium rhizogenes* has successfully produced hyoscyamine and scopolamine in vitro (Zhang et al. 2007; Ghorbanpour 2013; Zhao et al. 2020). Micropropagation is an asset for this purpose, involving screening, selection of germlines, tissue culture media standardization, and elicitation of secondary metabolite production by abiotic or biotic elicitors and metabolic engineering (Giri and Zaheer 2016). In vitro micropropagation introduces significant variations, potentially improving plant health and conserving endangered species. Micropropagation technologies have led to unique individual clones with high TA-yielding properties. Alkaloid production by *Datura* callus and suspension tissue culture was recorded (Chan and Staba 1965). Genetic variation in *H. niger* through intentional mutations (Shah et al. 2020) affects the alkaloid biosynthesis pathway genes putrescine N-methyltransferase (PMT) and 6-hydroxy hyoscyamine (H6H). This suggests that fostering genetic variety through targeted mutations can generate varieties with higher active chemical content. Until the discovery of the root pericycle-specific location of the PMT and H6H, the location of the TA synthesis site remained unclear (Hashimoto et al. 1993). Secondary metabolite accumulation in plants occurs due to abiotic stressors or interactions with microorganisms or their products (Gururaj et al. 2012; Akitha-Devi and Giridhar 2015; Akitha-Devi et al. 2018). Nitrogen and sucrose in culture media significantly influenced metabolite content variation in transformed root cultures of *Hyoscyamus muticus* (Oksman-Caldentey et al. 1987, 1994). Methyl jasmonate, salicylic acid, and acetylsalicylic acid are elicitor chemicals used in elicitation studies to increase alkaloid production (Giri and Zaheer 2016). The *PMT* gene is involved in tropane alkaloid production pathways. Methyl jasmonate may contribute to tropane alkaloid production (Kang et al. 2004a; Deng 2005; Zhang et al. 2007; Harfi et al. 2018). *Datura stramonium* suspension cultures have been enhanced with *Phytophthora megasperma* (Pmg) cell wall fragments, and the carbon-nitrogen ratio has been elevated 100 times (Ballica et al. 1993). Rhizobacterial strains *Pseudomonas putida*

Table 9.1 In vitro production of major TAs in various modes of production

Family	Plant	Type of cultures	Tropane alkaloid	References
Erythroxylaceae	*Erythroxylum coca* var. *coca*	Shoot cultures	Cocaine	Lydon et al. (1993)
	Erythroxylum coca	Callus cultures	Cocaine	Docimo et al. (2015)
Solanaceae	*Scopolia parviflora*	Shoot cultures	Hyoscyamine, scopolamine	Kang et al. (2004b)
	Datura innoxia	Suspension cultures	Scopolamine	Gontier et al. (1994)
	Datura stramonium	Root culture	Hyoscyamine and scopolamine increases	Sáenz-Carbonell et al. (1993)
	Hyoscyamus niger	Root culture	Scopolamine/hyoscyamine elevation of production	Woo et al. (1996)
	Scopolia parviflora	Adventitious root culture	Scopolamine and hyoscyamine rapidly increase	Min et al. (2007)
	Atropa belladonna	Root culture	Calystegines	Dräger et al. (1994)
		Transformed root cultures	Alkaloid content rises under salicylic acid treatment	Lee et al. (1998, 2001)
		Root cultures	Carbohydrates and hormone treatments alter alkaloid content	Rothe et al. (2001)
		Root cultures	A high carbohydrate supply combined with reduced mineral salts stimulates a specific alkaloid-	Rothe and Dräger (2002)
	Duboisia myoporoides	Cultured roots	Effect of oxygen on nicotine and tropane alkaloid production	Yukimune et al. (1994)
	Duboisia leichhardtii	Hairy root cultures (HRCs)	Scopolamine	Mano et al. (1989)
	Duboisia myoporoides	Shoot cultures	Scopolamine	Khanam et al. (2000)
	Duboisia Hybrid	Callus cultures	Scopolamine	Lin (1991) and Lin and Griffin (1992)
	Przewalskia tangutica	HRCs	Hyoscyamine, scopolamine content elevated	Lan et al. (2013)

(continued)

Table 9.1 (continued)

Family	Plant	Type of cultures	Tropane alkaloid	References
Solanaceae	*Scopolia lurida*	HRCs	Hyoscyamine	Zhao et al. (2017)
	Datura metel	HRCs	Atropine	Shakeran et al. (2015)
		HRCs	Scopolamine	Shakeran et al. (2017)
	H. reticulates, H. pusillus	HRCs	Scopolamine, hyoscyamine	Hedayati et al. (2020)
	Hyoscyamus albus	HRCs	Scopolamine, hyoscyamine	Sauerwein et al. (1992)
	Hyoscyamus muticus	HRCs	Growth and metabolites	Vanhala et al. (1998)
		HRCs	High scopolamine production	Zolala et al. (2007)
	Anisodus acutangulus	HRCs	Scopolamine, hyoscyamine and anisodine	Kai et al. (2012a, b)
	Anisodus luridus	HRCs	Scopolamine, hyoscyamine	Qin et al. (2014)
	Nicotiana tabacum	HRCs	Converted exogenously supplied hyoscyamine to scopolamine	Moyano et al. (2007)
	Atropa belladonna	HRCs	Increasing hyoscyamine/ scopolamine ratio	Chashmi et al. (2010)
		Transformed roots	on growth and tropane alkaloid production under the effect of Phytosulfokine-α	Sasaki et al. (2002)
			Chromium-induced production of tropane alkaloids and expression of the H6H gene	Vakili et al. (2012)
	Brugmansia candida	HRCs	Scopolamine and anisodamine scale-up and hairy root ecotype	Cardillo et al. (2010, 2013, 2017)
	H. niger	HRCs	TAs increased with changes in conditions like aeration and pH	Zhao (2014)
	H. niger	HRCs	Scopolamine, hyoscyamine, and cuscohygrine content increases	Jaremicz et al. (2014)
	Datura tatula,D. innoxia	HRCs	Hyoscyamine	Boitel-Conti et al. (1995, 2000) and Harfi et al. (2018)
	D. stramonium	HRCs	Hyoscyamine	Sáenz-Carbonell and Loyola-Vargas (1997)
		HRCs	Major tropane alkaloids under the effect of methyl jasmonate	Sun et al. (2013)

(PP) and *Pseudomonas fluorescens* (PF) significantly increased the accumulation of hyoscyamine and scopolamine in *Hyoscyamus niger* during water deficient stress (Ghorbanpour et al. 2013). TA was generated mainly in callus and cell suspension cultures and hairy root cultures (HRCs), as a result, HRCs are seen as a viable economic solution and have surpassed other in vitro TA production technologies in recent decades (Srivastava et al. 2017b). HRCs are not limited by seasonal or pathological restrictions, but they have options for resource sustainability along with the availability of biotransformation processes like hydroxylation, glycosylation, oxidoreduction, hydrolysis, and others. Enhanced levels of scopolamine production were accomplished in transgenic plants of *Atropa belladonna* by overexpressing *pmt* and *h6h* genes (Wang et al. 2011; Xia et al. 2016).

There are three biotransformation techniques with culture and substrate selection: precursor feeding, non-specific/exogenous molecule biotransformation, and co-culture (Srivastava et al. 2017a, b). Stirred tanks and plastic-lined vessel bioreactor cultivation of *H. muticus* suspension cultures led to fast TA accumulation. The concentration of scopolamine in *D. metel* callus cultures, derived from leaf, stem, and roots, varied with the explant type. Immobilization of *D. innoxia* cell suspension in Ca-alginate resulted in a nearly six-fold increase in hyoscyamine and scopolamine levels (Gontier et al. 1994). Exogenous tropic acid, tropinone, and tropanol supplementation improved the TA in *A. belladonna* suspensions and boosted hyoscyamine and scopolamine production (Srivastava et al. 2021). Even biotransformation of tropane alkaloids using recombinant *Escherichia coli* whole cells was reported (Cardillo et al. 2017). The effects of chitosan, acetic acid, gibberellic acid, citric acid, calcium chloride, hemicellulose, and other biotic and abiotic elicitors on both the growth and tropane alkaloid production in transformed roots of *Brugmansia candida* were explored (Pitta-Alvarez and Giulietti 1997, 1999; Pitta-Alvarez et al. 2000a, b, 2003). Similarly, calcium ion concentration in the medium on tropane alkaloid accumulation in *Datura stramonium* hairy roots was demonstrated (Piñol et al. 1999). The findings enhance our understanding of how these substances influence plant development and metabolite synthesis.

Metabolic engineering involves using specific culture systems for biotransformation, such as co-culture cultivation with complementary systems in one matrix. This method eliminates the need for the exogenous administration of the other system. An example is the efficient conversion of hyoscyamine into scopolamine using the co-culture system of *A. belladonna* shooty teratomas and HRCs (Subroto et al. 1996). Co-cultivation resulted in efficient conversion with excellent results compared to individual roots or shoot structures. *A. belladonna* is a popular plant for TAs production, and HRCs play a crucial role in various culture systems. HRCs have elevated major compound production, which enhances the desired chemical form of a compound and can be utilized for toxin remediation due to the enzymatic activity present in these systems (Srivastava et al. 2012; Mazaheri and Piri 2015; Habibi et al. 2018). HRCs have been standardized for *A. belladonna* and have been used in metabolic engineering tactics, such as the co-cultivation of *A. tumefaciens* modified with the *A. rhizogenes* rolC gene alone or with the rolC gene plus rolABC genes together, resulting in scopolamine and hyoscyamine elevation to 0.4% dry weight and 0.85% dry weight, respectively (Bonhomme et al. 2000). Biotransformation has also been explored in exogenous molecules (Srivastava et al. 2017a). Transforming TNT to ADNT with nitrogen reduction has been explored in *Datura innoxia* cell suspension cultures. An endemic Chinese plant, *Anisodus acutangulus*, has been explored in TAs production for a long time as a medicinal compound for a variety of purposes. HRCs have been utilized in this plant for TA production. Kai et al. (2012a) reported a transformed line TH53, which produced 4.293 mg/g TAs, 4.49-fold higher than control lines. Anisodine was elevated significantly by AaTRI plus *AaH6H* genes over-expression. In *Datura* multiple applications for the TAs production capabilities, particularly hyoscyamine, have been explored. *Datura* species majorly explored are D. *stramonium, D. tatula,* and *D. innoxia* which are widely used in HRCs production and enhancement of TA production (Harfi et al. 2018). The A4 strain of *A. rhizogenes* has been used to induce HRCs in cocultivation through infection with atropine-type plasmid. Cocaine in vitro has been introduced for the elevation of production with multiple approaches. Callus culture was introduced through stem and embryo for cocaine production in some reports (Lydon et al. 1993). Palazón et al. (2008) reviewed the application aspects of metabolic engineering on scopolamine production. Biotic and abiotic factors have negative and positive impacts on various TAs. Standard Murashige-Skoog media, supplemented with all micronutrients and macronutrients, carbon sources such as sucrose, level of ploidy, period of cultivation, type of vessel in culture, etc., are the properties that regulate in vitro production. Reports have shown an increase of around 200 times in concentration, especially for tetraploid tissue cultures in these plants (Madani et al. 2021). Srinivasan and Smolke (2019) provided valuable insights into the engineering of cellular metabolite transport, facilitating the biosynthesis of computationally predicted tropane alkaloid derivatives in yeast.

9.4 Industrial Production (Scale-Up) of TAs

Higher plants are rich in metabolites used in medications, and the global organic market favors natural goods. However, environmental, ecological, and climatic challenges hinder the synthesis of plant-derived compounds. Plant in vitro systems, established in the 1960s (Pavlov et al. 2009), can help address these challenges. However, issues like instability and yields hinder industrial deployment. Bioprocess engineering techniques can optimize plant in vitro system-based processes. Algorithms for media optimization, elicitation, bioreactor building adaption, environmental condition optimization, and other processes have been thoroughly detailed.

9.4.1 Approaches for Elevated TAs Production

Tissue culture approaches have been successful in multiple generations of TAs on a large scale. This can be achieved by efficiently providing precursors with targeted secondary metabolites, engineering pathways to enhancement, or suppressing genes or molecules to avoid divergence in nutrition (Srivastava et al. 2017b). Bioreactor up-scaling can also be used to obtain desired SM metabolite synthesis proportionately. Here are a few pertinent examples from such categories:

9.4.1.1 Precursor Feeding and Media Optimization

The precursor feeding technique enhances the synthesis of metabolites, including TAs, in tissue culture media. It involves adjusting the concentration of nitrate and ammonium in the culture medium, which can impact the production of secondary metabolites. Research has shown that increasing nitrate or ammonium concentrations raised the ratio of S/H in *A. belladonna* HRCs by two to three times (Bensaddek et al. 2001; Wei et al. 2018). In the roots of *H. niger*, deficiencies in Ca, Fe, and Mg lower hyoscyamine and increase scopolamine synthesis. Variations in the medium's exogenous magnesium and iron levels also show positive changes in scopolamine and hyoscyamine concentrations (Pudersell et al. 2012). Another example is investigating the influence of varied concentrations of glucose, sucrose, $CaCl_2$, KNO_3, and NH_4NO_3, as well as altering temperature, on *D. stramonium* cultures in suspension media (Iranbakhsh et al. 2007). Precursor feeding has been considered a significant choice for increasing TA production in certain genera, such as *Scopolia parviflora* callus cultures (Tabata et al. 1972). Feeding 4C-labeled phenylalanine and tropic acid to *Duboisia leichhardtii* resulted in radioactivity incorporation analyzed in the produced tropane alkaloids (Kitamura et al. 1992). The addition of the TA precursors L-phenylalanine or DL-phenyl lactic acid to *Datura innoxia*-converted roots increased hyoscyamine synthesis (Boitel-Conti et al. 2000). In one effort, tobacco cell cultures containing HmH6H (*H. muticus* H6H) grown from hairy roots produced scopolamine when treated with hyoscyamine (Moyano et al. 2007). Häkkinen et al. (2005) recorded increased secretion of tropane alkaloids in *Nicotiana tabacum* hairy roots expressing heterologous hyoscyamine-6-hydroxylase. Scopolamine build-up was observed after feeding hyoscyamine and 6-hydroxy hyoscyamine to non-TA-producing *Anisodus tanguiticus* suspension cultures (Chen et al. 2022a, b). *Calystegia sepium* produced the nortropane alkaloid calystegine after root cultures were given [15N] tropinone (Scholl et al. 2003). This approach has produced superior outcomes when combined with other yield improvement tactics such as metabolic engineering, elicitation, and bioreactor. Similarly, in vitro production of hyoscyamine and scopolamine from black henbane (*Hyoscyamus niger*) has been successfully demonstrated using shoot tip culture with a range of plant growth regulators and culture media (Ghorbanpour et al. 2003).

9.4.1.2 Elicitation

Elicitation is inducing or enhancing secondary metabolite (SM) production in plants by supplementing it with a physical, chemical, or biological elicitor (Namdeo et al. 2007). Studies have investigated elicitation in various in vitro systems, including root cultures, cell suspension cultures, and hairy root cultures (Giridhar and Parimalan 2010; Parimalan et al. 2011; Akitha-Devi and Giridhar 2015; Vaddadi and Giridhar 2015). This approach has led to the generation of various active phytotherapeutics, such as reserpine, scopolamine, hyoscyamine, podophyllotoxin, andrographolide, and tanshinone (Zabetakis et al. 1999; Srivastava et al. 2017a; Wen et al. 2023). Aluminium induces the production of tropane alkaloids and enhances the antioxidant system activity in micro propagated *Datura innoxia* plantlets (Karimi and Khataee 2012). Hong et al. (2012) studied the effect of elicitation on *Hyoscyamus niger* root cultures for the root growth and production of tropane alkaloids. *Datura stramonium* generated cultured, especially HRCs exposed to oligogalacturonide accumulated TAs (which are hyoscyamine, littorine, and scopolamine), methyl jasmonate (MeJA), and fungal elicitor (Abdel-Motaal et al. 2009). The sequence for exposure of cultures of *D. stramonium* for augmented levels of TAs was MeJa > fungal elicitor > oligogalacturonide (Zabetakis et al. 1999). Nano-sized titanium dioxide and bulk application on *Hyoscyamus niger* activated antioxidant enzymes, hyoscyamine, and scopolamine biosynthesis (Ghorbanpour et al. 2015). Sepehri et al. (2022) investigated the positive effects of iron oxide nanoparticles on the production of tropane alkaloids, *h6h* gene expression, and antioxidant enzyme activity in hairy roots of *Atropa belladonna*. These findings provide valuable insights into enhancing metabolic processes and promoting better growth in this important plant species. The influence of biotic and abiotic elicitors on the accumulation of hyoscyamine and scopolamine in root cultures of *Datura metal* was shown (Ajungla et al. 2009). The plants of *Datura* sp. (cultivated for their hyoscyamine content), *D. stramonium, D. tatula,* and *D. innoxia* were reared and then investigated for the hyoscyamine content following treatment with salicylic acid (SA) alongside acetylsalicylic acid (ASA). In this experiment, both elicitors (AS and ASA) increased hyoscyamine synthesis at optimum concentrations (0.1 mM). Tobamoviruses were used to elicit hyoscyamine production in *Datura stramonium* plants (Mihálik et al. 2022). Another study in *D. tatula* found the greatest content of TAs in HRCs treated with SA (Harfi et al. 2018). The treatment with ethanol, MeJa, and Ag + resulted in 1.51, 1.13-, and 1.08-fold increases in TA production after 24 hours. However, salicylic acid reduced the overall level of TAs. Ethanol treatment had an elicitation impact on the TA biosynthesis gene activation enhancement, particularly for hyoscyamine-6-hydroxylase (AaH6H). Similarly, Ag^+ showed a similar effect with putrescine N-methyltransferase I (AaPMT1) and methyl jasmonate with tropinone reductase I (AaTR1) (Kai et al. 2012a, b). Additionally, the elicitation impact of colchicine plus radiation (UV-B) treatment on scopolamine and hyoscyamine content of HRCs in *H. reticulata* has been reported (Zeynali et al. 2016). Water stress was also shown to have a positive effect on hyoscyamine and proline accumulation in *Hyoscyamus muticus* "hairy root" cultures (Halperin and

Flores 1997). MeJa-mediated productivity of scopolamine was enhanced in *H. niger* HRCs (Jaremicz et al. 2014). Sodium nitroprusside has been shown to enhance the production of tropane alkaloids and boost antioxidant enzyme activity in the hairy root culture of *Hyoscyamus reticulatus*. This suggests a promising avenue for improving the yield and efficacy of these important compounds through targeted growth conditions (Khezerluo et al. 2018). Two tropinone reductases with distinct stereo specificities that are associated with TAs biosynthesis are identified from cultured roots of *Hyoscyamus niger* (Hashimoto et al. 1992). UVB radiation affects *Anisodus luridus* hairy roots, causing elevated expression of PMT, TRI, CYP80F1, and H6H genes, increased TA production, and increased scopolamine accumulation. However, UVB therapy does not affect TA efflux in liquid media (Wang et al. 2011). The efficacy of yeast extract on nitrogen metabolism and secondary metabolism in *Atropa belladonna* hairy roots was shown (Guo et al. 2018). Hedayati et al. (2021) reported the effect of yeast extract on H6h gene expression and tropane alkaloid production in *Atropa belladonna* hairy roots. *D. metel* hairy root cultures produce atropine, an anticholinergic and parasympathetic drug. Methyl-β-Cyclodextrin and coronatine treatment for the hairy roots of *Atropa acuminata* and *Atropa belladonna* augmented tropane alkaloid biosynthesis (Fattahi et al. 2021). Biotic (*Bacillus cereus* and *Staphylococcus aureus*) and abiotic ($AgNO_3$ and nanosilver) elicitors applied showed nanosilver as the most effective elicitor for increasing atropine content in hairy roots (Shakeran et al. 2015). *Anisodus acutangulus* HRCs show 1.13 times greater TA accumulation after 24 h with Ag^+ stimulation (Kai et al. 2012b). Furthermore, iron oxide nanoparticles (FeNPs) increased scopolamine accumulation in *Hyoscyamus reticulatus* HRCs by 5 times at 900 and 450 mg/L for 24 and 48 h, likely converting hyoscyamine to scopolamine through Fe^{2+} availability with the help of H6H (Moharrami et al. 2017). A study demonstrated the significant influence of minerals and elicitation on the production of tropane alkaloids in *Datura stramonium*, as modelled through the in vitro biochemical response (Amdoun et al. 2009). Similarly, Amdoun et al. (2010) successfully optimized the culture medium composition, significantly enhancing hyoscyamine production in elicited *Datura stramonium* hairy roots through the application of Response Surface Methodology (RSM). The impact of nano-zinc oxide on tropane alkaloid production, H6h gene transcription, and antioxidant enzyme activity in *Hyoscyamus reticulatus* hairy roots is clearly established (Asl et al. 2019). Furthermore, the synergistic effects of polyploidization and elicitation on biomass and hyoscyamine content in the hairy roots of *Datura stramonium* are proven to be significant (Belabbassi et al. 2016). The bacterial elicitors were also found to enhance the production of scopolamine in the adventitious hairy root cultures of *Scopolia parviflora* (Jung et al. 2003). Improved root biomass and tropane alkaloid yields in transgenic hairy roots of *Datura* spp. were studied following the application of *Pseudomonas* spp. (Moussous et al. 2018). The effects of salt stress on selected hairy root lines of *Datura tatula* (LDT), *Datura innoxia* (LDI), and *Datura stramonium* (LDS) were studied to observe the accumulation of hyoscyamine using chemical elicitors such as calcium chloride ($CaCl_2$) and potassium chloride (KCl). The doses and exposure time of the elicitors significantly impacted the accumulation of hyoscyamine. $CaCl_2$ (1–2 gL^{-1}) led to an increase in hyoscyamine dry weight to 16.978 mg g^{-1}, 10.828 mg g^{-1}, and 12.697 mg g^{-1} for the LDT line, LDS line, and LDI line, respectively, representing a 2.07, 2.08, and 1.85-fold rise compared to the control. Similarly, KCl (2 g L^{-1}) showed increased hyoscyamine synthesis in the three HRCs (Harfi et al. 2016). Using hairy root cultures (in progress) indicates that silicon dioxide nanoparticles (SiO_2NPs) can enhance the synthesis of tropane alkaloids, specifically scopolamine, in *Hyoscyamus* species. The effects of SiO_2NPs (at concentrations of 0, 25, 50, 100, and 200 mg L^{-1}) were measured at the 24th and 48th hours on growth rate, total phenolic and flavonoid content, antioxidant enzyme activity, tropane alkaloid yield, and other parameters levels in hairy roots of two *Hyoscyamus* species (*H. reticulatus* and *H. pusillus*) was investigated by Hedayati et al. (2022). TPC and TFC measurements found that the maximum of total phenolic compounds (TPC) and total flavonoid compounds (TFC) was highest when induced by SiO_2 NPs (100 and 200 mg L^{-1}, respectively) after 24 h of treatment. Jasmonates cause an over-accumulation of methyl putrescine and conjugated polyamines in root cultures of *Hyoscyamus muticus* (Biondi et al. 2000). The production of tropane alkaloids in vitro and callus cultures of *Hyoscyamus aureus* is successfully demonstrated, along with a robust assessment of their genetic stability using ISSR markers (Besher et al. 2014). El Jaber-Vazdekis et al. (2008) demonstrated the effects of elicitors on tropane alkaloids and gene expression in *Atropa baetica* transgenic hairy roots. Luo et al. (2012) conducted a valuable study exploring how elicitation influences the expression of key enzyme genes and enhances the production of tropane alkaloids in *Anisodus acutangulus*. Their findings contribute significantly to our understanding of metabolic processes in this species.

9.4.1.3 Molecular Engineering

Metabolic engineering is a type of genetic engineering with regard to control over metabolite production via pathway alteration. It starts at the (a) Genetic level of pathway modulation (engineering of genes associated with the process to elevate the number of desired metabolites in plants.) and (b) Metabolic level pathway regulation systems (engineering of certain transcription factors that can impact gene expression in pathways), or a combination of both (Srivastava et al. 2017b). Furthermore, investigations have reported that the metabolic route does not conform to linear routes most of the time with multiple diversionary paths, which in turn contributes to the reduction in total metabolite production. To control these diversions, inhibition plays a significant role by blocking or silencing the expressions of specific genes, which plays a vital role (through approaches like RNAi). Most of these strategies are very effective at engineering SM metabolic pathways in commercially essential plants, resulting in the enhancement of many therapeutically important SMs, for example, camptothecin, reserpine, ajmaline, codeine, withanolides, silymarin, paclitaxel along with a few others (Mehrotra et al. 2010; Srivastava et al. 2017b). Most strategies in engineering are focused on significant and rate-limiting genes for the SM pathways, such as hyoscyamine-6-hydroxylase (H6H). This enzyme is responsible for catalysis of biosynthesis pathway oxidative processes that culminate in the conversion of hyoscyamine to scopolamine, a more pharmaceutically useful tropane alkaloid. The *H. niger* H6H gene was introduced into *A. belladonna* utilizing a binary vector system controlled by the cauliflower mosaic virus 35S promoter (Hashimoto et al. 1993). Engineering tropane biosynthetic pathway in *Hyoscyamus niger* hairy root cultures was elucidated (Zhang et al. 2004, 2005). The modified *A. belladonna* hairy roots had fivefold greater scopolamine contents than wild-type hairy roots.

MTA biosynthesis in *Atropa belladonna,* is contributed by ODC enzyme (AbODC) rather than arginine decarboxylase (AbADC). When the *A. belladonna* ODC gene overexpression was introduced in plants and HRCs, it increased the production of hyoscyamine and anisodamine significantly. Despite being at the start of the MTA biosynthetic pathway, ODC's involvement in increasing MTA synthesis suggests that it is a critical regulator driving ornithine toward alkaloid metabolism. Overexpression of ODC raises putrescine levels and provides sufficient metabolic flux to facilitate downstream processes of a biochemical nature (Zhao et al. 2020). However, such increased metabolic flow does not support the concept that ODC overexpression boosts MTA production. The alternative possibility is that ODC regulates polyamine metabolism (Nölke et al. 2008). Polyamines, for example, are plant growth regulators that participate in various cell physiologies. Because the ODC enzyme regulates polyamine biosynthesis, it may have complex synergistic effects on MTA production via modulating polyamine biosynthesis.

Littorine mutase (CYP80F1) converts littorine to hyoscyamine aldehyde, which can then be reduced by hyoscyamine dehydrogenase (Qiu et al. 2021). Plants have very little hyoscyamine aldehyde; either CYP80F1 is a rate-limiting enzyme, or hyoscyamine aldehyde is swiftly metabolized to produce other related alkaloids. When CYP80F1 was put for overexpression, it did not affect scopolamine and hyoscyamine synthesis in HRCs of *H. niger* (Li et al. 2006). This discovery of hyoscyamine dehydrogenase, also known as hyoscyamine aldehyde reductase and investigations on its significance in MTA engineering. When the HDH gene in *A. belladonna* was overexpressed in HRCs, the scopolamine and hyoscyamine production increased significantly, this demonstrated that HDH is an MTA biosynthesis gene that is functional for designing MTA production (Qiu et al. 2021).

After feeding the cultures with exogenous hyoscyamine, the *N. tabacum* and *H. muticus* hairy roots were examined to generate scopolamine and other tropane and nicotine alkaloids in a combinatorial effort that included gene engineering and precursor feeding. In *N. tabacum* roots and *H. muticus* roots for the production of nonhyoscyamine, the transformation was introduced through the CaMV 35S promoter-controlled H6H gene from *H. niger*. Scopolamine levels were much higher in *H. muticus* and *N. tabacum* HRCs carrying the 35S-H6H transgene, which also displayed excellent hyoscyamine absorption from the normal culture medium and a higher rate of hyoscyamine bioconversion to scopolamine. In transgenic HRCs of *Anisodus acutangulus,* the branch-controlling enzyme tropinone reductase I (AaTRI) along with downstream rate-limiting enzyme hyoscyamine-6-hydroxylase

(AaH6H) overexpression improved anisodine concentration (Kai et al. 2012a, b). In *A. belladonna* root lines, tropinone reductase, TRI, and TRII overexpression resulted in increased production of enzyme products, pseudo tropine and tropine, as a result of which tropine-derived alkaloids and pseudo-tropine-derived alkaloids were significantly modulated (Richter et al. 2005). Overexpression of the h6h (hyoscyamine 6-hydroxylase) and pmt. (putrescine N-methyltransferase) genes has been reported to result in higher scopolamine accumulation in grown plants of *A. belladonna* transgenic HRCs (Wang et al. 2011; Moradi et al. 2020). *A. belladonna* hairy root cultures showed an increase in TA production with overexpression of the pmt. and h6h genes (Yang et al. 2011). Metabolic engineering of *Scopolia parviflora* HRCs using putrescine N-methyl transferase (SpPMT) and hyoscyamine-6 hydroxylase (SpH6H) co-expression indicated increased scopolamine and hyoscyamine synthesis (Kang et al. 2011). In *Datura metel,* and *H. muticus* HRCs overexpression of *PMT* showed that both hyoscyamine and scopolamine, as well as simply hyoscyamine, were improved. Furthermore, the HmH6H expressed in a bioreactor (5 L-turbine stirred tank) with a tobacco cell line converts hyoscyamine to scopolamine with higher accumulation (Moyano et al. 2007).

9.4.1.4 Bioreactors

One of the key constraints is the low commercial output of TAs, an urgent need to establish a standardized culture system optimized for a higher amount of TA production. One notable endeavour to address this restriction of poor productivity is bioreactor upscaling (Mehrotra et al. 2018). Scaling up in vitro culture from lab-size bioprocesses to bigger culture containers on a commercial scale to generate higher volumes of product is part of the bioreactor approach (Mehrotra et al. 2018). Three types of bioreactors have been reported: liquid, gas, and hybrid (Kim et al. 2002). Bioreactor treatments can make expensive secondary metabolites like TA cheaper (Cardillo et al. 2010). Many bioreactors have implemented automated systems to scale up TA production without limitations regarding time or season (Gerth et al. 2007). Various cultures of plant-derived origins are cultivated in various bioreactors to produce secondary metabolites and TAs. Another example includes normal root cultures of *H. niger* alongside HRCs grown in culture systems of various types, such as stirred tanks, flasks, mist, and finally, unified mist systems showed elevated TA contents, including scopolamine, where an increase of more than 20 times within 25 days was noticed (Woo et al. 1996). Oxygen is employed as the primary component during cellular respiration and is one of the most important components during plant cell culture in bioreactors. A small-scale bubble column-type bioreactor was used to ensure sufficient oxygen supply, and it was demonstrated that TA synthesis increased when compared to control conditions. Furthermore, HRCs of *H. niger* grown in bioreactors with improved oxygen delivery by microbubbles and a hybrid bubble-column/spray approach produced several times more TA than controls (Jaremicz et al. 2014). *Datura stramonium* and *Datura innoxia* have also been cultured using bioreactor techniques (Monforte-González et al. 1992). *D. stramonium* root and HRCs treated with indole-butyric acid and Tween20, respectively, aided in the large-scale generation of TAs and enhanced their permeabilization in bioreactors. Compared to flask-type cultures, adventitious root cultures of *S. parviflora* cultivated in a bioreactor with a small-scale bubble column produced considerable scopolamine and hyoscyamine. The TA content of these cultures was enhanced after further elicitation with the bacteria *Staphylococcus aureus* (Min et al. 2007). These bioreactors may also be used as biotransformation platforms, such as the HRCs of H6H expressing tobacco cultivated in a 5 L bioreactor with a turbine-stirred tank reactor, which converts exogenously given hyoscyamine to scopolamine (Moyano et al. 2007). When treated with nitrate concentration, HRCs and in vitro cultivated *A. belladonna* in bioreactors of 100 mL flask size enhanced the synthesis of the hyoscyamine/scopolamine ratio (Chashmi et al. 2010). HRCs of *Przewalskia tangutica* and *B. candida* generated more scopolamine and hyoscyamine than control plants in various bioreactors of 1.5 L stirred tank size or 5-week growth in MS liquid medium (Cardillo et al. 2010; Lan and Hong 2010).

9.5 Future Perspectives

Tropane alkaloids are a class of naturally occurring compounds that have a variety of medicinal and commercial applications, including anticholinergic medications, analgesics, and the production of various chemicals. In vitro production refers to the development of plant cells, tissues, or organs in a controlled environment, generally utilizing cell culture techniques, to produce these vital chemicals. In vitro manufacture enables the investigation of lesser-known or uncommon tropane alkaloids with potentially novel characteristics or uses. This might lead to the discovery of new medications or compounds with useful applications. Production of tropane alkaloids in vitro, like any new technology, will certainly confront regulatory obstacles pertaining to safety, quality control, and clearance for diverse uses. Addressing these issues will be critical for the technology's mainstream adoption. Overall, the future of tropane alkaloids in vitro manufacturing shows potential for developing numerous sectors and applications. However, in pursuing these breakthroughs, it is critical to address ethical, environmental, and regulatory considerations.

9.6 Conclusion

In this review we have presented the current state and future prospects of tropane alkaloid synthesis using in vitro techniques. The significant insights covered are helpful for academics, biotechnologists, and enterprises interested in manufacturing and utilizing tropane alkaloids by critically evaluating existing approaches, problems, and prospects. The importance of in vitro production as a sustainable and scalable strategy is helpful in meeting the growing demand for tropane alkaloids in industries such as medicines, cosmetics, and agriculture. Furthermore, the current advances covered in molecular biology and genetic engineering provide prospective paths for increasing tropane alkaloid synthesis. The use of genetic engineering technologies such as metabolic engineering and synthetic biology to manipulate tropane alkaloid biosynthesis pathways is moderately reviewed, emphasizing their potential to increase yields and diversify alkaloid profiles.

References

Abdel-Motaal FF, Nassar MSM, El-Zayat SA, El-Sayed MA, Ito S (2009) Responses of fungi to tropane alkaloids produced by a medicinal plant *Hyoscyamus muticus* (Egyptian henbane). Folia Microbiol 54:207–212

Ajungla L, Patil PP, Barmukh RB, Nikam TD (2009) Influence of biotic and abiotic elicitors on accumulation of hyoscyamine and scopolamine in root cultures of *Datura metel* L. Indian J Biotechnol 8(3):317–322

Akitha-Devi MK, Giridhar P (2015) Variations in physiological response, lipid peroxidation, antioxidant enzyme activities, proline and isoflavones content in soybean varieties subjected to drought stress. Proc Natl Acad Sci India Sect B Biol Sci 85:35–44. https://doi.org/10.1007/s40011-013-0244-0

Akitha-Devi MK, Sravan Kumar S, Giridhar P (2018) LC–ESI–MS based characterisation of isoflavones in soybean (*Glycine max* (L.) Merr.) from India. J Food Sci Technol 55:5045–5054. https://doi.org/10.1007/s13197-018-3443-0

Amdoun R, Khelifi L, Khelifi-Slaoui M, Amroune S, Benyoussef EH, Thi DV et al (2009) Influence of minerals and elicitation on *Datura stramonium L.* tropane alkaloid production: modelization of the in vitro biochemical response. Plant Sci 177(2):81–87

Amdoun R, Khelifi L, Khelifi-Slaoui M, Amroune S, Asch M, Assaf-Ducrocq C, Gontier E (2010) Optimization of the culture medium composition to improve the production of hyoscyamine in elicited *Datura stramonium* L. hairy roots using the response surface methodology (RSM). Int J Mol Sci 11(11):4726–4740

Anonymous from Adroit Market Research (2022) Atropine market by type. URL https://www.adroitmarketresearch.com/industry-reports/atropine-market. Accessed 26 July 2023

Anonymous from Dataintelo (2021) Benztropine Mesylate market report [WWW Document]. URL https://dataintelo.com/report/benztropine-mesylate-market/. Accessed 26 July 2023

Anonymous from Dataintelo (2022) Hyoscyamine sulfate injection market report. URL https://dataintelo.com/report/global-hyoscyamine-sulfate-injection-market/. Accessed 26 July 2023

Anonymous from Growth Plus Reports (2023) Exclusive report by Growth Plus Reports URL https://www.globenewswire.com/news-release/2023/05/30/2678087/0/en/Belladonna-Alkaloids-Market-is-Projected-to-increase-at-a-CAGR-of-around-6-85-during-2023-to-2031-Exclusive-report-by-Growth-Plus-Reports.html. Accessed 26 July 2023

Anonymous from Polaris Market Research (2022) Anisole market share, size, trends, industry analysis report, by purity. URL https://www.polarismarketresearch.com/industry-analysis/anisole-market. Accessed 26 July 2023

Anonymous from Prescient & Strategic Intelligence (2023) Drug of abuse testing market outlook by product. URL https://www.psmarketresearch.com/market-analysis/drug-of-abuse-testing. Accessed 26 July 2023

Asl KR, Hosseini B, Sharafi A, Palazon J (2019) Influence of nano-zinc oxide on tropane alkaloid production, h6h gene transcription and antioxidant enzyme activity in *Hyoscyamus reticulatus* L. hairy roots. Eng Life Sci 19(1):73–89

Ballica R, Ryu DD, Kado CI (1993) Tropane alkaloid production in *Datura stramonium* suspension cultures: elicitor and precursor effects. Biotechnol Bioeng 41(11):1075–1081

Belabbassi O, Khelifi-Slaoui M, Zaoui D, Benyammi R, Khalfallah N, Malik S, Khelifi L (2016) Synergistic effects of polyploidization and elicitation on biomass and hyoscyamine content in hairy roots of *Datura stramonium*/Effets synergiques de la polyploïdisation et de l'élicitation sur la biomasse et le contenu en hyoscyamine des chevelus racinaires de *Datura stramonium*. Biotechnol Agron Soc Environ 20(3):408

Bensaddek L, Gillet F, Saucedo JEN, Fliniaux MA (2001) The effect of nitrate and ammonium concentrations on growth and alkaloid accumulation of *Atropa belladonna* hairy roots. J Biotechnol 85(1):35–40

Besher S, Al-Ammouri Y, Murshed R (2014) Production of tropane alkaloids in the in vitro and callus cultures of *Hyoscyamus aureus* and their genetic stability assessment using ISSR markers. Physiol Mol Biol Plants 20:343–349

Biondi S, Fornale S, Oksman-Caldentey KM, Eeva M, Agostani S, Bagni N (2000) Jasmonates induce over-accumulation of methylputrescine and conjugated polyamines in *Hyoscyamus muticus L.* root cultures. Plant Cell Rep 19:691–697

Boitel-Conti M, Gontier E, Laberche J, Ducrocq C, Sangwan-Norreel B (1995) Permeabilization of *Datura innoxia* hairy roots for release of stored tropane alkaloids. Planta Med 61:287–290

Boitel-Conti M, Laberche J-C, Lanoue A, Ducrocq C, Sangwan-Norreel BS (2000) Influence of feeding precursors on tropane alkaloid production during an abiotic stress in *Datura innoxia* transformed roots. Plant Cell Tissue Organ Cult 60:131–137

Bonhomme V, Laurain-Mattar D, Fliniaux MA (2000) Effects of the rol C gene on hairy root: induction development and tropane alkaloid production by *Atropa belladonna*. J Nat Prod 63:1249–1252

Cardillo AB, Otálvaro AÁM, Busto VD, Talou JR, Velásquez LME, Giulietti AM (2010) Scopolamine, anisodamine and hyoscyamine production by *Brugmansia candida* hairy root cultures in bioreactors. Process Biochem 45:1577–1581

Cardillo AB, Giulietti AM, Palazón J, Bonfill M (2013) Influence of hairy root ecotypes on production of tropane alkaloids in *Brugmansia candida*. Plant Cell Tissue Organ Cult 114:305–312

Cardillo AB, Perassolo M, Sartuqui M, Talou JR, Giulietti AM (2017) Production of tropane alkaloids by biotransformation using recombinant *Escherichia coli* whole cells. Biochem Eng J 125:180–189

Chan W, Staba EJ (1965) Alkaloid production by *Datura* callus and suspension tissue culture. Lloydia 28:55–62

Chashmi NA, Sharifi M, Karimi F, Rahnama H (2010) Differential production of tropane alkaloids in hairy roots and *in vitro* cultured two accessions of *Atropa belladonna* L. under nitrate treatments. Z Naturforsch C Biosci 65:373–379

Chen C, Li J, Xiong F, Wang B, Xiao Y, Zhou G (2022a) Multivariate statistical analysis of tropane alkaloids in *Anisodus tanguticus* (Maxim.) Pascher from different regions to trace geographical origins. Acta Chromatogr 34(4):422–429

Chen C, Wang B, Li J, Xiong F, Zhou G (2022b) Multivariate statistical analysis of metabolites in *Anisodus tanguticus* (Maxim.) Pascher to determine geographical origins and network pharmacology. Front. Plant Sci 13:927336

Chen K, Ma L, Chen C, Liu N, Wang B, Bao Y, Zhou G (2023) Long-term impact of N, P, K fertilizers in different rates on yield and quality of *Anisodus tanguticus* (Maxinowicz) Pascher. Plan Theory 12(11):2102

Deng F (2005) Effects of glyphosate, chlorsulfuron, and methyl jasmonate on growth and alkaloid biosynthesis of jimsonweed (*Datura stramonium* L.). Pestic Biochem Physiol 82(1):16–26

Docimo T, Davis AJ, Luck K, Fellenberg C, Reichelt M, Phillips M, Gershenzon J, D'Auria JC (2015) Influence of medium and elicitors on the production of cocaine, amino acids and phytohormones by *Erythroxylum coca* calli. Plant Cell Tissue Organ Cult 120:1061–1075

Dräger B, Funck C, Höhler A, Mrachatz G, Nahrstedt A, Portsteffen A, Schaal A, Schmidt R (1994) Calystegines as a new group of tropane alkaloids in Solanaceae. Plant Cell Tissue Organ Cult 38:235–240

El Jaber-Vazdekis N, Barres ML, Ravelo AG, Zárate R (2008) Effects of elicitors on tropane alkaloids and gene expression in *Atropa baetica* transgenic hairy roots. J Nat Prod 71(12):2026–2031

Endo T, Yamada Y (1985) Alkaloid production in cultured roots of three species of *Duboisia*. Phytochemistry 24(6):1233–1236

Fattahi F, Shojaeiyan A, Palazon J, Moyano E, Torras-Claveria L (2021) Methyl-β-cyclodextrin and coronatine as new elicitors of tropane alkaloid biosynthesis in *Atropa acuminata* and *Atropa belladonna* hairy root cultures. Physiol Plant 172(4):2098–2111

Gerth A, Schmidt D, Wilken D (2007) The production of plant secondary metabolites using bioreactors. Acta Hortic 764:95–103

Ghorbanpour M (2013) Role of plant growth promoting rhizobacteria on antioxidant enzyme activities and tropane alkaloids production of *Hyoscyamus niger* under water deficit stress. Turk J Biol. https://doi.org/10.3906/biy-1209-12

Ghorbanpour M, Omidi M, Etminan A, Hatami M, Shooshtari L (2003) In vitro hyoscyamine and scopolamine production of black henbane (*Hyoscyamus niger*) from shoot tip culture under various plant growth regulators and culture media. Trakia J Sci 2:125–134

Ghorbanpour M, Hatami M, Khavazi K (2013) Role of plant growth promoting *rhizobacteria* on antioxidant enzyme activities and tropane alkaloid production of *Hyoscyamus niger* under water deficit stress. Turk J Biol 37(3):350–360

Ghorbanpour M, Hatami M, Hatami M (2015) Activating antioxidant enzymes, hyoscyamine and scopolamine biosynthesis of *Hyoscyamus niger* L. plants with nano-sized titanium dioxide and bulk application. Acta Agric Slov 105(1):23–32

Giri CC, Zaheer M (2016) Chemical elicitors versus secondary metabolite production in vitro using plant cell, tissue and organ cultures: recent trends and a sky eye view appraisal. Plant Cell Tissue Organ Cult 126:1–18

Giridhar P, Parimalan R (2010) A biotechnological perspective towards the improvement of annatto colour production for value addition – the influence of biotic elicitors. Asia Pac J Mol Biol Biotechnol 18(1):77–79

Gontier E, Sangwan BS, Barbotin JN (1994) Effects of calcium, alginate, and calcium-alginate immobilization on growth and tropane alkaloid levels of a stable suspension cell line of *Datura innoxia* Mill. Plant Cell Rep 13. https://doi.org/10.1007/BF0023295

Grynkiewicz G, Gadzikowska M (2008) Tropane alkaloids as medicinally useful natural products and their synthetic derivatives as new drugs. Pharmacol Rep 60:439–463

Guo S, Yang Y, Liu X, Wu NB (2018) Mechanism exploration on nitrogen metabolism and secondary metabolism in *Atropa belladonna* hairy roots treated with yeast extract. Zhongguo Zhong yao za zhi Zhongguo Zhongyao Zazhi China J Chin Mater Med 43(8):1610–1617

Gururaj HB, Giridhar P, Ravishankar GA (2012) Laminarin as a potential non-conventional elicitor for enhancement of capsaicinoid metabolites. Asian J Plant Sci Res 2:490–495

Habibi P, Soccol CR, Grossi-de-Sa MF (2018) Hairy root-mediated biotransformation: recent advances and exciting prospects. In: Hairy roots: an effective tool of plant biotechnology. Springer, Singapore, pp 185–211

Häkkinen ST, Moyano E, Cusido RM, Palazon J, Pinol MT, Oksman-Caldentey KM (2005) Enhanced secretion of tropane alkaloids in *Nicotiana tabacum* hairy roots expressing heterologous hyoscyamine-6β-hydroxylase. J Exp Bot 56(420):2611–2618

Halperin SJ, Flores HE (1997) Hyoscyamine and proline accumulation in water-stressed *Hyoscyamus muticus* 'hairy root' cultures. In Vitro Cell Dev Biol - Plant 33:240–244

Harfi B, Khelifi-Slaoui M, Bekhouche M, Benyammi R, Hefferon K, Makhzoum A, Khelifi L (2016) Hyoscyamine production in hairy roots of three *Datura* species exposed to high-salt medium. In Vitro Cell Dev Biol - Plant 52:92–98. https://doi.org/10.1007/s11627-015-9725-6

Harfi B, Khelifi L, Khelifi-Slaoui M, Assaf-Ducrocq C, Gontier E (2018) Tropane alkaloids GC/MS analysis and low dose elicitors' effects on hyoscyamine biosynthetic pathway in hairy roots of Algerian *Datura* species. Sci Rep 8:4–11

Hashimoto T, Yamada Y (1987) Effects of culture conditions on tropane alkaloid formation in *Hyoscyamus niger* suspension cultures. Agric Biol Chem 51(10):2769–2774

Hashimoto T, Yukimune Y, Yamada Y (1986) Tropane alkaloid production in *Hyoscyamus* root cultures. J Plant Physiol 124(1–2):61–75

Hashimoto T, Nakajima K, Ongena G, Yamada Y (1992) Two tropinone reductases with distinct stereospecificities from cultured roots of *Hyoscyamus niger*. Plant Physiol 100(2):836–845

Hashimoto T, Yun DJ, Yamada Y (1993) Production of tropane alkaloids in genetically engineered root cultures. Phytochemistry 32:713–718. https://doi.org/10.1016/S0031-9422(00)95159-8

Hedayati A, Hosseini B, Palazon J, Maleki R (2020) Improved tropane alkaloid production and changes in gene expression in hairy root cultures of two *Hyoscyamus* species elicited by silicon dioxide nanoparticles. Plant Physiol Biochem 155:416–428. https://doi.org/10.1016/j.plaphy.2020.07.029

Hedayati A, Hemmaty S, Nourozi E, Amirsadeghi A (2021) Effect of yeast extract on h6h gene expression and tropane alkaloids production in *Atropa belladonna* L. hairy roots. Rus J Plant Physiol 68:102–109

Hedayati A, Maleki R, Palazon J (2022) Effects of silicon dioxide nanoparticles (SiO2 NPs) on polyphenolic compounds content and antioxidant capacity in *H. reticulatus* and *H. pusillus* hairy roots. J Plant Res (Iran J Biol) 35(1):84–98

Hong MLK, Bhatt A, Ping NS, Laikeng C (2012) Detection of elicitation effect on *Hyoscyamus niger* L. root cultures for the root growth and production of tropane alkaloids. Rom Biotechnol Lett 17:7340–7351

Iranbakhsh AR, Oshagi MA, Ebadi M (2007) Growth and production optimization of tropane alkaloids in *Datura stramonium* cell suspension culture. Pak J Biol Sci 10:1236–1242. https://doi.org/10.3923/pjbs.2007.1236.1242

Jaremicz Z, Luczkiewicz M, Kokotkiewicz A, Krolicka A, Sowinski P (2014) Production of tropane alkaloids in *Hyoscyamus niger* (black henbane) hairy roots grown in bubble-column and spray bioreactors. Biotechnol Lett 36:843–853. https://doi.org/10.1007/s10529-013-1426-9

Jung HY, Kang SM, Kang YM, Kang MJ, Yun DJ, Bahk JD et al (2003) Enhanced production of scopolamine by bacterial elicitors in adventitious hairy root cultures of *Scopolia parviflora*. Enzyme Microb Technol 33(7):987–990

Kai G, Yang S, Zhang Y, Luo X, Fu X, Zhang A, Xiao J (2012a) Effects of different elicitors on yield of tropane alkaloids in hairy roots of *Anisodus acutangulus*. Mol Biol Rep 39:1721–1729. https://doi.org/10.1007/s11033-011-0912-1

Kai G, Zhang A, Guo Y, Li L, Cui L, Luo X, Liu C, Xiao J (2012b) Enhancing the production of tropane alkaloids in transgenic *Anisodus acutangulus* hairy root cultures by over-expressing tropinone reductase I and hyoscyamine-6β-hydroxylase. Mol BioSyst 8:2883–2890. https://doi.org/10.1039/c2mb25208b

Kang SM, Jung HY, Kang YM, Yun DJ, Bahk JD, Yang JK, Choi MS (2004a) Effects of methyl jasmonate and salicylic acid on the production of tropane alkaloids and the expression of PMT and H6H in adventitious root cultures of *Scopolia parviflora*. Plant Sci 166(3):745–751

Kang YM, Min JY, Moon HS, Karigar CS, Prasad DT, Lee CH, Choi MS (2004b) Rapid in vitro adventitious shoot propagation of *Scopolia parviflora* through rhizome cultures for enhanced production of tropane alkaloids. Plant Cell Rep 23:128–133

Kang Y, Park D, Min J, Song H, Jeong M-J, Kim Y, Kang S, Karigar C, Choi M (2011) Enhanced production of tropane alkaloids in transgenic *Scopolia parviflora* hairy root cultures over-expressing putrescine N-methyl transferase (PMT) and hyoscyamine-6β-hydroxylase (H6H). In Vitro Cell Dev Biol - Plant 47:516–524

Karimi F, Khataee E (2012) Aluminum elicits tropane alkaloid production and antioxidant system activity in micropropagated *Datura innoxia* plantlets. Acta Physiol Plant 34:1035–1041

Khanam N, Khoo CAGK, Khan AG (2000) Effects of cytokinin/auxin combinations on organogenesis, shoot regeneration and tropane alkaloid production in *Duboisia myoporoides*. Plant Cell Tissue Organ Cult 62:125–133

Khezerluo M, Hosseini B, Amiri J (2018) Sodium nitroprusside stimulated production of tropane alkaloids and antioxidant enzymes activity in hairy root culture of *Hyoscyamus reticulatus* L. Acta Biol Hung 69(4):437–448

Kim Y, Wyslouzil BE, Weathers PJ (2002) Secondary metabolism of hairy root cultures in bioreactors. In Vitro Cell Dev Biol - Plant 38:1–10

Kitamura Y (1988) *Duboisia* spp.: In vitro regeneration, and the production of tropane and pyridine alkaloids. In: Medicinal and aromatic plants, I. Springer, Berlin, Heidelberg, pp 419–436

Kitamura Y, Miura H, Sugii M (1985) Alkaloid composition and atropine esterase activity in callus and differentiated tissues of *Duboisia myoporoides* R. Br Chem Pharm Bull 33(12):5445–5448

Kitamura Y, Misuzu S, Hiroshi M (1992) Differences of atropine esterase activity between intact roots and cultured roots of various tropane alkaloid-producing plants. Phytochemistry 31(4):1191–1194

Lan X, Hong Q (2010) Hairy root culture of *Przewalskia tangutica* for enhanced production of pharmaceutical tropane alkaloids. J Med Plants 4:1477–1481

Lan X, Chang K, Zeng L, Liu X, Qiu F, Zheng W, Quan H, Liao Z, Chen M, Huang W, Liu W, Wang Q (2013) Engineering salidroside biosynthetic pathway in hairy root cultures of *Rhodiola crenulata* based on metabolic characterization of tyrosine decarboxylase. PLoS One 8. https://doi.org/10.1371/journal.pone.0075459

Lee KT, Yamakawa T, Kodama T, Shimomura K (1998) Effects of chemicals on alkaloid production by transformed roots of *belladonna*. Phytochemistry 49(8):2343–2347

Lee KT, Hirano H, Yamakawa T, Kodama T, Igarashi Y, Shimomura K (2001) Responses of transformed root culture of *Atropa belladonna* to salicylic acid stress. J Biosci Bioeng 91(6):586–589

Li R, Reed DW, Liu E, Nowak J, Pelcher LE, Page JE, Covello PS (2006) Functional genomic analysis of alkaloid biosynthesis in *Hyoscyamus niger* reveals a cytochrome P450 involved in littorine rearrangement. Chem Biol 13:513–520. https://doi.org/10.1016/j.chembiol.2006.03.005

Lin GD (1991) Tissue culture of a *Duboisia* hybrid (*D. leichhardtii x D. myoporoides*) and the production of tropane alkaloids. Ph.D. thesis, School of Pharmacy, The University of Queensland, Brisbane

Lin GD, Griffin WJ (1992) Scopolamine content of a *Duboisia* hybrid in callus cultures. Phytochemistry 31(12):4151–4153

Luo X, Weng S, Ni X, Wang X, Fu X, Xiao J, Kai G (2012) The effects of elicitation on the expression of key enzyme genes and on production of tropane alkaloids in *Anisodus acutangulus* plant. Biologia 67(2):352–359

Lydon J, Zimmerman RH, Fordham IM, Lusby WR (1993) Tissue culture and alkaloid production of *Erthroxylum coca* var. *coca*. Int J Geogr Inf Syst 2:3–14. https://doi.org/10.1300/J044v02n01_02

Madani H, Escrich A, Hosseini B, Sanchez-Muñoz R, Khojasteh A, Palazon J (2021) Effect of polyploidy induction on natural metabolite production in medicinal plants. Biomol Ther 11(6):899

Mano Y, Ohkawa H, Yamada Y (1989) Production of tropane alkaloids by hairy root cultures of *Duboisia leichhardtii* transformed by *Agrobacterium rhizogenes*. Plant Sci 59(2):191–201

Mazaheri H, Piri K (2015) Removal of phenol by *A. belladonna* L. hairy root. Int J Phytoremediation 17(12):1212–1219

Mehrotra S, Rahman LU, Kukreja AK (2010) An extensive case study of hairy-root cultures for enhanced secondary-metabolite production through metabolic-pathway engineering. Biotechnol Appl Biochem 56:161–172. https://doi.org/10.1042/BA20100171

Mehrotra S, Mishra S, Srivastava V (2018) Bioreactor technology for hairy roots cultivation. In: Bioprocessing of plant in vitro systems, Reference series in phytochemistry, pp 483–506. https://doi.org/10.1007/978-3-319-54600-1_10

Mihálik D, Hančinský R, Kaňuková Š, Mrkvová M, Kraic J (2022) Elicitation of hyoscyamine production in *Datura stramonium* L. plants using tobamoviruses. Plan Theory 11(23):3319

Min JY, Jung HY, Kang SM, Kim YD, Kang YM, Park DJ, Prasad DT, Choi MS (2007) Production of tropane alkaloids by small-scale bubble column bioreactor cultures of *Scopolia parviflora* adventitious roots. Bioresour Technol 98:1748–1753

Moharrami F, Hosseini B, Sharafi A, Farjaminezhad M (2017) Enhanced production of hyoscyamine and scopolamine from genetically transformed root culture of *Hyoscyamus reticulatus* L. elicited by iron oxide nanoparticles. In Vitro Cell Dev Biol - Plant 53:104–111

Monforte-González M, Ayora-Talavera T, Maldonado-Mendoza IE, Loyola-Vargas VM (1992) Quantitative analysis of serpentine and ajmalicine in plant tissues of *Catharanthus roseus* and hyoscyamine and scopolamine in root tissues of *Datura stramonium* by thin layer chromatography-densitometry. Phytochem Anal 3(3):117–121

Moradi A, Sharifi M, Mousavi A (2020) Induced production of tropane alkaloids, and expression of hyoscyamine 6β-hydroxylase (h6h) and putrescine N-methyl transferase (pmt2) genes in hairy roots and propagated plantlets of *Atropa belladonna* L. elicited by methyl jasmonate. South Afr J Bot 131:328–334. https://doi.org/10.1016/j.sajb.2020.01.042

Moussous A, Paris C, Khelifi-Slaoui M, Bekhouche M, Zaoui D, Rosloski SM, Khelifi L (2018) *Pseudomonas spp.* increases root biomass and tropane alkaloid yields in transgenic hairy roots of *Datura* spp. In Vitro Cell Dev Biol - Plant 54:117–126

Moyano E, Palazón J, Bonfill M, Osuna L, Cusidó RM, Oksman-Caldentey K-M, Piñol MT (2007) Biotransformation of hyoscyamine into scopolamine in transgenic tobacco cell cultures. J Plant Physiol 164:521–524. https://doi.org/10.1016/j.jplph.2006.06.012

Namdeo AG, Jadhav TA, Rai PK, Gavali S, Mahadik K (2007) Precursor feeding for enhanced production of Secondary metabolites: a review. Pharmacogn Rev 1:227–231

National Center for Biotechnology Information (2023) [WWW Document]. PubChem Compound Summary for CID 79037, Tropane. https://pubchem.ncbi.nlm.nih.gov/compound/Tropane. Accessed 7 Dec 2023

Nölke G, Schneider B, Agdour S, Drossard J, Fischer R, Schillberg S (2008) Modulation of polyamine biosynthesis in transformed tobacco plants by targeting ornithine decarboxylase to an atypical subcellular compartment. Open Biotechnol J 2:183–189

Oksman-Caldentey K-M, Vuorela H, Strauss A, Hiltunen R (1987) Variation in the tropane alkaloid content of *Hyoscyamus muticus* plants and cell culture clones. Planta Med 53:349–354

Oksman-Caldentey KM, Sevón N, Vanhala L, Hiltunen R (1994) Effect of nitrogen and sucrose on the primary and secondary metabolism of transformed root cultures of *Hyoscyamus muticus*. Plant Cell Tissue Organ Cult 38:263–272

Palazón J, Navarro-Ocaña A, Hernandez-Vazquez L, Mirjalili MH (2008) Application of metabolic engineering to the production of scopolamine. Molecules 13:1722–1742

Parimalan R, Mahendranath G, Giridhar P, Ravishankar GA (2011) Abiotic elicitor mediated augmentation of annatto pigment production in standing crop of *Bixa orellana* L. Indian J Fundam Appl Life Sci 1(4):229–236

Pavlov AI, Georgiev VG, Marchev AS, Berkov SH (2009) Nutrient medium optimization for hyoscyamine production in diploid and tetraploid Datura stramonium L. hairy root cultures. World J Microbiol Biotechnol 25:2239–2245

Piñol MT, Palazón J, Cusidó RM, Ribó M (1999) Influence of calcium ion-concentration in the medium on tropane alkaloid accumulation in *Datura stramonium* hairy roots. Plant Sci 141(1):41–49

Pitta-Alvarez SI, Giulietti AM (1997) Effects of gibberellin GA 7 on kinetics of growth and tropane alkaloid accumulation in hairy roots of *Brugmansia candida*. In Vitro Cell Dev Biol - Plant 33:147–153

Pitta-Alvarez SI, Giulietti AM (1999) Influence of chitosan, acetic acid and citric acid on growth and tropane alkaloid production in transformed roots of *Brugmansia candida* Effect of medium pH and growth phase. Plant Cell Tissue Organ Cult 59:31–38

Pitta-Alvarez SI, Spollansky TC, Giulietti AM (2000a) Scopolamine and hyoscyamine production by hairy root cultures of *Brugmansia candida*: influence of calcium chloride, hemicellulase and theophylline. Biotechnol Lett 22:1653–1656

Pitta-Alvarez SI, Spollansky TC, Giulietti AM (2000b) The influence of different biotic and abiotic elicitors on the production and profile of tropane alkaloids in hairy root cultures of *Brugmansia candida*. Enzyme Microb Technol 26(2–4):252–258

Pitta-Alvarez S, Marconi PL, Giulietti A (2003) Comparison of the influence of different elicitors on hyoscyamine and scopolamine content in hairy root cultures of *Brugmansia candida*. In Vitro Cell Dev Biol - Plant 39:640–644

Pudersell K, Vardja T, Vardja R, Matto V, Arak E, Raal A (2012) Inorganic ions in the medium modify tropane alkaloids and riboflavin output in *Hyoscyamus niger* root cultures. Pharmacogn Mag 8:73–77. https://doi.org/10.4103/0973-1296.93330

Qin B, Ma L, Wang Y, Chen M, Lan X, Wu N, Liao Z (2014) Effects of acetylsalicylic acid and UV-B on gene expression and tropane alkaloid biosynthesis in hairy root cultures of *Anisodus luridus*. Plant Cell Tissue Organ Cult 117:483–490

Qiu F, Yan Y, Zeng J, Huang JP, Zeng L, Zhong W, Lan X, Chen M, Huang SX, Liao Z (2021) Biochemical and metabolic insights into hyoscyamine dehydrogenase. ACS Catal 11:2912–2924

Richter U, Rothe G, Fabian A-K, Rahfeld B, Dräger B (2005) Overexpression of tropinone reductases alters alkaloid composition in *Atropa belladonna* root cultures. J Exp Bot 56:645–652. https://doi.org/10.1093/jxb/eri067

Rothe G, Dräger B (2002) Tropane alkaloids—metabolic response to carbohydrate signal in root cultures of *Atropa belladonna*. Plant Sci 163(5):979–985

Rothe G, Garske U, Dräger B (2001) Calystegines in root cultures of *Atropa belladonna* respond to sucrose, not to elicitation. Plant Sci 160(5):1043–1053

Sáenz-Carbonell L, Loyola-Vargas VM (1997) *Datura stramonium* hairy roots tropane alkaloid content as a response to changes in Gamborg's B 5 medium. Appl Biochem Biotechnol 61:321–337

Sáenz-Carbonell LA, Maldonado-Mendoza IE, Moreno-Valenzula O, Clau-Uitz R, Lopez-Meyer M, Oropeza C, Loyola-Vargas VM (1993) Effect of the medium pH on the release of secondary metabolites from roots of *Datura stramonium*, *Catharanthus roseus*, and *Tagetes patula* cultured in vitro. Appl Biochem Biotechnol 38:257–267

Sasaki K, Ishise T, Shimomura K, Kobayashi T, Matsubayashi Y, Sakagami Y et al (2002) Effects of phytosulfokine-α on growth and tropane alkaloid production in transformed roots of *Atropa belladonna*. Plant Growth Regul 36:87–90

Sauerwein M, Wink M, Shimomura K (1992) Influence of light and phytohormones on alkaloid production in transformed root cultures of *Hyoscyamus albus*. J Plant Physiol 140(2):147–152

Savita, Srivastava A, Jain R, Nagpal AK, Pati PK (2021) Plant tropane alkaloids (TAs): chemotaxonomic significance, biogenesis, and biotechnological interventions. In: Tropane alkaloids: pathways, potential and biotechnological applications. Springer, Singapore, pp 3–19

Scholl Y, Schneider B, Dräger B (2003) Biosynthesis of calystegines: 15N NMR and kinetics of formation in root cultures of *Calystegia sepium*. Phytochemistry 62(3):325–332

Sepehri E, Hosseini B, Hedayati A (2022) The effect of iron oxide nano-particles on the production of tropane alkaloids, h6h gene expression and antioxidant enzyme activity in *Atropa belladonna* hairy roots. Rus J Plant Physiol 69(6):122

Shah D, Kamili AN, Wani AA, Majeed U, Wani ZA, Sajjad N, Ahmad P (2020) Promoting the accumulation of scopolamine and hyoscyamine in *Hyoscyamus niger* L. through EMS based mutagenesis. PLoS One 15(5):e0231355

Shakeran Z, Keyhanfar M, Asghari G, Ghanadian M (2015) Improvement of atropine production by different biotic and abiotic elicitors in hairy root cultures of *Datura metel*. Turk J Biol 39:111–118. https://doi.org/10.3906/biy-1405-25

Shakeran Z, Keyhanfar M, Ghanadian M (2017) Biotic elicitation for scopolamine production by hairy root cultures of *Datura metel*. Mol Biol Res Commun 6(4):169

Shi Z, Zou W, Zhu Z, Xiong Z, Li S, Dong P, Zhu Z (2022) Tropane alkaloids (hyoscyamine, scopolamine and atropine) from genus *Datura*: Extractions, contents, syntheses and effects. Ind Crop Prod 186:115283

Srinivasan P, Smolke CD (2019) Engineering a microbial biosynthesis platform for de novo production of tropane alkaloids. Nat Commun 10(1):3634

Srivastava V, Negi AS, Ajayakumar PV, Khan SA, Banerjee S (2012) *Atropa belladonna* hairy roots: orchestration of concurrent oxidation and reduction reactions for biotransformation of carbonyl compounds. Appl Biochem Biotechnol 166:1401–1408

Srivastava V, Mehrotra S, Mishra S (2017a) Biotransformation through hairy roots: perspectives, outcomes, and major challenges. In: Jha S (ed) Transgenesis and Secondary Metabolism. Reference Series in Phytochemistry. Springer, Cham. https://doi.org/10.1007/978-3-319-27490-4_10-1

Srivastava V, Mehrotra S, Verma PK (2017b) Biotechnological interventions for production of therapeutic secondary metabolites using hairy root cultures of medicinal plants. In: Current developments in biotechnology and bioengineering: crop modification, nutrition, and food production. Elsevier B.V. https://doi.org/10.1016/B978-0-444-63661-4.00012-8

Srivastava V, Mishra S, Lhamo S, Chowdhary AA, Mehrotra S (2021) Involvement of various biotechnological contrivances for tropane alkaloid biosynthesis and applications of tropane alkaloid-bearing in vitro cultures. In: Srivastava V, Srivastava S, Mishra S (eds) Tropane alkaloids: pathways, potential and biotechnological applications. Springer, Singapore, pp 69–88

Subroto MA, Hamill JD, Doran PM (1996) Development of shooty teratomas from several solanaceous plants: growth kinetics, stoichiometry and alkaloid production. J Biotechnol 45(1):45–57

Sun JW, Zhang H, Wang FY, Sun YM, Sun M (2013) Effects of methyl jasmonate on accumulation and release of main tropane alkaloids in liquid cultures of *Datura stramonium* hairy root. Zhongguo Zhong yao za zhi Zhongguo Zhongyao Zazhi China J Chin Mater Med 38(11):1712–1718

Tabata M, Yamamoto H, Hiraoka N, Konoshima M (1972) Organization and alkaloid production in tissue cultures of *Scopolia parwxora*. Phytochemistry 11:949–955

Vaddadi S, Giridhar P (2015) Impacts of biotic and abiotic stress on major quality attributing metabolites of coffee beans. J Environ Biol 36:377–387

Vakili B, Karimi F, Sharifi M, Behmanesh M (2012) Chromium-induced tropane alkaloid production and H6H gene expression in *Atropa belladonna L.* (Solanaceae) in vitro-propagated plantlets. Plant Physiol Biochem 52:98–103

Vanhala L, Eeva M, Lapinjoki S, Hiltunen R, Oksman-Caldentey KM (1998) Effect of growth regulators on transformed root cultures of *Hyoscyamus muticus*. J Plant Physiol 153(3–4):475–481

Wang X, Chen M, Yang C, Liu X, Zhang L, Lan X, Tang K, Liao Z (2011) Enhancing the scopolamine production in transgenic plants of *Atropa belladonna* by overexpressing pmt and h6h genes. Physiol Plant 143:309–315

Wei Y, Zhang CP, Guo S, Lu KH, Liu X, Yang Y, Wu NB (2018) Effects of nitrogen form on accumulation of alkaloids and expression of relative genes in *Atropa belladonna*. Zhongguo Zhong yao za zhi Zhongguo Zhongyao Zazhi China J Chin Mater Med 43(1):72–78

Wen Y, Liao Y, Tang Y, Zhang H, Zhang J, Liao Z (2023) Metabolic effects of elicitors on the biosynthesis of tropane alkaloids in medicinal plants. Plan Theory 12(17):3050

Woo SH, Park JM, Yang JW (1996) Root culture using a mist culture system and estimation of scale-up feasibility. J Chem Tech Biotechnol 65:355–362

World Health Organization (2019) World Health Organization model list of essential medicines. In: Mental and holistic health: some international perspectives, vol 21. WHO, Geneva, pp 119–134

Xia K, Liu X, Zhang Q, Qiang W, Guo J, Lan X, Liao Z (2016) Promoting scopolamine biosynthesis in transgenic *Atropa belladonna* plants with pmt and h6h overexpression under field conditions. Plant Physiol Biochem 106:46–53

Yang C, Chen M, Zeng L, Zhang L, Liu X, Lan X, Tang K, Liao Z (2011) Improvement of tropane alkaloids production in hairy root cultures of *Atropa belladonna* by overexpressing pmt and h6h genes. Plant Omics 4:29–33

Yukimune Y, Yamagata H, Hara Y, Yamada Y (1994) Effects of oxygen on nicotine and tropane alkaloid production in cultured roots of *Duboisia myoporoides*. Biosci Biotechnol Biochem 58(10):1824–1827

Zabetakis I, Edwards R, O'Hagan D (1999) Elicitation of tropane alkaloid biosynthesis in transformed root cultures of *Datura stramonium*. Phytochemistry 50(1):53–56

Zeynali Z, Hosseini B, Ehsan R (2016) Effect of elicitation on antioxidant activity and production of tropane alkaloids in *Hyoscyamus reticulatus* hairy root cultures. Res J Pharmacogn 3:43–53

Zhang L, Ding R, Chai Y, Bonfill M, Moyano E, Oksman-Caldentey KM, Tang K (2004) Engineering tropane biosynthetic pathway in *Hyoscyamus niger* hairy root cultures. Proc Natl Acad Sci 101(17):6786–6791

Zhang L, Kai GY, Lu BB, Zhang HM, Tang KX, Jiang JH, Chen WS (2005) Metabolic engineering of tropane alkaloid biosynthesis in plants. J Integr Plant Biol 47(2):136–143

Zhang L, Yang B, Lu B, Kai G, Wang Z, Xia Y, Ding R, Zhang H, Sun X, Chen W, Tang K (2007) Tropane alkaloids production in transgenic *Hyoscyamus niger* hairy root cultures over-expressing Putrescine N-methyltransferase is methyl jasmonate-dependent. Planta 225:887–896. https://doi.org/10.1007/s00425-006-0402-1

Zhao B (2014) Alkaloid production by hairy root cultures. Utah State University, Logan, pp 17–107

Zhao K, Zeng J, Zhao T, Zhang H, Qiu F, Yang C, Liao Z (2017) Enhancing tropane alkaloid production based on the functional identification of tropine-forming reductase in *Scopolia lurida,* a Tibetan medicinal plant. Front Plant Sci 8:1745

Zhao T, Li S, Wang J, Zhou Q, Yang C, Bai F, Lan X, Chen M, Liao Z (2020) Engineering tropane alkaloid production based on metabolic characterization of ornithine decarboxylase in *Atropa belladonna*. ACS Synth Biol 9:437–448

Zhou LG, Wu JY (2006) Development and application of medicinal plant tissue cultures for production of drugs and herbal medicinals in China. Nat Prod Rep 23(5):789–810

Zolala J, Farsi M, Gordan HR, Mahmoodnia M (2007) Producing a high scopolamine hairy root clone in *Hyoscyamus muticus* through transformation by *Agrobacterium rhizogenes*. J Agric Sci Technol 9(4):327–339

Chapter 10
Production of Important Pharmaceutical Compounds, Tropane Alkaloids, Through Metabolic Engineering

Arun Sam Chacko, P. V. Kavya, and T. Dennis Thomas

Abstract Tropane alkaloids (TAs) are a class of specialized metabolites with significant pharmaceutical importance, including compounds like hyoscyamine, anisodamine, and scopolamine. These molecules, primarily derived from Solanaceae plants, exhibit potent biological activities, including anticholinergic, analgesic, and sedative properties. However, the natural production of TAs is often constrained by low yields from plants, limited availability of source plants, and environmental variability. Metabolic engineering provides a promising avenue to address these challenges by enhancing TA biosynthesis in plants or heterologous systems. This chapter highlights recent advances in metabolic engineering, focusing on boosting the production of specific metabolites by targeting bottleneck steps or blocking competing pathways. Furthermore, it discusses the strategies employed to engineer primary metabolic flux and enhance the expression of pathway-specific genes.

Keywords Metabolic engineering · Hairy root cultures · Scopolamine · Hyoscyamine · Anisodamine · Solanaceae · Plant secondary metabolites

10.1 Introduction

Being sessile organisms, plants synthesize several natural products, also known as "specialized metabolites," as a chemical response to biotic and abiotic stimuli. Plants utilize specialized metabolites to interact, communicate, and defend with other plants and organisms in their surrounding environment (Kim et al. 2016). The production of secondary metabolites forms part of the defence system of plants, for example, the constitutive production of antifeedants and phytoanticipins, and the inducible production of phytoalexins (Verpoorte and Memelink 2002). Plants produce many organic molecules that are not involved in their growth and development

A. S. Chacko · P. V. Kavya · T. D. Thomas (✉)
Department of Plant Science, Central University of Kerala, Kasaragod, Kerala, India
e-mail: den_thuruthiyil@cukerala.ac.in

T. Pullaiah et al. (eds.), *Tropane Alkaloids*,
https://doi.org/10.1007/978-981-96-5805-3_10

directly. These substances are called natural products, secondary metabolites, or secondary products. Secondary metabolites differ from primary metabolites in having restricted distribution within the plant kingdom.

Plant secondary metabolism supports a variety of functions during a plant's life cycle. They influence the food's quality in terms of color, taste, aroma, and flower color and the scent of ornamental plants. Hence, the Plant Secondary Metabolites (PSMs) are also of interest. Recently, research has revealed that PSMs have several of health-improving and disease-preventing qualities, including anti-oxidative and cholesterol-lowering capabilities. Apart from these features, some PSMs separated from plants are sold as fine chemicals in the pharmaceutical, flavoring, fragrance, and dye industries. Because of their low synthesis in naturally occurring plants, several phytochemicals are highly costly (Zhang et al. 2005).

Nowadays, bioactive chemicals isolated from plants are used as food additives, food colouring, pesticides, cosmetics, and fragrances, in addition to being significant therapeutic and restorative items. These substances are grouped as secondary metabolites because plants synthesize various secondary chemicals and the primary metabolites (carbohydrates, proteins, lipids, and nucleic acids), which are not essential for plants to survive (Palazón et al. 2006).

Although about 100,000 plant secondary metabolites are already known, only a small percentage of all plant species have been studied to some extent for the presence of secondary metabolites. However, nearly 4000 new ones are uncovered each year (Zhang et al. 2005; Verpoorte et al. 2000a). Since the beginning of recorded history, people have used a variety of specialized metabolites, which are pharmacologically active, for both medicinal and recreational purposes.

Alkaloids, in particular, are metabolites generated from amino acids that contain cyclic nitrogen, which are known to have pharmacological effects (Nocquet and Opatz 2016). These naturally occurring compounds typically contain one or more nitrogen atoms in their molecular structure. Alkaloids derive their name from the word "alkaline-like," which describes the alkaline characteristics of many alkaloid substances. Since not all compounds containing nitrogen fall under this category, defining alkaloids is complicated. For example, biogenic amines, or amino sugars, are natural plant products and are N-containing but do not belong to the alkaloids group (Kohnen-Johannsen and Kayser 2019).

The primary structure of plant-derived tropane alkaloids (TAs) is the 8-azabicyclo [3.2.1] octane nucleus (Fig. 10.1). Similar to other alkaloids, TAs exhibit a wide range of strong pharmacological properties, including narcotic analgesics,

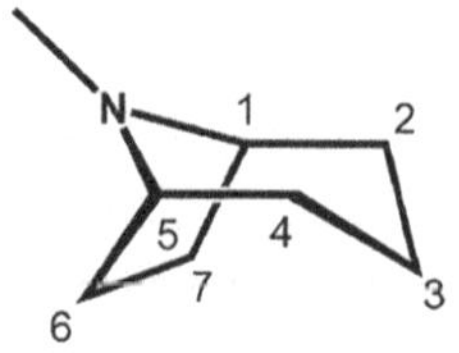

Fig. 10.1 The tropane core skeleton

deliriants, and anticholinergic agents. Humans have been exploiting the pharmacological properties of TAs since the beginning of history (Jirschitzka et al. 2013).

These metabolites are among the important secondary plant metabolites which are predominantly found in high concentrations in the Solanaceae and Erythroxylaceae families, and they can be divided into three major groups based on their chemical structure and occurrence: hyoscyamine and scopolamine, cocaine and calystegines (Kohnen-Johannsen and Kayser 2019). Depending on their structural features and biosynthetic relationship, TAs are divided into six main classes (Fig. 10.2). They are tropine (4, 3a-tropanol) and its derivatives (e.g., hyoscyamine, anisodamine, scopolamine), pseudotropine (6, 3b-tropanol) and its derivatives, calystegines, ecgonine and its derivatives, 4α-benzyltropane and its derivatives (core structure: ferrugine), pyranotropane and its derivatives (core structure: strobiline; Huang et al. 2021).

Scopolamine and hyoscyamine are two important medicinal tropane alkaloids (MTAs) used as anticholinergic medications. Clinically, they are used to treat a variety of conditions, including Parkinson's syndrome, gastrointestinal problems,

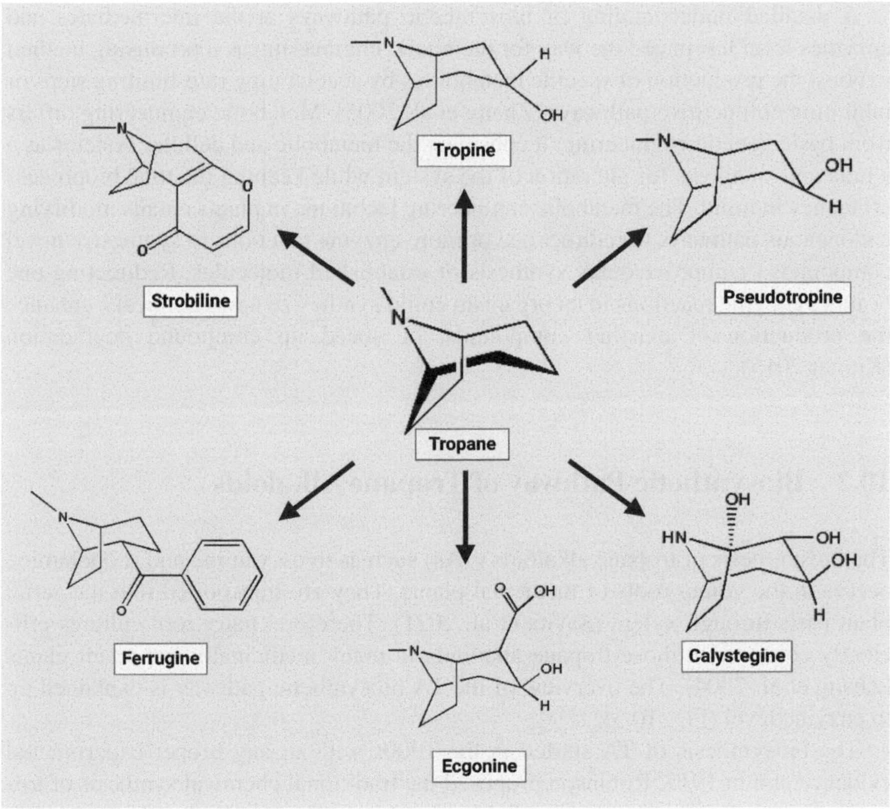

Fig. 10.2 Classes of tropane alkaloids

motion sickness, asthma, and pain etc. (Gong et al. 2022). Furthermore, hyoscyamine is the key compound for the production of ipratropium bromide, and scopolamine for tiotropium bromide, which are essential drugs in the treatment of chronic obstructive pulmonary disease (COPD; Griffin et al. 2008). Anisodamine is commonly used to improve and maintain blood circulation during shock treatment. It works by relaxing blood vessels and reducing the effects of certain nerve signals, making it effective for treating conditions like septic shock and other circulation problems. Because of these properties, anisodamine is an important ingredient in medicines that support blood flow and manage cardiovascular and inflammatory issues (Kohnen-Johannsen and Kayser 2019).

This increasing commercial importance of secondary metabolites has recently increased interest in secondary metabolism. It resulted in the potential use of genetic engineering to change how secondary metabolites are produced (Verpoorte and Memelink 2002). In this regard, molecular breeding—which uses genetic engineering—is a viable strategy. Even though many studies have been conducted in this area over the last 10 years (Verpoorte et al. 2000b), a significant constraint has been the lack of information about the plant's secondary metabolic pathways at the level of biosynthetic intermediates and enzymes.

A detailed understanding of biosynthetic pathways at the intermediates and enzymes level has paved the way for metabolic engineering as a promising method to boost the production of specific metabolites by accelerating rate-limiting steps or inhibiting competitive pathways (Zhang et al. 2005). Metabolic engineering differs from basic genetic engineering, it considers the metabolic and cellular systems as a whole and so allows for alteration of the system while keeping the total bioprocess efficiency in mind. The metabolic engineering technique in plants entails modifying endogenous pathways to redirect one or more enzyme reactions to synthesize novel compounds or improve/delay synthesis of established molecules. Redirecting one or more enzyme reactions in an organism could synthesize new chemicals, enhance the production of existing compounds, or speed up compound degradation (Kumar 2015).

10.2 Biosynthetic Pathway of Tropane Alkaloids

The biosynthesis of tropane alkaloids (TAs) such as hyoscyamine and scopolamine occurs in the young roots of medicinal plants. They are transported into the aerial plant parts through xylem (Savita et al. 2021). Therefore, hairy root cultures efficiently concentrate those tropane alkaloids in many medicinally important plants (Zhang et al. 2004). The overview of the TA biosynthetic pathway is explained up to enzyme level (Fig. 10.3).

The biosynthesis of TA started in the 1900s without any proper experimental evidence and in 1917, Robinson proposed the traditional chemical synthesis of tropinone in the Solanaceae family. Subsequently stable radiolabeled isotopes were incorporated in the biosynthetic pathway from 1950 to the twentieth century (Huang

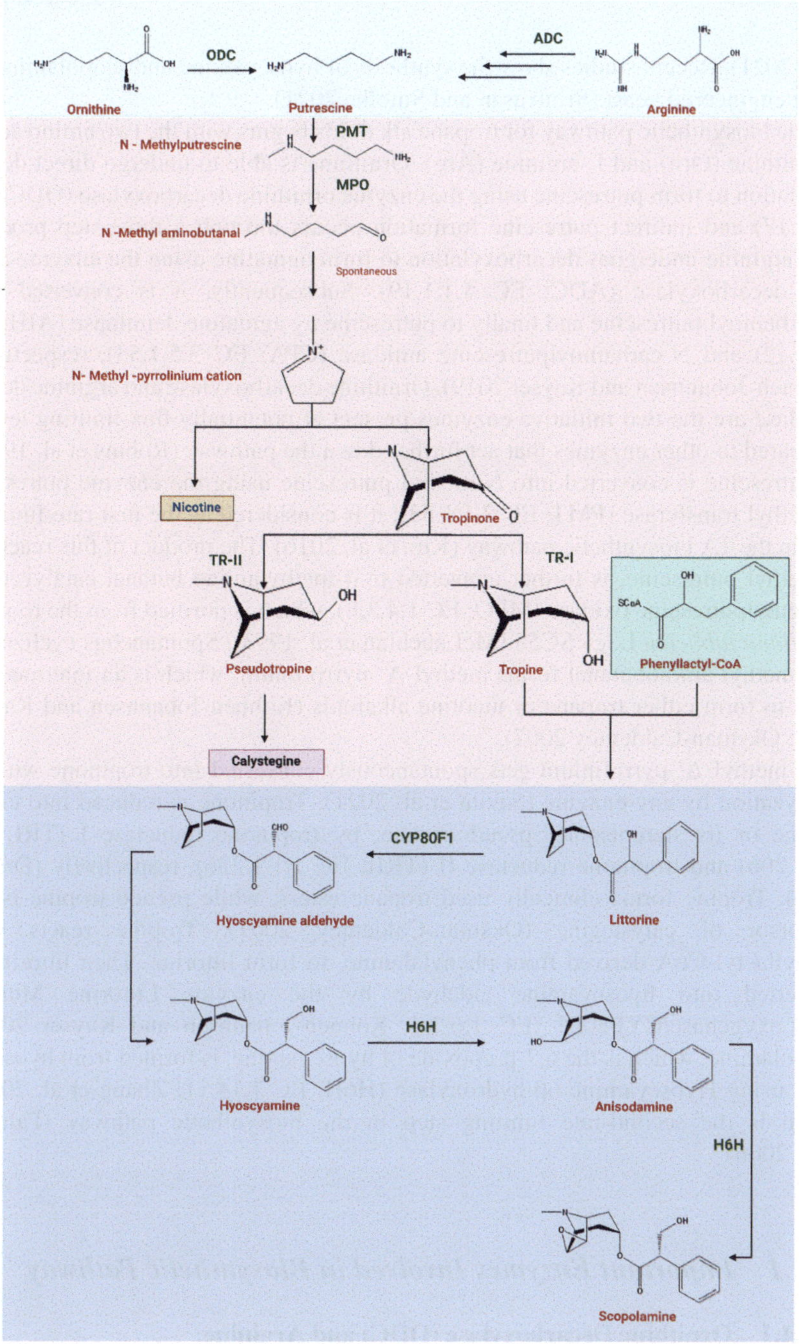

Fig. 10.3 Overview of the biosynthetic pathway of tropane alkaloids. Abbreviations: *ODC* ornithine decarboxylase, *ADC* arginine decarboxylase, *PMT* Putrescine N- methyltransferase, *MPO* N-methyl putrescine oxidase, *TRI* Tropinone reductase I, *TRII* Tropinone reductase II, *CYP80F1* cytochrome p450, *H6H* Hyoscyamine 6β-hydroxylase

et al. 2021). Recent studies show the synthesis of hyoscyamine and scopolamine by using engineered yeast (Srinivasan and Smolke 2021).

The biosynthetic pathway for tropane alkaloids begins with the two amino acids, L-ornithine (Orn) and L-arginine (Arg). Ornithine is able to undergo direct decarboxylation to form putrescine using the enzyme ornithine decarboxylase (ODC; EC 4.1.1.17) and indirect putrescine formation occurs through a three-step process. First arginine undergoes decarboxylation to form agmatine using the enzyme arginine decarboxylase (ADC; EC 4.1.1.19). Subsequently, it is converted into N-carbamoyl putrescine and finally to putrescine by agmatine deiminase (AIH; EC 3.5.3.12) and N-carbamoylputrescine amidase (CPA; EC 3.5.1.53), respectively (Kohnen-Johannsen and Kayser 2019). Ornithine decarboxylase and arginine decarboxylase are the two initiative enzymes present at potentially flux-limiting levels, compared to other enzymes that act further down the pathway (Robins et al. 1990).

Putrescine is converted into N-methyl putrescine using the enzyme putrescine N-methyl transferase (PMT; EC 2.1.1.53); it is considered as the first rate-limiting step in the TA biosynthetic pathway (Kim et al. 2016). The product of this reaction, N-methyl putrescine, is further converted to 4-methylamino butanal catalyzed by N-methylputrescine oxidase (MPO; EC 1.4.3.6), which is purified from the roots of *Nicotiana tabacum* L. cv SC58 (McLauchlan et al. 1993). Spontaneous cyclization of 4-methyl aminobutanal forms methyl-Δ^1-pyrrolinium, which is an intermediate stage to form either tropane or nicotine alkaloids (Kohnen-Johannsen and Kayser 2019; Oksman-Caldentey 2007).

N-methyl-Δ^1-pyrrolinium gets spontaneously converted into tropinone without catalyzation by any enzyme (Savita et al. 2021). Tropinone is reduced into either tropine or its stereoisomer pseudotropine, by tropinone reductase I (TRI; EC 1.1.1.206) and tropinone reductase II (TRII; EC 1.1.1.236), respectively (Dräger 2006). Tropine forms clinically used tropane esters, while pseudo tropine is the precursor of calystegine (Oksman-Caldentey 2007). Tropine reacts with phenyllactyl-CoA derived from phenylalanine, to form littorine. Then littorine is converted into hyoscyamine aldehyde by the enzyme Littorine Mutase/Monooxygenase/CYP80F1 (EC 1.6.2.4; Kohnen-Johannsen and Kayser 2019). Scopolamine, which is the 6,7-β-epoxide of hyoscyamine, is formed from hyoscyamine using Hyoscyamine 6β-hydroxylase (H6H; EC 1.14.11; Zhang et al. 2005), which is the second-rate limiting step in the biosynthetic pathway (Palazón et al. 2006).

10.2.1 Important Enzymes Involved in Biosynthetic Pathway

10.2.1.1 Ornithine Decarboxylase (ODC) and Arginine Decarboxylase (ADC)

Ornithine decarboxylase (ODC) and arginine decarboxylase (ADC) are the starting enzymes in TA biosynthetic pathways. The former forms putrescine by direct decarboxylation of amino acid ornithine, and the latter is involved in the conversion of arginine to putrescine through a three-step process. ODC plays an important role in the TA biosynthesis as compared to ADC. By overexpressing ODC gene of *Atropa belladonna* L. (AbODC) shows an increment in hyoscyamine, anisodamine production, so AbODC has a major role in TA biosynthesis production (Zhao et al. 2020).

10.2.1.2 Putrescine N-Methyltransferase (PMT)

Putrescine N-methyltransferase is the first committed enzyme in TA biosynthetic pathway. It is an upstream enzyme that catalyzes the N-methylation of diamine to form N-methylputrescine by removal of putrescine from the polyamine pool (Zhang et al. 2005). In solanaceous plants, high PMT activities occur in the in-branch roots and cultured roots. Partial purification and characterization of PMT was carried out from the cultured roots of *Hyoscyamus albus* L. which primarily produce that hyoscyamine as its main alkaloid (Hibi et al. 1992).

10.2.1.3 Tropinone Reductase I and Tropinone Reductase II

Tropinone reductase I and Tropinone reductase II are NADPH dependent enzyme belonging to the SDR family (short-chain dehydrogenase/ reductase) having an amino acid sequence, which constitutes a branching point in the metabolism. TR-I and TR-II catalyze the conversion of tropinone into tropine and pseudotropine, respectively (Kohnen-Johannsen and Kayser 2019). Tropine is the precursor for many pharmaceutically important tropane esters, while pseudotropine is the precursor of calystegines and also forms some rare esters, such as tigloidine (Oksman-Caldentey 2007). Both tropinone reductases have 50% similarity and show that it originates from a common ancestor (Dräger 2006). Immunoblot analysis results show that concentration of both TRs was high in the lateral roots of *Hyoscyamus niger* L. in its entire development. In cultured roots, TR protein accumulation shows in the basal region not in the root apex (Nakajima and Hashimoto 1999).

10.2.1.4 Hyoscyamine 6 β-Hydroxylase (H6H)

Hyoscyamine 6 β-hydroxylase is 2-oxoglutarate-dependent dioxygenase (EC 1.14.11), which is extracted from the roots of *H. niger* L. (Hashimoto and Yamada 1986). H6H is the final enzyme in the TA biosynthetic pathway, which converts hyoscyamine to scopolamine (Hashimoto et al.1992) by 2-step reaction. In the first step, L-hyoscyamine undergoes hydroxylation to form 6-hydroxy hyoscyamine (Anisodamine) and in the second step 6-hydroxy hyoscyamine converted to scopolamine (6,7-β-epoxide of hyoscyamine) by epoxidation (Palazón et al. 2006; Kohnen-Johannsen and Kayser 2019). The L-isomer of hyoscyamine is a substrate in this reaction; D-hyoscyamine is nearly inactive (Hashimoto and Yamada 1986). Several alkaloids structurally similar to L-hyoscyamine have also undergone hydroxylation at the C-6 position of the tropane group by the H6H enzyme (Kohnen-Johannsen and Kayser 2019). It is a prominent target enzyme, and its overexpression enhances the scopolamine levels in hyoscyamine-producing tissues of the transgenic plants or roots (Palazón et al. 2006).

10.3 Metabolic Engineering to Produce Tropane Alkaloids

Several investigations have been conducted on the genetic engineering of the pharmaceutically significant tropane alkaloids (Verpoorte and Memelink 2002).

10.3.1 Alteration of Biosynthetic Pathway by Genetic Engineering

The in-depth understanding of biosynthetic pathways and the increasing number of cloned genes involved in biosynthesis open up the opportunities of metabolic engineering as a potentially a practical approach to increasing the yield of specific metabolites by enhancing rate-limiting steps or blocking competitive pathways (Zhang et al. 2004).

Hyoscyamine and scopolamine, the Medicinally important Tropane Alkaloids (MTA), are primarily synthesized in young root cells and translocated to the aerial parts of the plant (Hashimoto et al. 1992). Therefore, high quantities of these metabolites can build up in cultured roots.

10.3.1.1 Engineering of Single Enzyme

Hyoscyamine 6 β-Hydroxylase (H6H)

Scopolamine is synthesized from hyoscyamine via 6β-hydroxy hyoscyamine (Hashimoto and Yamada 1986). H6H is a promising target enzyme that would raise the amounts of scopolamine in the transformants if it was highly expressed in plants that accumulate hyoscyamine. When the H6H gene is expressed ectopically in tissues other than the biosynthetic site, the H6H enzyme may be able to reach its alkaloid substrates, hyoscyamine and 6β-hydroxy hyoscyamine at the time of the alkaloids' translocation in the plant's aerial parts or during its storage (Yun et al. 1992).

The h6h gene of *H. niger* was inserted under the regulation of the 35S promoter in a binary T-DNA vector and transferred into two different *Agrobacterium rhizogenes* strains, LBA9402 and 15,834. It produced the hairy root clones in liquid cultures (Vanhala et al. 1995). Even among the best clones produced, hyoscyamine was the main alkaloid, with scopolamine accounting for around 10% of the morphological, chemical, and genetic content of hyoscyamine. KB7 was the best clone that produces scopolamine, over 100 times more than the control clones. However, the conversion of hyoscyamine to scopolamine was still incomplete (Jouhikainen et al. 1999).

The best transgenic clone of *Hyoscyamus muticus* L. hairy root cultures, overexpressing h6h, had a 100-fold increase in the level of scopolamine. Furthermore, this clone produced hyoscyamine as the primary alkaloid in similar amounts to controls. The transgenic plants exhibited almost exclusive production of scopolamine in their leaves and stems, making them valuable for medicinal use. Compared to diploid clones, tetraploid hairy root cultures of *H. muticus* exhibited higher levels of h6h transgene expression and scopolamine production. Thus, polyploidy can be a beneficial parameter in metabolic engineering for increasing alkaloid production. *A. rhizogenes* strains, LBA9402, carrying pLAL21 plasmid, containing the h6h gene from *H. niger* under the control of CaMV 35S promoter (Jouhikainen et al. 1999) used for plant transformation. In *H. muticus*, overexpression of the h6h gene significantly increased scopolamine production, with the best clone producing over 100 times more scopolamine than control clones (Dehghan et al. 2017).

The Hnh6h (*H. niger* h6h) gene was transformed into the genome of *Scopolia parviflora* (Dunn) Nakai by using the *A. rhizogenes* binary vector system. Constitutive expression of the H6H enzyme was detected in transgenic hairy root lines which resulted in the enhanced production of hyoscyamine and scopolamine. The best transgenic line (#5) produced scopolamine, three times more than wild type root (Kang et al. 2005).

To evaluate the impact of H6H on enhancing the production of tropane alkaloids (TAs), the H6H genes from *Scopolia lurida* (Link) Dunal (*SlH6H*) and *Hyoscyamus niger* (*HnH6H*) were cloned and overexpressed in hairy root cultures of *S. lurida*. This led to a significant increase in scopolamine and anisodamine levels in the transformed root cultures, while the hyoscyamine content showed a slight decrease compared to the control group (Lan et al. 2018).

The roots of *Datura innoxia* Mill. were used to clone the cDNA encoding the H6H gene. The protein sequence of *D. innoxia* H6H (DiH6H) and H6Hs from other plants shows a high degree of similarity. The recombinant DiH6H converts hyoscyamine to scopolamine. The cDNA of DiH6H was overexpressed in *D. innoxia* hairy roots, and scopolamine production was observed in the positive transformants (Li et al. 2020).

Under the control of the cauliflower mosaic virus 35S promoter, the H6H gene from *H. niger* was introduced into *A. belladonna* L., a hyoscyamine-rich plant by the use of an *Agrobacterium*-mediated transformation system. Kanamycin resistance and immunoscreening for the hydroxylase enzyme were used for selecting transgenic plants. The primary transformant and its progeny exhibited a significant shift in alkaloid composition. Major alkaloids seen in the leaves and stems were scopolamine. This indicates that the introduced gene was effectively expressed and functional in transgenic plants, leading to the desired metabolic change (Yun et al. 1992). Several transgenic root clones showed five-fold higher concentrations of scopolamine than the wild-type hairy roots (Sharma and Agnihotri 2021).

Putrescine N-Methyltransferase (PMT)

The first key step in the production of tropane alkaloids involves N-methylputrescine, which is formed from putrescine through the action of the enzyme PMT. This enzyme plays a crucial role in controlling the metabolic flow and is therefore regarded as the initial committed step in the biosynthesis of tropane alkaloids (Zhang et al. 2004).

A binary vector system combining the T-DNA of the Ri plasmid and the tobacco pmt. gene under the control of the CaMV 35S promoter was developed and introduced into the genomes of *Datura metel* L. and *H. muticus*. This approach was aimed to enhance the production of tropane alkaloids, such as hyoscyamine and scopolamine, in hairy root cultures. In *D. metel*, overexpression of the pmt. gene in transgenic hairy roots resulted in significantly increased production of both alkaloids. Additionally, the transgenic roots exhibited faster aging and accumulated higher levels of these alkaloids compared to non-transgenic (control) roots. In the case of *H. muticus,* pmt. gene overexpression resulted in increased hyoscyamine production but did not significantly enhance scopolamine levels. This indicates a limited ability to convert hyoscyamine into scopolamine in this species (Moyano et al. 2003).

The Ntpmt gene, under the control of CaMV 35S promoter, was integrated into the genome of *A. belladonna*. In transgenic plants and root cultures, the levels of alkaloids like hyoscyamine, scopolamine, tropinone, tropine, pseudotropine, and calystegines were found to be either unchanged or slightly reduced compared to control plants. Treatment of root cultures with 5% sucrose increased calystegine levels in control roots, but no such change was observed in roots overexpressing the pmt. gene. While the 1 μM auxin treatment reduced calystegine levels in control roots, while all alkaloid levels in pmt-overexpressing roots remained unaltered (Rothe et al. 2003). The overexpression of PMT does not significantly alter the levels of key alkaloids, indicating that other factors may be involved in regulating alkaloid biosynthesis in this plant.

The tobacco pmt. gene was successfully introduced into the genome of *S. parviflora* using a binary vector system involving disarmed *A. rhizogenes*. In the resulting kanamycin-resistant hairy root lines, the pmt. gene, along with its mRNA and protein, was overexpressed by at least twofold. This increase in pmt. expression enhanced PMT enzyme activity, significantly boosting the production of scopolamine and hyoscyamine in the transgenic lines (Lee et al. 2005).

The NtPMT gene was introduced into the genome of *H. niger* using the disarmed *A. tumefaciens* strain C58C1, which also contained the *A. rhizogenes* Ri plasmid pRiA4 and was regulated by the CaMV 35S promoter. Hairy root lines transformed with pmt. presented fivefold higher PMT activity than the control, and the methyl putrescine (MPUT) levels of the resulting engineered hairy roots increased four to fivefold compared to the control and wild-type roots, but there was no significant increase in tropane alkaloids (Zhang et al. 2007).

The *A. tumefaciens* strain C58C1, carrying the pRiA4 plasmid and the NtPMT gene controlled by the CaMV 35S promoter, was used to transform *Hyoscyamus senecionis* Willd. In the transformed plants, the pmt. gene was overexpressed, resulting in a significant increase in hyoscyamine production. However, h6h-overexpressing clones did not accumulate higher levels of scopolamine than the leaves of intact plants, indicating species-specific responses to genetic modifications (Dehghan et al. 2017).

Tropinone Reductases (TRs)
Tropinone reductases (TRs) are crucial enzymes at the branch point of TA metabolism, reducing tropinone to either tropine or pseudotropine. This reduction determines the flow of metabolites into different alkaloid groups (Nakajima and Hashimoto 1999).

In *A. belladonna*, root cultures overexpressed with tropinone reductases have been shown to alter tropane alkaloid composition significantly. Root cultures with strong overexpression of tropine-forming reductase showed a five-fold increase in scopolamine and a three-fold increase in hyoscyamine, while the levels of calystegine were decreased by 30–90%, as compared to control roots. High levels of pseudotropine were associated with an increased accumulation of calystegines in the roots (Richter et al. 2005).

The gene SlTRI (*S. lurida* Tropinone Reductases I), identified as a short-chain dehydrogenase/reductase (SDR), was isolated and characterized from *S. lurida*. Overexpression of SlTRI in hairy root cultures of *S. lurida* resulted in a significant increase in hyoscyamine and scopolamine contents, demonstrating the potential of SlTRI in enhancing TA production (Zhao et al. 2017).

Ornithine Decarboxylase (ODC)
ODC is a crucial enzyme in the TA biosynthetic pathway and a valuable target for metabolic engineering. Binary vector pMVR1ODC of *A. tumefaciens* strain LBA4404, containing mouse odc gene with CaMV 35S promoter and hygromycin phosphotransferase (hpt) gene, was used for transformation into *D. innoxia* genome. The odc gene overexpressed in *D. innoxia* transgenic plants, to modulate the metabolic flux downstream and eventually achieved higher accumulation of scopolamine. One line, O10 among six independent transformed lines, produced scopolamine almost six times higher than that produced by control plants (Singh et al. 2011).

Suppressing AbODC in *A. belladonna* root cultures led to a decrease in the production of putrescine, N-methyl putrescine, and tropane alkaloids (TAs). In contrast, overexpressing AbODC significantly boosted the biosynthesis of hyoscyamine, putrescine, N-methyl putrescine, and amidoamine. Transgenic *A. belladonna* plants with AbODC overexpression produced much higher levels of anisodamine and hyoscyamine compared to control plants. The physiological and metabolic roles of AbODC, particularly in TA production, were explored using metabolic engineering strategies. AbODC converts ornithine to putrescine more efficiently than other plant ODCs (Zhao et al. 2020).

10.3.1.2 Engineering of Two Enzymes

Putrescine N-Methyltransferase (PMT) and Hyoscyamine 6 β-Hydroxylase (H6H): Co-expression

In *H. niger*, putrescine *N*-methyltransferase (PMT) and hyoscyamine 6 β-hydroxylase (H6H) genes were introduced and overexpressed. The hairy root formation is induced in it. Scopolamine synthesis was considerably greater in transgenic hairy root lines that expressed both pmt. and h6h than in wild-type and transgenic lines that only carried one gene (pmt or h6h). The best line (T_3) produced scopolamine, which was over nine times more than that in the wild type and more than twice the amount in the highest scopolamine-producing *h6h* single-gene transgenic line H_{11} (Zhang et al. 2004).

The same gene combination was simultaneously introduced and overexpressed in transgenic hairy root cultures of *A. belladonna*. All nine transgenic hairy root lines overexpressed both pmt. and h6h and produced tropane alkaloids at higher levels (Wang et al. 2011). The best line T3 produced a higher amount of hyoscyamine, which was about 24 times more than that in the wild type (intact *A. belladonna*) and 11 times more than that in non-transgenic hairy root cultures. In addition to this, T3 produced more scopolamine, which was five times more than that in untransgenic hairy root cultures and four times more than that in the wild type (Yang et al. 2011).

With the help of *A. rhizogenes*, both the two genes PMT and H6H cDNAs were co-expressed to metabolically modify the adventitious roots of *S. parviflora*. The co-expression of these essential enzyme genes increased the production of alkaloids such as scopolamine and hyoscyamine (Kang et al. 2011).

Tropinone Reductase I (TRI) and Hyoscyamine 6 β-Hydroxylase (H6H): Co-expression

Using *Agrobacterium*-mediated gene transfer, the tropinone reductase I (TRI) and hyoscyamine-6β-hydroxylase (H6H) genes were successfully introduced into *Anisodus acutangulus* C.Y. Wu & C. Chen. The resulting transgenic hairy root lines, which expressed both TRI and H6H (referred to as TH lines), showed significantly higher levels of tropane alkaloids (TAs) compared to the control and lines transformed with only one of the genes (T or H lines). Among these, the TH53 line stood out, producing approximately 4.49 times more TAs than the control lines. The simultaneous introduction of TRI and H6H genes improved the content of anisodine. The elevated amounts of scopolamine and anisodamine were about 12-fold and five-fold, respectively. The hyoscyamine content was significantly higher than that of the control (Kai et al. 2012).

Putrescine N-Methyltransferase (PMT) and Tropinone Reductase I (TRI): Co-expression

The simultaneous introduction and overexpression of AaPMT and AaTRI, under the control of CaMV 35S promoter, were separately introduced into *A. acutangulus* using disarmed *A. tumefaciens* (C58C1 strain). The contents of Anisodamine,

Hyoscyamine, Anisodine, and Scopolamine were increased than single gene expressed transformants and control (Kai et al. 2011).

10.3.2 Enhancing Alkaloid Production

10.3.2.1 CRISPR/Cas9 System

CRISPR (Clustered regularly interspaced short palindromic repeats)/Cas9 (CRISPR-associated) is an endonuclease system which is considered a powerful RNA-guided genome editing tool. CRISPR/Cas9 has been well-studied in model plant species for targeted genome editing (Alagoz et al. 2016). Recent advancements in molecular biology, such as the CRISPR/Cas9 technology used in high-yielding lines, have been developed for improved production of functional and efficient molecules, promising to food, cosmetic, and pharmaceutical industries (Biswas et al. 2023).

The combination of CRISPR-Cas9, hairy root cultures, and metabolic engineering is a promising strategy for enhancing the production of secondary metabolites in plants. This approach leverages the precision of CRISPR-Cas9 for genome editing, the stability and high growth rate of hairy root cultures, and the targeted pathway modifications of metabolic engineering.

A. belladonna produces hyoscyamine or atropine, and also produces scopolamine and anisodamine. In wild plants, hyoscyamine content is very low, and it is difficult and too expensive to isolate hyoscyamine independently from the tropane alkaloids mixture. Using the CRISPR/Cas9 system to disrupt the gene encoding hyoscyamine 6β-hydroxylase (AbH6H) in *A. belladonna* resulted in plants with high hyoscyamine yields and no anisodamine or scopolamine production. This approach is promising for producing hyoscyamine at a lower cost by eliminating the need to separate it from other alkaloids (Zeng et al. 2021).

10.3.2.2 Methyl Jasmonate (MJ) Elicitation

For the elevated production of hyoscyamine and scopolamine, methyl jasmonate (MJ) is applied to hairy root cultures and propagated plantlets of *A. belladonna.* MJ enhanced the biosynthesis of these alkaloids, by upregulating the expression of PMT and H6H genes, although MJ decreased the growth parameters in hairy root cultures and in plantlets also. Moreover, in the plantlets, MJ upregulated gene expression of pmt2 and h6h in the root, which is the site of biosynthesis of hyoscyamine and scopolamine. Nevertheless, the scopolamine content decreased by increasing the MJ concentration as it also showed a reverse pattern compared to the h6h gene expression, presumably due to the negative effect of hyoscyamine accumulation on H6H enzyme activity (Moradi et al. 2020).

Application of methyl jasmonate in *H. senecionis* increased pmt. expression and decreased tropinone reductase II (trII) expression, leading to higher biosynthesis of hyoscyamine and total alkaloids. This indicates that elicitation can be an effective strategy to boost alkaloid production (Dehghan et al. 2017).

Metabolic engineering has proven to be a useful tool for boosting the production of tropane alkaloids, which are important for making medicines. By changing the way one or more key enzymes in the production pathway work—either by overexpressing them individually or together—scientists have been able to increase the yields (Table 10.1) significantly. These strategies help to overcome natural limits in the pathway, making production more efficient. With advancements like CRISPR technology and synthetic biology, researchers can fine-tune these processes even further. Combining these methods with better process design and system-level understanding will make producing these valuable compounds at larger scales easier. Overall, metabolic engineering offers a promising way to meet the growing demand for tropane alkaloids in healthcare.

10.3.2.3 Targeted Expression of *Vitreoscilla* Hemoglobin (VHb)

The bacteria *Vitreoscilla stercoraria* express a kind of hemoglobin, —*Vitreoscilla* hemoglobin (VHb) during hypoxic conditions, by promoting oxygen delivery enhances respiration and energy metabolism. It has been shown to improve growth, protein secretion, metabolite productivity, and stress resistance of hosts, thus rendering the protein promising in metabolic engineering, especially in plant metabolism optimization (Yu et al. 2021). Expression of VHb in plastids of *H. niger* hairy roots enhanced the production of hyoscyamine and scopolamine. This strategy improved oxygen delivery to the biosynthetic pathway, thereby increasing the overall yield of tropane alkaloids. Specifically, hyoscyamine levels were 1.25 times higher than in non-targeted VHb expression lines and 3.66 times higher than in wild-type (WT) lines. Scopolamine content increased by 2.20 times compared to VHb transgenic lines and 4.70 times compared to WT lines (Guo et al. 2018).

10.3.2.4 Hairy Root Culture: A Revolutionary Tool for Boosting Secondary Metabolite Yield

"Hairy roots" are the mirrors to the corresponding root of the parent plant which has the same phytochemical pattern with higher growth rate and productivity, are therefore, hairy root cultures are extensively studied as an effective alternative for the in vitro production of metabolites (Biswas et al. 2023).

A. rhizogenes-mediated gene transfer is a method used to produce secondary metabolites through hairy root cultures. During infection, the T-DNA from *A. rhizogenes* integrates into the plant's genome, causing a "hairy root." These hairy roots are often chosen as a model system for producing bioactive compounds because they have special advantages, such as higher production of secondary metabolites,

Table 10.1 Engineering of medicinally important plants for the tropane alkaloids production through overexpression of gene(s)

Name of the plant	Enzyme(s) altered	Gene altered	Effect			References
			Scopolamine	Hyoscyamine	Anisodamine	
Hyoscyamus niger L. (Henbane)	Hyoscyamine 6-hydroxylase	Hn H6H	+	0	0	Jouhikainen et al. (1999)
Hyoscyamus muticus L. (Egyptian henbane)	Hyoscyamine 6-hydroxylase	Hn H6H	+	0	0	Jouhikainen et al. (1999) Dehghan et al. (2017)
Scopolia parviflora (Dunn) Nakai	Hyoscyamine 6-hydroxylase	Hn H6H	+	+	0	Kang et al. (2005)
Scopolia lurida (Link) Dunal	Hyoscyamine 6-hydroxylase	HnH6H and SlH6H	+	–	+	Lan et al. (2018)
Datura innoxia Mill (Angel's Trumpet)	Hyoscyamine 6-hydroxylase	*Di*H6H	+	0	0	Li et al. (2020)
Atropa belladonna L. (deadly nightshade)	Hyoscyamine 6-hydroxylase	Hn H6H	+	0	0	Sharma and Agnihotri (2021)
Datura metel L. (Devil's trumpet)	Putrescine N-methyltransferase (PMT)	NtPMT	+	+	0	Moyano et al. (2003)
Hyoscyamus muticus L.	Putrescine N-methyltransferase (PMT)	NtPMT	0	+	0	Moyano et al. (2003)
Atropa belladonna L. (deadly nightshade)	Putrescine N-methyltransferase (PMT)	NtPMT	0	0	0	Rothe et al. (2003)
Scopolia parviflora (Dunn) Nakai	Putrescine N-methyltransferase (PMT)	NtPMT	+	+	0	Lee et al. (2005)
Hyoscyamus niger L. (Henbane)	Putrescine N-methyltransferase (PMT)	NtPMT	0	0	0	Zhang et al. (2007)
Hyoscyamus senecionis Willd.	Putrescine N-methyltransferase (PMT)	NtPMT	0	+	0	Dehghan et al. (2017)
Atropa belladonna L. (deadly nightshade)	Tropinone reductases (TRs)	DsTRI	+	+	0	Richter et al. (2005)

(continued)

Table 10.1 (continued)

Name of the plant	Enzyme(s) altered	Gene altered	Effect			References
			Scopolamine	Hyoscyamine	Anisodamine	
Scopolia lurida (link) Dunal	Tropinone reductases (TRs)	SlTRI	+	+	0	Zhao et al. (2017)
Datura innoxia Mill (Angel's Trumpet)	Ornithine decarboxylase	Mouse ODC	+	0	0	Singh et al. (2011)
Atropa belladonna L. (deadly nightshade)	Ornithine decarboxylase	AbODC	+	0	+	Zhao et al. (2020)
Hyoscyamus niger L. (Henbane)	Putrescine N-methyltransferase and hyoscyamine 6β-hydroxylase	NtPMT and HnH6H	+	0	0	Zhang et al. (2004)
Atropa belladonna L. (deadly nightshade)	Putrescine N-methyltransferase and hyoscyamine 6β-hydroxylase (H6H)	NtPMT and HnH6H	+	+	0	Yang et al. (2011)
Scopolia parviflora (Dunn) Nakai	Putrescine N-methyltransferase and hyoscyamine 6β-hydroxylase	SpPMT and SpH6H	+	+	0	Kang et al. (2011)
Anisodus acutangulus C.Y.Wu & C.Chen.	Tropinone reductase I and hyoscyamine 6 β-hydroxylase	AaTRI and AaH6H	+	+	+	Kai et al. (2012)
Anisodus acutangulus C.Y.Wu & C.Chen.	Putrescine N-methyltransferase and tropinone reductase I	AaPMT and AaTRI	+	+	+	Kai et al. (2011)

faster growth rates, and stable genetic, biochemical, and physiological properties compared to naturally growing plants. The hairy roots are selected as a model system for generating bioactive chemicals over naturally growing plants because of their unique characteristics for increased secondary metabolite production, higher growth rate, and genetic, biochemical, and physiological stability (Dubey et al. 2021).

10.4 Conclusions

The byproducts and intermediates of physiological metabolism are called metabolites (Biswas et al. 2023). In order to defend themselves against biotic and abiotic stress, plants have an almost infinite capacity to generate a vast array of secondary metabolic chemicals in various organs, such as leaves, shoots, roots, and flowers, via various biochemical pathways (Wang et al. 2022). These bioproducts are tremendously important as agrochemicals, dyes, insecticides, food additives, and pharmaceuticals. Producing pharmacologically and commercially important compounds has gained tremendous attention over the past decades.

Among these, tropane alkaloid is a major anticholinergic drug whose biosynthesis takes place in plant roots. Due to its limited production by plants, alternative methods are essential for its effective large-scale production. Hairy root culture and targeted metabolic modifications are considered as the effective methods for the enhanced production of valuable plant metabolites. The integration of CRISPR-Cas9 and elicitor treatments are also used for enhanced production of pharmaceutically important plant compounds both quantitatively and qualitatively.

10.5 Future Perspectives

Metabolic engineering is a promising tool used to improve plant secondary metabolite production by altering the biosynthetic pathway. It is essential in the pharmaceutical industry in order to produce valuable medicinal compounds. Tropane alkaloids are one of the prominent alkaloids that have many medicinal properties. Hairy root culture is a proven method for the enhanced production of scopolamine and hyoscyamine by genetic transformation using *A. tumefaciens* (Savita et al. 2021). We can use microbial hosts such as yeast for the synthesis of plant natural products (PNPs) and use computational tools for gene discovery and expanding the heterologous biosynthetic diversity. Recent year's studies show that the elicitor treatments can also accumulate TA production. Further investigations are essential for using this effective method properly for commercial production (Wen et al. 2023). Microbial co-culture and transport engineering have enabled the generation as well as enhanced production of various important metabolites (Yamada et al. 2024).

Acknowledgments ASC and KPV acknowledge the Chemdraw pro 8.0 software for all chemical structures and Biorender.com for Figs. 10.1, 10.2, and 10.3, the financial assistance from the Council of Scientific & Industrial Research in the form of JRF and SRF to carry out this chapter, and also acknowledge research scholars (Anju T, Kaviraj R and Ashifa Nizam) of Plant Science Department of Central University of Kerala, Kasaragod, for their outstanding support.

References

Alagoz Y, Gurkok T, Zhang B, Unver T (2016) Manipulating the biosynthesis of bioactive compound alkaloids for next-generation metabolic engineering in Opium poppy using CRISPR-Cas 9 genome editing technology. Sci Rep 6:30910

Biswas D, Chakraborty A, Mukherjee S, Ghosh B (2023) Hairy root culture: a potent method for improved secondary metabolite production of Solanaceous plants. Front Plant Sci 14:1197555

Dehghan E, Reed D, Covello P, Hasanpour Z, Palazón J, Oksman-Caldentey K, Ahmadi F (2017) Genetically engineered hairy root cultures of *Hyoscyamus senecionis* and *H. muticus*: ploidy as a promising parameter in the metabolic engineering of tropane alkaloids. Plant Cell Rep 36:1615–1626

Dräger B (2006) Tropinone reductases, enzymes at the branch point of tropane alkaloid metabolism. Phytochem 67:327–337

Dubey NK, Singh P, Singh A, Yadav SK (2021) Current progress in tropane alkaloid production by transgenic and heterologous gene expression approaches. In: Srivastava V, Mehrotra S, Mishra S (eds) Tropane alkaloids: pathways, potential and biotechnological applications. Springer, Singapore, pp 113–131

Gong H, He P, Lan X, Zeng L, Liao Z (2022) Biotechnological approaches on engineering medicinal tropane alkaloid production in plants. Front Plant Sci 13:924413

Griffin J, Lee S, Caiado M, Kesten S, Price D (2008) Comparison of tiotropium bromide and combined ipratropium/salbutamol for the treatment of COPD: a UK General Practice Research Database 12-month follow-up study. NPJ Prim Care Respir Med 17:104–110

Guo Z, Tan H, Lv Z, Ji Q, Huang Y, Liu J, Chen D, Diao Y, Si J, Zhang L (2018) Targeted expression of Vitreoscilla hemoglobin improves the production of tropane alkaloids in *Hyoscyamus niger* hairy roots. Sci Rep 8:17969

Hashimoto T, Yamada Y (1986) Hyoscyamine 6β-hydroxylase, a 2-oxoglutarate-dependent dioxygenase, in alkaloid-producing root cultures. J Plant Physiol 81:619–625

Hashimoto T, Nakajima K, Ongena G, Yamada Y (1992) Two tropinone reductases with distinct stereospecificities from cultured roots of *Hyoscyamus niger*. J Plant Physiol 100:836–845

Hibi N, Fujita T, Hatano M, Hashimoto T, Yamada Y (1992) Putrescine N-methyltransferase in cultured roots of *Hyoscyamus albus*: n-butylamine as a potent inhibitor of the transferase both in vitro and in vivo. J Plant Physiol 100:826–835

Huang JP, Wang YJ, Tian T, Wang L, Yan Y, Huang SX (2021) Tropane alkaloid biosynthesis: a centennial review. Nat Prod Rep 38:1634–1658

Jirschitzka J, Dolke F, D'Auria JC (2013) Increasing the pace of new discoveries in tropane alkaloid biosynthesis. In: Giglioli-Guivarc'h N (ed) Advances in botanical research. Academic, London, pp 39–72

Jouhikainen K, Lindgren L, Jokelainen T, Hiltunen R, Teeri TH, Oksman-Caldentey KM (1999) Enhancement of scopolamine production in *Hyoscyamus muticus* L. hairy root cultures by genetic engineering. Planta 208:545–551

Kai G, Yang S, Luo X, Zhou W, Fu X, Zhang A, Xiao J (2011) Co-expression of AaPMT and AaTRI effectively enhances the yields of tropane alkaloids in *Anisodus acutangulus* hairy roots. BMC Biotechnol 11:1–12

Kai G, Zhang A, Guo Y, Li L, Cui L, Luo X, Xiao J (2012) Enhancing the production of tropane alkaloids in transgenic *Anisodus acutangulus* hairy root cultures by over-expressing tropinone reductase I and hyoscyamine-6β-hydroxylase. Mol Biosyst 8:2883–2890

Kang YM, Lee OS, Jung HY, Kang SM, Lee BH, Karigar C, Choi MS (2005) Overexpression of hyoscyamine 6β-Hydroxylase (h6h) gene and enhanced production of tropane alkaloids in *Scopolia parviflora* hairy root lines. JMB 15:91–98

Kang YM, Park DJ, Min JY, Song HJ, Jeong MJ, Kim YD, Choi MS (2011) Enhanced production of tropane alkaloids in transgenic *Scopolia parviflora* hairy root cultures over-expressing putrescine N-methyltransferase (PMT) and hyoscyamine-6β-hydroxylase (H6H). In Vitro Cell Dev Biol – Plant 47:516–524

Kim N, Estrada O, Chavez B, Stewart C Jr, D'Auria JC (2016) Tropane and granatane alkaloid biosynthesis: a systematic analysis. Molecules 21(11):1510

Kohnen-Johannsen KL, Kayser O (2019) Tropane alkaloids: chemistry, pharmacology, biosynthesis and production. Molecules 24:796

Kumar A (2015) Metabolic engineering in plants. In: Bahadur B, Rajam MV, Sahijram L, Krishnamurthy KV (eds) Plant biology and biotechnology, Plant genomics and biotechnology. Springer, New Delhi, pp 517–526

Lan X, Zeng J, Liu K, Zhang F, Bai G, Chen M, Huang L (2018) Comparison of two hyoscyamine 6β-hydroxylases in engineering scopolamine biosynthesis in root cultures of *Scopolia lurida*. Biochem Biophys Res Commun 497:25–31

Lee OS, Kang YM, Jung HY, Min JY, Kang SM, Karigar CS, Choi MS (2005) Enhanced production of tropane alkaloids in *Scopolia parviflora* by introducing the PMT (putrescine N-methyltransferase) gene. In vitro Cell Dev Biol – Plant 41:167–172

Li Q, Zhu T, Zhang R, Bu Q, Yin J, Zhang L, Chen W (2020) Molecular cloning and functional analysis of hyoscyamine 6β-hydroxylase (H6H) in the poisonous and medicinal plant *Datura innoxia* Mill. PPB 153:11–19

McLauchlan WR, McKee RA, Evans DM (1993) The purification and immunocharacterisation of N-methylputrescine oxidase from transformed root cultures of *Nicotiana tabacum* L. cv SC58. Planta 191:440–445

Moradi A, Sharifi M, Mousavi A (2020) Induced production of tropane alkaloids, and expression of hyoscyamine 6β-hydroxylase (h6h) and putrescine N-methyl transferase (pmt2) genes in hairy roots and propagated plantlets of *Atropa belladonna* L. elicited by methyl jasmonate. S Afr J Bot 131:328–334

Moyano E, Jouhikainen K, Tammela P, Palazón J, Cusido R, Piñol M, Teeri T, Oksman-Caldentey K (2003) Effect of pmt. gene overexpression on tropane alkaloid production in transformed root cultures of *Datura metel* and *Hyoscyamus muticus*. J Exp Bot 54:203–211

Nakajima K, Hashimoto T (1999) Two tropinone reductases, that catalyze opposite stereospecific reductions in tropane alkaloid biosynthesis, are localized in plant root with different cell-specific patterns. PCP 40:1099–1107

Nocquet PA, Opatz T (2016) Total synthesis of (±)-scopolamine: challenges of the tropane ring. Eur J Org Chem 2016:1156–1164

Oksman-Caldentey KM (2007) Tropane and nicotine alkaloid biosynthesis-novel approaches towards biotechnological production of plant-derived pharmaceuticals. Curr Pharm Biotechnol 8:203–210

Palazón J, Moyano E, Bonfill M, Cusidó RM, Piñol MT, da Silva JT (2006) Tropane alkaloids in plants and genetic engineering of their biosynthesis. FOB 2:219

Richter U, Rothe G, Fabian A, Rahfeld B, Dräger B (2005) Overexpression of tropinone reductases alters alkaloid composition in *Atropa belladonna* root cultures. J Exp Bot 56:645–652

Robins RJ, Parr AJ, Payne J, Walton NJ, Rhodes MJ (1990) Factors regulating tropane-alkaloid production in a transformed root culture of a *Datura candida*× *D. aurea* hybrid. Planta 181:414–422

Rothe G, Hachiya A, Yamada Y, Hashimoto T, Dräger B (2003) Alkaloids in plants and root cultures of *Atropa belladonna* overexpressing putrescine N-methyltransferase. J Exp Bot 54:2065–2070

Savita SA, Jain R, Nagpal AK, Pati PK (2021) Plant tropane alkaloids (TAs): chemotaxonomic significance, biogenesis, and biotechnological interventions. In: Srivastava V, Mehrotra S, Mishra S (eds) Tropane alkaloids: pathways, potential and biotechnological applications. Springer, Singapore, pp 3–19

Sharma SK, Agnihotri AK (2021) Hairy root cultures and plant tropane alkaloids: production matrix and product in biotechnological perspective. In: Srivastava V, Mehrotra S, Mishra S (eds) Tropane alkaloids: pathways, potential and biotechnological applications. Springer, Singapore, pp 89–104

Singh A, Nirala NK, Das S, Narula A, Rajam MV, Srivastava PS (2011) Overexpression of odc (ornithine decarboxylase) in *Datura innoxia* enhances the yield of scopolamine. Acta Physiol Plant 33:2453–2459

Srinivasan P, Smolke DC (2021) Engineering cellular metabolite transport for biosynthesis of computationally predicted tropane alkaloid derivatives in yeast. Proc Natl Acad Sci 118:e2104460118

Vanhala L, Hiltunen R, Oksman-Caldentey KM (1995) Virulence of different *Agrobacterium* strains on hairy root formation of *Hyoscyamus muticus*. Plant Cell Rep 14:236–240

Verpoorte R, Memelink J (2002) Engineering secondary metabolite production in plants. Curr Opin Biotechnol 13:181–187

Verpoorte R, van der Heijden R, Memelink J (2000a) Engineering the plant cell factory for secondary metabolite production. Transgenic Res 9:323–343

Verpoorte R, Van der Heijden R, Memelink J (2000b) General strategies. In: Verpoorte R, Alfermann AW (eds) Metabolic engineering of plant secondary metabolism. Springer, Dordrecht, pp 31–50

Wang X, Chen M, Yang C, Liu X, Zhang L, Lan X, Tang K, Liao Z (2011) Enhancing the scopolamine production in transgenic plants of *Atropa belladonna* by overexpressing pmt. and h6h genes. Physiol Plant 143:309–315

Wang L, Ning C, Pan T, Cai K (2022) Role of silica nanoparticles in abiotic and biotic stress tolerance in plants: a review. Int J Mol Sci 23:1947

Wen Y, Liao Y, Tang Y, Zhang H, Zhang J, Liao Z (2023) Metabolic effects of elicitors on the biosynthesis of tropane alkaloids in medicinal plants. Plants 12:3050

Yamada Y, Urui M, Shitan N (2024) Integration of co-culture and transport engineering for enhanced metabolite production. Plant Biotechnol J 41:195–202

Yang C, Chen M, Zeng L, Zhang L, Liu X, Lan X, Tang K, Liao Z (2011) Improvement of tropane alkaloids production in hairy root cultures of *Atropa belladonna* by overexpressing pmt. and h6h genes. Plant Omics 4:29–33

Yu F, Zhao X, Wang Z, Liu L, Yi L, Zhou J, Du G (2021) Recent advances in the physicochemical properties and biotechnological application of Vitreoscilla hemoglobin. Microorganisms 9:1455

Yun DJ, Hashimoto T, Yamada Y (1992) Metabolic engineering of medicinal plants: transgenic *Atropa belladonna* with an improved alkaloid composition. PNAS 89(11799–11):803

Zeng L, Zhang Q, Jiang C, Zheng Y, Zuo Y, Qin J, Liao Z, Deng H (2021) Development of *Atropa belladonna* L. plants with high-yield hyoscyamine and without its derivatives using the CRISPR/Cas9 system. Int J Mol Sci 22:1731

Zhang L, Ding R, Chai Y, Bonfill M, Moyano E, Oksman-Caldentey KM, Tang K (2004) Engineering tropane biosynthetic pathway in *Hyoscyamus niger* hairy root cultures. PNAS 101(17):6786–6791

Zhang L, Kai GY, Lu BB, Zhang HM, Tang KX, Jiang JH, Chen WS (2005) Metabolic engineering of tropane alkaloid biosynthesis in plants. J Integr Plant Biol 47:136–143

Zhang L, Yang B, Lu B, Kai G, Wang Z, Xia Y, Tang K (2007) Tropane alkaloids production in transgenic *Hyoscyamus niger* hairy root cultures over-expressing putrescine N-methyltransferase is methyl jasmonate-dependent. Planta 225:887–896

Zhao K, Zeng JunLan ZJ, Zhao TengFei ZT, Zhang HaoXing ZH, Qiu Fei QF, Yang ChunXian YC, Liao ZhiHua LZ (2017) Enhancing tropane alkaloid production based on the functional identification of tropine-forming reductase in *Scopolia lurida*, a Tibetan medicinal plant. Front Plant Sci 8:1745

Zhao T, Li S, Wang J, Zhou Q, Yang C, Bai F, Lan X, Chen M, Liao Z (2020) Engineering tropane alkaloid production based on metabolic characterization of ornithine decarboxylase in *Atropa belladonna*. ACS Synth Biol 9:437–448

Chapter 11
Cultivation of Tropane Alkaloid Yielding Plants

T. Pullaiah

Abstract Different species of the genera *Duboisia, Atropa, Scopolia, Anisodus,* and *Hyoscyamus* are propagated worldwide for the large-scale production of scopolamine, hyoscyamine, and other related Tropane alkaloids (TAs). In the present chapter, cultivation methods of Tropane alkaloid-yielding plants are given.

Keywords Tropane alkaloids · Propagation · *Atropa* · *Hyoscyamus* · *Datura* · *Duboisia*

11.1 Introduction

The global demand for tropane alkaloids, a class of naturally occurring compounds with significant medicinal properties, has been on a consistent upward trajectory. However, the rampant and unsustainable harvesting of plants known for their tropane alkaloid content has resulted in a notable depletion of these vital wild resources. This depletion poses significant challenges for industries reliant on these compounds, making it increasingly difficult to ensure a reliable supply as production levels soar to keep pace with growing consumer needs.

In response to this pressing issue, the cultivation of tropane alkaloid-yielding plants has emerged as the foremost strategy to rectify the imbalance between supply and demand. This approach not only seeks to meet the industrial need for these important compounds but also plays a crucial role in conserving the vanishing natural plant populations that produce them. Despite the progress made in synthetic production methods, it is noteworthy that most of the tropane alkaloids utilized by pharmaceutical companies are still sourced from the traditional cultivation of these specialized plants. Maintaining this cultivation is essential for ensuring a sustainable and consistent supply of these valuable alkaloids, safeguarding the industry and the natural ecosystems from which they originate (Table 11.1).

T. Pullaiah (✉)
Department of Botany, Sri Krishnadevaraya University, Anantapur, Andhra Pradesh, India

T. Pullaiah et al. (eds.), *Tropane Alkaloids*,
https://doi.org/10.1007/978-981-96-5805-3_11

Table 11.1 Distribution of commercially grown TA-producing plant species and their total alkaloid yields (Mehrotra et al. 2021)

Species	Distribution	Alkaloid yields
Atropa belladonna, A. baetica	Europe, N America, and West Asia	0.34–0.68% DW
Hyoscyamus muticus, H. niger, H. albus	North Africa, West Asia, Europe, India	0.04–0 0.72% DW
Datura metel, D. innoxia, D. leichhardtii, D. stramonium	America, SE Asia, Australia	1.2–1.8% DW
Duboisia myoporoides, D. leichhardtii hybrid	Australia, Japan	Up to 7%
Scopolia japonica	Europe and Asia	
Erythroxylum coca	Brazil, Bolivia, and Peru	0.3–1.5%

Different species of the genera *Atropa, Datura, Duboisia, Hyoscyamus,* and *Scopolia* are propagated round the world for production of scopolamine, hyoscyamine, and other related TAs on a large-scale. The Australian hybrid of *Duboisia myoporoides* and *D. leichhardtii* is cultivated in Southwest Queensland, where it grows naturally. In recent years, *D. myoporoides* has not only seen limited cultivation, but it has also emerged as an undesirable source of alkaloids due to its pyridine base content, which significantly hinders the efficiency of scopolamine isolation. Consequently, a hybrid of *D. leichhardtii* and *D. myoporoides* is now being actively cultivated. This hybrid's leaves boast an impressive 1.54% scopolamine and 0.1% hyoscyamine (Griffin and Lin 2000). Moreover, *Duboisia* interspecific hybrids have consistently shown superior alkaloid yields, especially when *D. myoporoides* serves as the female parent (Ikenga et al. 1979).

The alkaloid content of plants depends on several factors, including the plant's age, humidity, temperature, sunlight intensity, shade duration, and seasonal changes (Ikenaga et al. 1985). The presence of weeds and insects also affects the productivity of these plants. Environmental conditions, soil quality, and the use of fertilizers play a significant role in how plants develop and produce alkaloids. For example, *Datura* plants that receive fertilizer show higher levels of hyoscyamine and scopolamine. Certain *Hyoscyamus* species demonstrate a clear connection between adding nitrogen (N), phosphorus (P), and potassium (K) to the soil and their levels of hyoscyamine and scopolamine. However, *Duboisia* plants grown in the field do not show a relationship between alkaloid content and fertilizer use (Al-Humaid 2004; Alaghemand et al. 2013).

More effort is necessary to improve the large-scale production of plant-derived tertiary metabolites (TAs) through traditional propagation methods. The first steps should involve selecting and screening high-yielding clones and conserving their germplasm. Controlled trials are needed to grow elite TA-producing plants in consistent conditions. This research will help us understand how different environmental factors affect plant biomass and alkaloid yield. The knowledge gained can help optimize productivity in field-grown plants.

11.2 *Atropa belladonna*

Propagation: Belladonna thrives in temperate climates with well-defined seasons. It prefers cool temperatures and requires a winter chilling period for optimal growth. It grows well in slightly acidic, deep, fertile soils of medium texture, which are rich in humus. The soil should be well-draining and slightly acidic, with a pH ranging from 5.5 to 6.5. A loamy soil enriched with organic matter is ideal for belladonna cultivation.

Cultivating Belladonna: A Constructive Approach
When considering cultivating Belladonna, it is essential to select an appropriate soil type. Avoid heavy clay soils that are prone to waterlogging, as they can hinder plant growth. In subtropical regions, Belladonna is best suited for winter cropping, but keep in mind that it behaves like an annual, often dying off in the summer and resulting in lower yields.

Land Preparation
To prepare the land effectively, start by enhancing the soil's structure. Aim for a fine tilth and incorporate organic matter at a rate of 25 tons per hectare. This foundational step will enrich the soil, creating a conducive environment for Belladonna growth. You can propagate Belladonna using seeds, cuttings, or rootstocks; however, establishing a nursery generally yields better results than direct sowing.

Nursery Cultivation
Establish your nursery between the second week of May and autumn (September to October), or during summer months with adequate shading. Prior to setting up nursery beds, thoroughly plough the land to achieve the desired tilth. Construct raised beds measuring 3 × 1 m, ensuring they are surrounded by efficient drainage and irrigation channels. Incorporate well-decomposed farmyard manure (FYM) or sheep manure into the top 10 cm of the soil to enhance fertility.

Seeding Requirements
For optimal seedling production, approximately 4 kg of seeds will suffice for 1 ha, while about 20 kg is necessary for broadcast sowing. Following sowing, water the beds immediately with a watering can and maintain daily watering to support germination. Expect seeds to germinate within 3 weeks, and the seedlings will be ready for transplanting when they reach a height of 15–20 cm after 8–12 weeks.

Transplanting Techniques
Timing is crucial for successful transplanting; aim for early spring (March–April) or autumn (October–November) for best results. To protect seedlings from disease, treat them with a fungicide, such as Agallol, prior to planting. Maintain a spacing of 45–60 cm between seedlings to facilitate growth. Using raised beds with 1-m-wide strips or ridges will help prevent waterlogging and ease irrigation.

Optimal Growing Conditions

Belladonna thrives in partial shade to full sunlight. In hotter climates, providing afternoon shade can significantly reduce stress on the plants. Immediately after transplanting, ensure proper irrigation. A balanced fertilizer containing nitrogen, phosphorus, and potassium will promote healthy growth. Apply a basal dose of 25–30 kg of nitrogen, 50–60 kg of phosphorus, and 40–50 kg of potassium per hectare before planting.

Fertilization Strategy

After harvest, consider applying an additional 60–80 kg of nitrogen divided into 3–4 split doses as top dressing at monthly intervals. Implementing heavier nitrogen doses will support robust leaf development and enhance the alkaloid content. From the first year onward, repeat the basal applications of nitrogen and phosphorus each spring, followed by top-dressing with nitrogen.

Watering and Maintenance

Belladonna has a high water requirement, necessitating irrigation every 10–15 days during dry spells, with 6–7 irrigations recommended in dry months. To avoid waterlogging, monitor soil moisture closely. Incorporating a layer of mulch around the plants will help retain moisture, suppress weeds, and regulate soil temperature effectively.

Weed management is vital for plant health. Engage in hand weeding when plants are young, aiming for a total of 3–4 weeding and hoeing sessions throughout the growing season. Leaves are typically richest in alkaloids at the end of June or in July, and providing ample sunlight often results in more potent leaves than those grown in shade.

Harvesting and Post-harvest Care

Plants approximately 3 years old will yield a good supply of leaves. If roots are part of your harvest plan, it's recommended to replant every third year. You can collect two or more crops of leaves annually. To maintain quality, dry the harvested leaves immediately and store them properly.

By following these constructive guidelines, you can successfully cultivate Belladonna while ensuring optimal growth and yield.

Three months after planting, the first leaves can be harvested as soon as the plants begin to flower, when alkaloid content is highest. Use pruning scissors to cut the leaves, and dry them immediately in the shade, sunlight, or with artificial heat while ensuring good air circulation by turning them frequently. Belladonna grown in sunny areas yields more leaves, and applying dehydration stress can moderately increase alkaloid concentration. After 3 years, the roots can be harvested. Wash the roots, cut them into 10 cm lengths, and split them if thick before drying. In the first year, average yields are 1000 kg of dry herb per hectare, increasing to 1500 kg in the second and third years. Dry root yields range from 170 to 335 kg per hectare.

Baricevic et al. (1999) studied the effects of nitrogen (N) fertilization and water stress on hyoscyamine and scopolamine content in the roots of *Atropa belladonna*. Using seeds from a Slovene population, plants were grown under various water regimes (35% to 95% soil water depletion) and nitrogen supplies (0.37 to 1.60 g per pot) in a greenhouse.

The optimal yield of tropane alkaloids, with 54 mg of hyoscyamine and 7 mg of scopolamine per plant, was achieved with 35% soil water depletion and 0.37 g N per pot. Conversely, the highest alkaloid content was recorded at 95% water depletion with 1.60 g N per pot. Increasing nitrogen is generally associated with higher alkaloid production. For example, a tenfold increase in nitrogen for *A. acuminata* raised the levels of hyoscyamine and scopolamine in the whole plant and stems, while leaf and root contents remained unchanged.

Genotype and developmental stage also significantly affect alkaloid levels in *A. belladonna*. The lowest alkaloid percentage (0.38% dry weight) was found at developmental stage 3 (50% flowering), while the highest (0.67% dry weight) occurred at stage 9 (130 days after the first flower).

11.3 *Duboisia*

One of the little known trees that is now a plant that is a commercially viable crop are *Duboisia* species. These small native trees reach only a few meters in height, attractive green foliage and small white star-shaped flowers. It is *Duboisia hopwoodii* that is the most sought after, and this is a shrubby plant found in Western Australia, Southern Australia, and other arid areas in Queensland and New South Wales. They have been traditionally used for medicinal purposes where the *Duboisa hopwoodii* leaves are harvested also known as Pituri. The plant does have attractive flowers, trumpet-shaped, white and veined. Foliage is a mid to deep green, narrow and lance-shaped. One of the interesting features is that this plant varies greatly in chemical composition from one population to another. It is only selected plants that were used for medicinal purposes, some were regarded as poisons. Today the trees are grown in plantations for the leaves. The leaves are harvested and sent overseas for processing in commercially produced pharmaceutical products.

Duboisia myoporoides is commercially grown for two potent anticholinergic and antispasmodic alkaloids, scopolamine and hyoscyamine. Global demand for scopolamine is expected to be ten times higher than that for hyoscyamine and atropine (Palazón et al. 2008). Production is significantly lower than demand, causing prices to reach AUD 15,000 per kilogram, with no alternatives available. (Wang et al. 2011). Scopolamine synthesis is possible via yeast (Srinivasan and Smolke 2020); however, the process is costly and complicated. *Duboisa, Datura, Hyoscyamus, Atropa*, and *Scopolia* are the main sources of scopolamine (Cardillo et al. 2010). *Duboisia* is the most productive, yielding 6% of total alkaloids by leaf dry weight, while *Atropa* and *Datura* only produce 0.2% to 0.8% (Palazón et al. 2008). This fast-growing plant reaches harvest in 10 months and is resistant to pests and diseases (Loftus-Hills and Kelenyi 1946; Singh et al. 1985), making it ideal for commercial cultivation. Scopolamine production relies heavily on *Duboisia* propagation and efficient leaf drying and alkaloid extraction processes.

Due to the increasing demand for scopolamine, in the 1950s, large-scale cultivation of *Duboisia* species for leaf production was initiated in Queensland, Australia (Foley 2006). Today, Australia is the world leader in scopolamine production, owning 70% of the world supply (Foley 2006, Ullrich et al. 2017). Since 1992, Australia has produced 994,678 kg of *Duboisia* leaf decennially, with a gross value of AUD 6,604,439, exported internationally for alkaloid extraction (Foley 2006). Due to the economic value of this species, other regions such as India (Williams 2013), S America (Williams 2013), Brazil, and Ecuador (Palazón et al. 2008, Machado et al. 2014) have also initiated its cultivation.

Duboisia species are commercially propagated on a large scale, as they are the main source of scopolamine. Hyoscyamine and scopolamine are produced from *Duboisia* plants growing on massive plantations in Queensland, Australia (Figs. 11.1, 11.2, 11.3 and 11.4) (Grynkiewicz and Gadzikowska 2008).

Climate change and the resulting new biotic and abiotic factors make it harder for the pharmaceutical industry to produce massive amounts of scopolamine continuously.

Duboisia has emerged as a highly significant crop in the agricultural sector, primarily due to its production of two valuable alkaloids: scopolamine and hyoscyamine. With the global demand for scopolamine expected to surpass that for hyoscyamine and its derivative, atropine, by a factor of ten, there is substantial market potential for this crop. However, scopolamine production currently falls short of meeting this rising demand, resulting in impressive market prices that can reach

Fig. 11.1 Australian *Duboisia* planting material. (Source: https://alkaloids.org/duboisia-plantations/)

Fig. 11.2 Aerial view of cultivated *Duboisia* in Queensland Australia. (Source: www.alamy.com)

Fig. 11.3 *Duboisia* plantation in Queensland, Australia (ABC Southern Queensland: Jon Daly) https://www.abc.net.au/news/2021-11-13/sleepy-town-secretive-drug-centre-big-pharma-criminal-cartel/100572628)

Fig. 11.4 *Duboisia* plantation in Queensland, Australia (ABC Southern Queensland: Jon Daly) https://www.abc.net.au/news/2021-11-13/sleepy-town-secretive-drug-centre-big-pharma-criminal-cartel/100572628)

AUD 15,000 per kilogram, particularly given the absence of viable substitutes. This presents a unique opportunity for *Duboisia* growers, as the crop stands out with a remarkable 6% yield of total alkaloids by leaf dry weight—significantly higher than the 0.2–0.8% produced by related species such as *Atropa* and *Datura*. By enhancing production methods and exploring sustainable cultivation practices, the *Duboisia* industry could boost its output to meet global demand and solidify its position as a leader in alkaloid production. (Palazón et al. 2008).

Given the scattered distribution in Australia of the two *Duboisia* spp., *D. myoporoides* and *D. leichhardtii*, the leaves of which are a source of alkaloids, cultivation of these drug plants may prove advantageous. Three years of trials were carried out in the subtropical climate of Nambour, southeast Queensland, and in the continental climate of Canberra. The results show that satisfactory yields are obtained at both localities, conditions at Nambour possibly favoring the formation of hyoscine and at Canberra that of hyoscyamine. At Canberra, at a spacing of 4 ft. by 2 ft., 1-year-old *D. leichhardtii* yielded 3, 839 lb. of dry leaf per acre and *D. myoporoides* 3, 540 lb. The largest leaf area was obtained by a system of cutting the trees down to waist height at intervals of 3 months in the sub-tropics and 1 year in a continental climate. Progeny tests, though in progress only for a short time, indicate that selection will be profitable of strains containing the alkaloids in a high concentration and in a relatively pure form irrespective of environmental conditions.

11.3.1 Duboisia myoporoides

Industrial production of scopolamine is based extensively on field cultivation of hybrids of *Duboisia myoporoides* and *Duboisia leichhardtii* mainly in Australia. *Duboisia myoporoides* is a small to medium-sized tree and is known as Corkwood. It can reach 5 m in good condition. It is widespread on the east coast from Queensland to Southern New south Wales. Shukla (1996) provides valuable information on cultivating *Duboisia myoporoides* in India. This species thrives in cool, dry climates with an ideal annual rainfall of around 76 cm and can grow in diverse soil types, including red clay and dark brown loams, with a pH between 4.5 and 5.5. Good soil drainage is essential to avoid issues with water-logging.

For successful cultivation, selecting warm sites with proper air drainage and protection from strong winds is important. Propagation can be done through root cuttings or seeds, with treatments including GA_3 at 250 ppm for 24 h at approximately 40 °C. After thorough washing, seeds are planted in polythene bags after 6 weeks. Germination typically takes 3 to 4 weeks, leading to field planting 10 to 12 weeks later when seedlings reach 15 to 22 cm. Planting should be done at spacings of 2 × 1 m, 2 × 2 m, or 3 × 3 m, with pre-plant applications of farmyard manure and N, P, and K fertilizers. Weed control during the first year is crucial, and strategies such as discing can be effective. The use of trifluralin and parquat supports weed management while avoiding deep ploughing and spraying young tree trunks until bark formation is advisable. Typically, three harvests can be obtained within 2 years, starting 7–10 months after planting and with further intervals of 7–8 months. For better regrowth, it is recommended to leave 20–30% of the leaves on the trees during harvest. Yields can vary, with average leaf production around 450 kg per hectare for 740–980 trees planted per hectare.

11.3.2 Duboisia *Hybrids*

The leaves of Australian *Duboisia* hybrids stand out as the premier source of hyoscine (scopolamine), distinctly native to the Kingaroy district of Australia. Australian Alkaloids effectively harness the potent qualities of these specially developed hybrids, which have been commercially cultivated for decades due to their exceptional content of tropane alkaloids, with about half comprising scopolamine (Foley 2006).

Environmental factors, such as daily sunlight exposure and soil composition, can obscure genetic characteristics. Furthermore, it is well-documented that alkaloid content experiences a decline during autumn and winter, primarily due to cold weather (Luanratana and Griffin 1980). This variability presents significant challenges in achieving consistent and reliable results from field trials. To address these challenges, Australian Alkaloids employs strictly controlled cultivation in climate chambers. This method effectively minimizes seasonal influences and reduces interactions with other organisms. Additionally, understanding whether alkaloid production is age-dependent is critical for optimizing the timing of harvesting or sampling during the breeding process.

***Duboisia hopwoodii* × *myoporoides*.**
This fast-growing hybrid requires well-drained soil and full sun to slight shade. It is drought-tolerant and adaptable to the wet and humid conditions of the Australian Sunshine Coast.

11.4 *Datura*

Datura can be effectively cultivated in a variety of soil types, but it shows optimal growth in rich clay-loam soils. For the best results, focus on alkaline or neutral clay loams that have saline-alkaline reactions and are rich in organic matter. This combination creates the perfect environment for robust growth. *Datura* thrives in regions with low rainfall, particularly those where winter and monsoon rains are followed by extended dry periods. Aim for areas with an annual rainfall of less than 100 cm and mean winter temperatures between 10 to 15 °C, along with warmer temperatures of 27 to 28 °C during May and June. Since *Datura* is sensitive to frost, it is crucial to select sheltered areas with plenty of sunlight for successful cultivation. Propagation can be achieved using seeds or vegetative cuttings. To boost seed germination, it is beneficial to soak the seeds overnight in water and rinse them a few times with fresh water before planting. Another effective method is to alternate exposure to freezing and thawing, which helps weaken the seed coat and accelerates germination. When sowing seeds, maintain a spacing of 1 m between rows, or broadcast them over prepared beds and cover with sand and farmyard manure. Keeping the beds moist through spraying ensures that seeds germinate within 15–30 days. Be mindful that germination may be low and inconsistent due to the presence of inhibitors. For vegetative propagation, take cuttings from plants during the pre-flowering phase and allow them to root. Treating the cuttings with auxins (at IBA 5000 ppm) can significantly improve the rooting percentage. Once the seedlings reach a height of 8–12 cm and have developed four leaves, they can be transplanted. Space them 30–45 cm apart in rows that are 45–60 cm apart for optimal growth. Regular irrigation is essential for healthy plants. If you are broadcasting seeds, ensure irrigation is provided if there hasn't been rainfall within a week. Follow up with additional watering every 8–10 days, adjusting according to weather conditions, until the final harvest. This thoughtful approach will help ensure a successful *Datura* cultivation experience.

11.5 *Hyoscyamus niger*

It is an endangered plant of the cold desert region of trans-Himalaya. *H. niger* is propagated only through seeds which exhibit dormancy. Seed pre-treatments with $NaNO_3$, KNO_3, NaN_3, sodium nitroprusside, cold stratification and gibberellic acid proved effective in initiating germination and caused 34, 42, 62, 68, 88, and 96% germination, respectively (Sharma 2024).

The plant can be propagated from seeds. The seeds are first raised in the nursery during October. Five kilograms of seeds is sufficient for 1 ha. Seeds germinate within a week after sowing in nursery beds and can be transplanted after about a month. When the seedlings attain 4–5 leaves, seedlings are planted with a spacing of 30–45 cm. The crop requires 3–4 irrigations. When plants reach the flowering stage, they are cut at the ground level. The entire herbage is spread in the shade and dried. From 1 ha, the yield of leaves is about 15 quintals, while stalks, pedicels, and leaves are about 30q/ha.

11.6 *Anisodus*

Anisodus tanguticus thrives in mountainous terrain and prefers moderate moisture, benefiting from consistent, weekly watering while being drought-tolerant. Its water needs increase during the growing season to support growth. Propagation can be done through seeds and cuttings. For seeds, stratify them at 4 °C for 4–6 weeks, then sow in well-draining soil. Keep the seed bed moist at 20–25 °C until germination. For cuttings, select healthy stems, apply rooting hormone, and plant in a moist, sterile mix, maintaining high humidity until roots are established. This species prefers well-draining, organic-rich soil and a location with full sun or partial shade. Prepare the area by weeding and turning the soil, adding compost for extra nutrients. When planting, dig a hole twice the width of the root ball and water generously afterwards. (https://www.picturethisai.com/care/Anisodus.html).

Anisodus is relatively resilient to temperature fluctuations, tolerating cold down to −15 °C (5 °F) and heat up to 30 °C (86 °F). This adaptability impacts its care requirements significantly: during cold extremes, it may need mulching or insulation to prevent root damage, and in high heat, ensuring adequate hydration is vital. Symptoms of stress from cold include leaf wilting and discoloration, while overheating can cause leaf scorch and drooping. Countermeasures include moving the plant to a sheltered location during frosts, providing shade, and increased watering during heatwaves. Overall, *Anisodus*'s moderate temperature tolerance makes it suitable for various environments, with some seasonal adjustments. Effects of different planting densities and harvesting periods on the growth and major alkaloids of *Anisodus tanguticus* on the Qinghai–Tibetan Plateau were studied by Liu et al. (2022). They reported that the mass per plant was higher at low density, while the yield per unit area of the underground parts was greater at high density, and the mass of the aboveground parts was higher at low density.

The implementation of suitable cultivation methods has the potential to enhance both the quality and yield of *A. tanguticus*. In a commendable study conducted by Liu et al. (2024), a comprehensive two-year field experiment was carried out from 2021 to 2023, utilizing a single-factor randomized complete block design with three replicates. This research thoughtfully investigated the influence of various nutrient (NPK) levels on the growth, primary alkaloid content, and alkaloid yield of *A. tanguticus* at different growth stages, focusing on both roots and aboveground portions. The findings suggest that nutrient levels play a significant role in influencing the growth and alkaloid accumulation in *A. tanguticus*. Notably, higher nitrogen levels (375 kg/ha) were associated with increases in both root and aboveground biomass, while phosphorus appeared to have a lesser impact, particularly on aboveground biomass. In terms of alkaloid content—including scopolamine, anisodamine, anisodine, and atropine—a moderate nitrogen level (225 kg/ha) demonstrated the most effective results, followed by lower potassium (75 kg/ha), with phosphorus showing limited effects. Interestingly, increased levels of phosphorus were linked to a decrease in scopolamine content. During the T-Growing period, the introduction of moderate nitrogen (225 kg/ha) resulted in the highest alkaloid levels per unit area, amounting to 205.79 kg/ha. In contrast, during the T-Wilting period, low potassium (75 kg/ha) and low phosphorus (750 kg/ha) contributed to alkaloid levels of 146.91 kg/ha and 142.18 kg/ha, respectively. These insights indicate that nitrogen exerts the most pronounced effect on alkaloid accumulation, with potassium and phosphorus following in importance. In summary, the results point to the T-Wilting period as the optimal harvesting time for *A. tanguticus*, with recommended nitrogen addition at 225 kg/ha, potassium at 75 kg/ha, and phosphorus at no more than 600 kg/ha. This research provides valuable guidance for future cultivation practices.

11.7 *Brugmansia*

Brugmansia species are known for their valuable alkaloid content, particularly scopolamine and hyoscyamine, which play a crucial role in clinical applications. As a result, these plants have been cultivated extensively in South America for the production of these important alkaloids. *Brugmansia*, characterized by its striking large, pendulous flowers resembling trumpets, can thrive as large shrubs or small trees. Notably, *B. sanguinea* has gained attention due to its diverse range of tropane alkaloids and is commercially cultivated in Ecuador, achieving yields of approximately 0.8% scopolamine. These plantations typically have an economically productive lifespan of about 10 years. Furthermore, *Brugmansia* species have been successfully introduced to Europe since the mid-1990s, where they can also yield around 0.8% scopolamine from dried leaves (Levy 1977).

For optimal growth, it is advisable to plant *Brugmansia* in mid-spring, ensuring that nighttime temperatures are consistently above 10 °C. This fast-growing plant can reach heights of 60–90 cm each year and blooms beautifully every summer. It's important to be aware that all parts of the *Brugmansia* plant are toxic to humans and pets, so care should be taken when cultivating it. This perennial plant is relatively low-maintenance and flourishes in moist, well-draining soil with full sun exposure, requiring at least 6 h of direct sunlight daily. To maintain healthy growth, ensure that the plant is watered whenever the top layer of soil feels dry or when it starts to wilt. Optimal growth conditions include well-draining soil, which is essential for successful *Brugmansia* propagation.

11.8 *Schizanthus*

Schizanthus is a genus of 15 species, all native to Chile and Argentina. It is cultivated in the USA and Europe. *Schizanthus* plants are bushy, half-hardy annuals that range from 30 cm to 1.2 m (1–4 ft.) in height. *Schizanthus* species are cultivated in the horticulture trade and widely available as ornamental plants for gardens. Propagation of *S. hookeri*, and *S. porrigens* through seeds is challenging due to low and uneven germination percentages. Moreno et al. (2024) investigated the morpho-anatomical characteristics, dormancy, and germination requirements of the seeds of these species. Germination tests showed that the seeds of these species exhibit physiological dormancy. Imbibition in GA_3 (200 ppm) proved an effective treatment to promote germination. Cold stratification and after-ripening of seeds also improved germination in *S. hookeri* seeds. The optimal temperatures for seed germination were 19 °C for *S. hookeri*, and 23 °C for *S. porrigens*.

Schizanthus seeds take about 2 or 3 weeks to germinate in the dark at a temperature of 15–23 °C. *Schizanthus* seedlings are transplanted outdoors in the middle of spring, a couple of weeks after the last chance of a frost. *Schizanthus* thrives in cool, moist environments with well-draining soil. Partial shade helps to promote longevity of their delicate blooms during warmer months. Regular watering maintains a lush appearance. Deadhead spent flowers to encourage a robust second bloom. Provide consistent moisture and avoiding waterlogged roots are key to a vibrant display. Be watchful for slugs.

Schizanthus species are grown as garden ornamentals but not for tropane alkaloid extraction. Efforts should be made to cultivate and extract tropane alkaloids from *Schizanthus* species on a large scale.

11.9 *Erythroxylum*

Restrepo et al. (2019) provided an overview of the genus *Erythroxylum*, focusing on two significant species: *Erythroxylum coca* and *Erythroxylum novogranatense*. These species have been cultivated for thousands of years by South American societies and are now sources for cocaine production, often diverted into the global illicit trade. *E. coca* can be divided into two varieties: *E. coca* var. *coca*, or Huánuco/Bolivian coca, primarily found in the eastern Andes of Peru and Bolivia, and *E. coca* var. *ipadu*, or Ipadú coca, grown in the Amazon region. *E. novogranatense* includes Colombian coca (var. *novogranatense*), found in Colombia's inter-Andean valleys, and Trujillo coca (var. *truxillense*), which thrives in Peru's coastal deserts. Coca is one of the oldest cultivated medicinal plants, with evidence of use dating back at least 8000 years (Plowman 1984, 1986). Traditional consumption remains strong, with Bolivia's coca-using population estimated at three million in 2012 and Peru's at 3.5 million in 2013 (Restrepo et al. 2019). Significant populations are also present in Argentina, Brazil, Chile, and Colombia. *Erythroxylum coca* leaves are notable for containing significant amounts of cocaine and other tropanes (Johnson 1996). The alkaloid content in these leaves varies depending on the specific variety and the environmental conditions they are exposed to. For example, *E. coca* var. *coca* typically has an average cocaine content of 0.63% by dry weight (Plowman and Rivier 1983), while Amazonian coca shows a range of 0.11–0.41%.

Importantly, research indicates that there are no significant differences in alkaloid production between plants grown in laboratories and those cultivated in natural field conditions. This observation suggests that the processes involved in biosynthesis are likely regulated at the genetic level. Among common coca cultivars, *E. novogranatense* varieties stand out, with *E. novogranatense* var. *novogranatense* averaging 0.77% cocaine, and *E. novogranatense* var. *truxillense* averaging 0.72%.

Additionally, studies by Acock et al. (1996) have highlighted the importance of temperature on alkaloid production. While light intensity appears to have a minimal effect, lower temperatures can negatively influence the overall alkaloid content. Understanding these factors could help optimize cultivation practices and enhance the alkaloid yield from coca plants.

References

Acock MC, Lydon J, Johnson E, Collins R (1996) Effects of temperature and light levels on leaf yield and cocaine content in two *Erythroxylum* species. Ann Bot 78:49–53

Alaghemand A, Ghorbanpour M, Asli DE, Moghaddasian B (2013) Influence of urea fertilization on tropane alkaloids content of Henbane (*Hyoscyamus niger* L.) under hydroponic culture conditions. Adv Environ Biol 7(2):301–307

Al-Humaid AI (2004) Effects of compound fertilization on growth and alkaloids of *Datura* plants. J Plant Nutr 27(12):2203–2219

Baricevic D, Umek A, Kreft S, Maticic B, Zupancic A (1999) Effect of water stress and nitrogen fertilization on the content of hyoscyamine and scopolamine in the roots of deadly nightshade (*Atropa belladonna*). Environ Exp Bot 42:17–24

Cardillo AB, Otálvaro AÁM, Busto VD, Talou JR, Velásquez LME, Giulietti AM (2010) Scopolamine, anisodamine and hyoscyamine production by *Brugmansia candida* hairy root cultures in bioreactors. Process Biochem 45:1577–1581

Foley P (2006) *Duboisia myoporoides*: the medical career of a native Australian plant. Hist Rec Aust Sci 17:31–69

Griffin WJ, Lin GD (2000) Chemotaxonomy and geographical distribution of tropane alkaloids. Phytochemistry 53:623–637

Grynkiewicz G, Gadzikowska M (2008) Tropane alkaloids as medicinally useful natural products and their synthetic derivatives as new drugs. Pharmacol Rep 60:439

https://www.agrownet.com/contents/en-us/p623_Belladonna_Cultivation_agrownet_Belladonna_crop_management.html

https://www.picturethisai.com/care/Anisodus.html

Ikenaga T, Hama K, Takemoto S, Ohashi H (1985) Seasonal variations of tropane alkaloids content in *Duboisia*. Jpn J Trop Agric 29(4):229–230

Ikenga T, Abe M, Itaka A, Ohashi H (1979) Planta Med 35:51

Johnson EL (1996) Alkaloid content in *Erythroxylum coca* tissue during reproductive development. Phytochemistry 42:35–38

Levy R (1977) Jimson seed poisoning – a new hallucinogen on the horizon. J Am Coll Emerg Physicians 6(2):58–61

Liu N, Chen C, Wang B, Chen KY, Feng SH, Zhang DS, Zhou GY (2022) Effects of different planting densities and harvesting periods on the growth and major alkaloids of *Anisodus tanguticus* (Maxim.) Pascher on the Qinghai-Tibetan Plateau. Agriculture-Basel 12(11):1881

Liu N, Chen C, Wang B, Zang D, Zhou G (2024) Exogenous regulation of macronutrients promotes the accumulation of alkaloid yield in *Anisodus tanguticus* (Maxim.) Pascher. BMC Plant Biol 24:602. https://doi.org/10.1186/s12870-024-05299-8

Loftus-Hills K, Kelenyi GP (1946) A preliminary report upon the cultivation of *Duboisia* spp. CABI digital Library.org, pp 359–375

Luanratana O, Griffin WJ (1980) Cultivation of a *Duboisia* hybrid. Part A. Nutritional requirements and effects of growth regulators on alkaloid content. J Nat Prod 43(5):546–551

Machado ACZ, Dorigo OF, Boss A, Tironi P (2014) First report of *Meloidogyne javanica* parasitizing *Duboisia* sp. in Paraná State, Brazil. Plant Dis 98:1745

Mehrotra S, Mishra S, Srivastava V (2021) Plant Tropane alkaloids: Commercial stature and production developments. In Srivastava V, Mehrotra S, Mishra S (eds.) Tropane alkaloids, Springer

Moreno J, Gómez M, Contreras S (2024) Seed morpho-anatomy, dormancy and germination requirements in three *Schizanthus* species (Solanaceae) with ornamental potential. Horticulturae 10(8):867. https://doi.org/10.3390/horticulturae10080867

Palazón J, Navarro-Ocaña A, Hernandez-Vazquez L, Mirjalili MH (2008) Application of metabolic engineering to the production of scopolamine. Molecules 13:1722–1742

Plowman T (1984) The ethnobotany of coca (*Erythroxylum* spp., Erythroxylaceae). Repos Inst CEDRO

Plowman T (1986) Coca chewing and the botanical origins of coca (*Erythroxylum* spp.). In: Stark J, Pacini D, Franquemont C (eds) Proceedings of the coca and cocaine: effects on people and policy in Latin America, vol 30. Cultural Survival, Inc. and LASP (Cornell University), Cambridge, p 235

Plowman T, Rivier L (1983) Cocaine and cinnamoylcocaine content of *Erythroxylum* species. Ann Bot 51:641–659

Restrepo DA, Saenz E, Jara-Muñoz OA, Calixto-Botía IF, Rodríguez-Suárez S, Zuleta P, Chavez BG, Sanchez JA, D'Auria JC (2019) *Erythroxylum* in focus: an interdisciplinary review of an overlooked genus. Molecules 24(20):3788. https://doi.org/10.3390/molecules24203788

Sharma RK (2024) Harvesting seeds before dormancy induction in *Hyoscyamus niger*, a threatened medicinal herb of trans Himalaya: implication for propagation without seed pre-treatments. 1: article no. 13

Shukla YN (1996) Cultivation of *Duboisia* spp. in India. In: Handa SS, Kaul MK (eds) Supplement to cultivation and utilization of medicinal plants. RRL, Jammu, pp 207–212

Singh A, Singh DV, Rao MR, Shukla YN, Hussain A (1985) Cultivation of *Duboisia myoporoides* R. Brown as source of tropane alkaloids in India. Indian J Pharmaceut Sci 47(3):120–121

Srinivasan P, Smolke CD (2020) Biosynthesis of medicinal tropane alkaloids in yeast. Nature 585:614–619

Ullrich SF, Hagels H, Kayser O (2017) Scopolamine: a journey from the field to clinics. Phytochem Rev 16:333–353

Wang X, Chen M, Yang C, Liu X, Zhang L, Lan X, Tang K, Liao Z (2011) Enhancing the scopolamine production in transgenic plants of *Atropa belladonna* by overexpressing Pmt and H6h genes. Physiol Plant 143:309–315

Williams C (2013) Medicinal plants in Australia volume 4: an antipodean apothecary. Rosenberg Publishing, Kenthurst

Chapter 12
Micropropagation of Tropane Alkaloid Yielding Plants

T. Pullaiah and Mallappa Kumara Swamy

Abstract Tropane alkaloids, recognized for their significant therapeutic applications, are primarily derived from plant species such as *Datura, Duboisia, Atropa, Hyoscyamus, Schizanthus*, and *Anisodus*. These alkaloids are extensively used in the pharmaceutical and herbal industries for their anticholinergic and other medicinal properties. However, the large-scale cultivation of these plants to meet current demand is hindered by substantial variability in alkaloid content across harvested materials, posing challenges to consistent production. The growing global demand for tropane alkaloids and the difficulties of achieving uniform yields have underscored the need for alternative approaches. Biotechnological advancements, particularly micropropagation, offer promising solutions for the sustainable production of these valuable compounds. This chapter provides a detailed exploration of micropropagation techniques tailored for major tropane alkaloid-yielding plants. It elaborates on the selection and preparation of explants, highlighting sterilization protocols to ensure contamination-free cultures. Comprehensive information on the composition of culture media, the role of plant growth regulators, and optimized concentrations for effective shoot and root induction is presented. The chapter also discusses the combination of plant growth regulators and their impact on regeneration efficiency. Additionally, procedures for acclimatizing in vitro-propagated plantlets to ex vitro conditions are described. By addressing these critical aspects, this chapter emphasizes the potential of micropropagation to enhance alkaloid production and contribute to the conservation and large-scale cultivation of these medicinally important plant species.

Keywords Micropropagation · Tropane alkaloids · Hyoscyamine · Scopolomine · *Duboisia* · *Datura* · *Atropa. Hyoscyamus* · *Anisodus*

T. Pullaiah (✉)
Department of Botany, Sri Krishnadevaraya University, Anantapur, Andhra Pradesh, India

M. K. Swamy
Department of Biotechnology, East-West First Grade College, Bangalore, Karnataka, India

T. Pullaiah et al. (eds.), *Tropane Alkaloids*,
https://doi.org/10.1007/978-981-96-5805-3_12

Abbreviations

BA	N6-Benzyladenine
BAP	N6-Benzyl aminopurine
GA_3	Gibberellic acid
IAA	Indole 3-acetic acid
IBA	Indole 3-butyric acid
Kn	Kinetin
MS	Murashige and Skoog
NAA	Naphthalene acetic acid
TDZ	Thiadiazuron
WPM	Woody Plant *Medium*
2,4-D	2,4-Dichlorophenoxy acetic acid

12.1 Introduction

Tropane alkaloids represent an important class of compounds characterized by their unique structure, including both a pyrrole ring and a piperidine ring, forming the foundational framework of tropane. These alkaloids play a significant role in pharmacology due to their anticholinergic properties; they effectively inhibit the action of the neurotransmitter acetylcholine in both the central and peripheral nervous systems by binding to muscarinic and/or nicotinic receptors (Lockery et al. 2021). These compounds are predominantly sourced from plants within various families, including Convolvulaceae, Brassicaceae, and Solanaceae, with the highest concentrations in the Solanaceae family. Notable tropane alkaloids, such as anisodine, anisodamine, scopolamine, and atropine, have established therapeutic uses. They are employed in treating a variety of conditions, including muscle spasms, visual impairments, gastrointestinal and renal colic, as well as in anesthesia, shock management, and antidotes for poisoning by organophosphorus compounds (Fan et al. 2016; Xu and Zhao 2017; Liu et al. 2019; Yu et al. 2021).

The increasing recognition of tropane alkaloids' therapeutic potential is reflected in the growing market for associated medications. It was projected that by 2020, the market for antispasmodic drugs in China would surpass 3 billion yuan (Zhou et al. 2023). However, it is important to note that there are currently no commercially feasible methods for chemically synthesizing these compounds. Instead, they are primarily extracted from medicinal plants, highlighting an area for future research and development that could lead to more sustainable and efficient production methods.

Tropane alkaloids are primarily sourced from *Datura, Duboisia, Atropa, Hyoscyamus*, *Schizanthus,* and *Anisodus*. While large-scale cultivation generally meets current demand, the herbal industry struggles with inconsistent alkaloid content in these plants. This variability, along with the rising demand for these therapeutic agents, has led to the exploration of alternative sources and biotechnological methods for sustainable production. Propagation of the best commercial varieties of these herbs, in addition to their preservation with genetic stability, is one of the important aspects of the pharmaceutical industry. In general, seeds are used for commercial cultivation of these crops; however, they are not preferred because seeds may show genetic variability with that of their parents because of outcrossing nature. Genetic diversity is not favorable as these plants accompany tropane alkaloids. In addition, seeds exhibit a deprived germination rate with sluggish growth of seedlings. For instance, the production of *Duboisia* seedlings is critically restricted along with the short-term seed viability (Luanratana 1997; Xue et al. 2023a). Alternatively, the conventional propagation of medicinal plants relies upon vegetative propagation approaches, predominantly via stem cuttings. This approach promises genetic uniformity amongst propagules; nevertheless, it faces numerous limits, such as lower rooting rates, inconsistent rooting success, and prolonged root initiation times. Moreover, cutting propagation demands a consistent and abundant supply of cuttings in a rhizogenic state, which can restrict production volumes and necessitate additional plantation maintenance to gather suitable cuttings (Xue et al. 2023a, b).

Micropropagation is a highly effective and economically viable method for producing large volumes of genetically uniform plants. This technique enables the rapid and large-scale production of clones that are genetically identical to the elite mother plant, provided that the tissue culture media and growth conditions are carefully optimized (Xue et al. 2023b). By employing controlled environments and carefully formulated media compositions, micropropagation facilitates the production of high-quality, genetically consistent plants, which are crucial for both commercial applications and conservation efforts. Tissue culture plays a crucial role beyond propagation, serving as a prerequisite for the genetic modification of plants, germplasm preservation, plant breeding, and the production of secondary metabolites (Xue et al. 2023a). Plant tissue culture studies have been conducted to preserve medicinal plant resources and enhance the production of pharmaceutically significant secondary metabolites. Specifically, micropropagation techniques for tropane alkaloids yielding plants have been developed, which significantly improve plant propagation efficiency compared to traditional methods such as seedling. These tissue culture methods not only ensure uniform growth in the field but also result in higher alkaloid content in the propagated plants (Yoshimatsu 2008; Lemma et al. 2020).

This book chapter aims to provide an in-depth exploration of micropropagation techniques for plants that produce major tropane alkaloids, which are known for their significant pharmacological properties. The chapter focuses on the efficient propagation of these specialized medicinal plants to ensure sustainable production. It discusses various protocols and methodologies used to achieve high-yield, genetically uniform plantlets through in vitro culture techniques. This includes examining the specific conditions and media compositions necessary for successful micropropagation, as well as the challenges and solutions associated with these techniques.

12.2 Micropropagation of *Datura*

In 1973, research on the in vitro culturing of *Datura* species explored using various growth regulators, including 2,4-D, Kn, and GA_3, to promote the growth of regenerated shootlets from callus cultures. This study laid the groundwork for understanding how these chemicals influence plant tissue culture (Engvild 1973).

Zayed et al. (2006) studied in vitro organogenesis and alkaloid accumulation in *Datura innoxia*. They established undifferentiated callus from stem explants using MS medium supplemented with 1 mg/l BAP and 0.5 mg/l IAA for 6 weeks. Callus differentiation occurred after subculturing onto hormone-free solid MS medium for over 10 months, with shoots forming after 4 weeks. Further subculturing without growth regulators led to root development after 6 weeks.

The alkaloid analysis revealed that the callus produced only trace amounts, while re-differentiated shoots were mainly composed of scopolamine, and re-differentiated roots predominantly biosynthesized hyoscyamine. A micropropagation method for *Datura innoxia* using nodal and shoot tip explants was described by Waoo et al. (2013). In this method, multiple shoots were induced in vitro on MS medium with BAP, either alone or in combination with NAA and Kn. Nodal explants produced an average of three shoots per culture, each about 4.5 cm long, while shoot tip explants yielded six to seven shoots per culture, averaging 4 cm in length. The nodal segments, due to their pre-existing meristems, are ideal for clonal propagation because they are easily handled, have a high proliferation rate, and maintain clonal fidelity. Sangekar et al. (2019) described a micropropagation method for *Datura innoxia* using leaf and stem explants from in vitro-grown plants. Cultured on MS medium with different plant growth regulators, 2,4-D and NAA, alone or combined with BAP and Kn, induced callus formation that later exhibited somatic embryogenesis and shoot multiplication.

Figueiredo et al. (1987, 1988, 1989) studied the micropropagation of *Datura insignis*, a vegetatively propagated species with no observed seeds. They induced a compact, non-homogeneous callus from internodal explants on B5 medium supplemented with 1 mg/l BAP and either 0.25 mg/l or 1 mg/l NAA. Histological analysis revealed that only the tissues between the epidermis and central cylinder formed callogenic tissue, while xylematic elements and dispersed meristematic centers developed within the callus. On the medium with 1 mg/l BAP and 0.25 mg/l NAA, globular structures appeared, showing cells in periclinal division arranged radially, along with multiple meristematic centers. However, somatic embryos could not be identified, leading to the conclusion that the callus was not embryogenic.

Dos Santos et al. (1990) described a method for in vitro propagation of *D. insignis*. Nodal explants were cultured on MS medium with BAP, either alone or combined with 2,4-D or IAA, achieving the best results with 1.0 mg/l of BAP. Individual shoots were then excised and rooted in a growth regulator-free medium. Additional multiplication occurred through single-node culture using explants from the rooted shoots, which were later successfully grown in soil. Sangekar and Pandhure (2019) outlined methods for micropropagating *D. innoxia*, using cotyledons as explants.

They tested various concentrations of auxins and cytokinins in MS medium. The highest proliferation occurred with 2 mg/l of 2,4-D, both alone and in combination with 2 mg/l of BAP. Naphthaleneacetic acid (NAA) at 0.2 to 0.4 mg/l also enhanced proliferation. Muthukumar et al. (2000) outlined a micropropagation protocol for *Datura metel* using hypocotyl explants. The highest shoot proliferation was achieved with 1.5 mg/L BAP on MS medium. Regenerated shoots elongated on MS basal medium and rooted best at 2.0 mg/l IBA. Once hardened, the plantlets were transferred to soil. In vitro plant regeneration was achieved by Muthukumar et al. (2004) in *D. metel* from nodal explants collected from both in vitro germinated seedlings and field grown plants. The explants were cultured on MS medium supplemented with different concentrations of BAP and NAA. The nodal explants isolated from in vivo source exhibited a greater number of healthy multiple shoots than in vitro. BAP at 3 mg/l with NAA at 0.5 mg/l was found to be optimal for shoot regeneration.

In another approach, nodal explants from both in vitro germinated seedlings and field-grown plants were cultured on MS medium with BAP (0.5–3.0 mg/l) and 0.5 mg/l NAA. In vivo nodal explants produced more healthy shoots, with the optimal combination being 3 mg/l BAP and 0.5 mg/l NAA. Nithiya and Arockiasamy (2007) developed somatic embryos from root explants on MS medium with 4 mg/l BAP. These embryos grew into plants while attached to the explants. For fast shoot elongation and root growth, a medium containing 2 mg/l BAP, 1 mg/l GA_3, and 1 mg/l IBA was used. After hardening in a soil mixture, the plants had a survival rate of 90% when transferred to the soil. Jha and Pandey (2012) carried out a study to analyze the effects of growth regulators on callus formation of *Datura metel*. Leaf and other explants were cultured in MS medium supplemented with different concentrations of 2–4 D, NAA, Kn, BAP and IAA (from 1 mg/l to 5 mg/l). The most suitable medium for callus induction from leaf and anther were 2-4D (1.5 mg/l) alone and 2–4 D (1.5 mg/l) + Kn (2 mg/l) + IAA (1 mg/l) in combinations.

Thomas (2018) reported the effects of plant growth regulators on the formation of calli and multiple shoots in *D. metel* using terminal buds and nodal segments as explants. They used MS basal media consisting of four different growth regulators, viz., 2,4-D, Kn, IAA, and BAP. According to their research outcome, profuse callus formation was observed on MS media supplemented with Kn (2 mg/l) and IAA (1 mg/l) in combination. Also, nodal segments produced calli on MS medium having combinations of Kn (1.5 mg/l) and BAP (2 mg/l). Nodal segments showed multiple shoot formation on the media supplemented with the combinations of Kn (1 mg/l) and BAP (2 mg/l). In a pioneering study by Sharma et al. (1993), successfully achieved in vitro plant regeneration through embryogenesis in suspension cultures of *Datura innoxia*. This was accomplished by inducing embryogenesis in suspension cultures derived from callus tissue that had an androgenetic origin. The researchers utilized an LS liquid medium enriched with a precise concentration of 0.22 mg/l of 2,4-D, a plant growth regulator that played a crucial role in the formation of embryos. Throughout the culture period, the number of embryos produced varied significantly, demonstrating the dynamic nature of the embryogenesis process. Histological examinations revealed that the developing embryos showed clear signs of differentiation and maturation while suspended in the liquid medium.

Remarkably, once these embryos were transferred to a semisolid MS medium devoid of additional growth substances, they successfully germinated, giving rise to healthy plantlets.

Additionally, the regenerated plants exhibited a haploid chromosome number, indicating that the embryogenesis process preserved the genetic integrity characteristic of the androgenic callus from which they originated. This method illustrates a significant advancement in plant tissue culture, offering insights into the potential for the regeneration of haploid plants. Amiri and Kazemitabar (2011) established an efficient tissue culture cycle for *Datura stramonium* involving callus induction and plant regeneration by adjusting carbon sources and concentrations. They cultured embryo explants with six carbon sources (sucrose, glucose, fructose, galactose, maltose, and lactose) at four concentrations (1%, 2%, 3%, and 4%) for callus induction. The study found that 2% lactose was best for callus induction, while 3% glucose yielded the highest regeneration. Amiri et al. (2011) described in vitro propagation and whole-plant regeneration from callus in *Datura stramonium*. They cultured leaf explants and mature embryos on MS basal medium supplemented with various combinations of 2,4-D and Kn. The best treatments for callus induction were 2 mg/l 2,4-D + 0.5 mg/l Kn for leaf explants and 2 mg/l 2,4-D for embryos. For shoot regeneration, the optimal hormonal treatments were 3 mg/l BAP + 1 mg/l NAA for leaf calli and 2 mg/l BAP + 1 mg/l NAA for embryo calli. Rooting of regenerated shoots was best achieved with 0.5 mg/l IBA. Thapa et al. (2022) also used MS medium with BAP, Kn, and IBA for shoot induction from leaf explants of *D. stramonium*. They found that transferring healthy in vitro shoots (2–3 cm) to MS media with IBA effectively induced rooting.

12.3 Micropropagation of *Duboisia*

Duboisia, an Australian native shrub, is the sole commercial source for extracting scopolamine, a crucial alkaloid utilized as an antimuscarinic medication. This industry is highly lucrative, commanding an impressive 70% of the global market for scopolamine. However, one of the primary challenges confronting this sector is satisfying the increasing demand for elite *Duboisia* selections. As an outcrossing species, *Duboisia*'s seeds do not produce plants that are true to the parent material, rendering them unsuitable for plantation use. Additionally, the germination of these seeds is often problematic, resulting in low germination rates and subpar growth vigour (Singh et al. 1985). Currently, cutting propagation is the most reliable method for clonal propagation; however, it is fraught with inefficiencies. The method frequently experiences poor and inconsistent rooting, particularly with older mother plants (Luanratana 1997). Furthermore, the volume of plant material necessary for cutting propagation is significant and can lead to considerable economic burdens. Kitamura (1988) has provided an essential review of in vitro regeneration and the production of tropane and pyridine alkaloids in *Duboisia*.

Kukreja and Mathur (1985) demonstrated high-frequency plant regeneration from nodal explants of *Duboisia myoporoides*. For optimal axillary bud growth and multiplication, 1 mg/l IAA combined with 2 mg/l Kn or BAP was essential. Other auxins caused callus formation. Rooting was achieved with 0.5 mg/l NAA in a static liquid MS medium. No synergistic effects were found between cytokinins and adenine sulfate, and GA_3 inhibited shoot growth. The plantlets were successfully grown to maturity in the field. Additionally, Lin (1991) conducted a tissue culture of a *Duboisia* hybrid (*D. leichhardtii* x *D. myoporoides*) and produced tropane alkaloids. Khanam et al. (2000) investigated the effects of cytokinin/auxin combinations on organogenesis, shoot regeneration, and tropane alkaloid production in *Duboisia myoporoides*. Two types of callus were induced, leading to the regeneration of non-rooted shoots from various combinations in both the regeneration and callus induction mediums. The shoot-bud regeneration ability, number of leaves per shoot, and tropane alkaloid production of the non-rooted shoots were influenced by the cytokinin/auxin combinations used during callus induction. Successful shoot regeneration occurred in a liquid medium with 10 μM BA and 0.1 μM IBA; however, no tropane alkaloids were detected in the 4-week-old shoots cultured in this medium.

Xue et al. (2023a) explored micropropagation methods for *Duboisia myoporoides, D. leichhardtii,* and *D. hopwoodii*, testing various nutrient media, hormone combinations, and incubation conditions. The study found that MS (Murashige and Skoog) medium was superior to QL (Quoirin and Lepoivre) and WPM (Woody Plant Medium) for promoting shoot growth in *Duboisia* species and hybrids. The highest rates of meristem induction and multiplication were achieved for *D. myoporoides* and *D. leichhardtii* using BAP as the cytokinin. Species-specific rooting protocols resulted in 100% rooting for *D. myoporoides* and *D. leichhardtii* and 70% for *D. hopwoodii*. However, a 5-second IBA pulse treatment did not improve rooting in a solid agar medium. The addition of activated charcoal enhanced root formation but had varying effects across species; for instance, it improved rooting in *D. myoporoides,* while reducing it in *D. leichhardtii* and *D. hopwoodii*. An optimized acclimatization protocol led to 100% survival for *D. myoporoides* and *D. leichhardtii*, and 80% for *D. hopwoodii.*

Mason et al. (2022) presented methods for the micropropagation of the *Duboisia* hybrid, which are outlined below. For seed germination, various concentrations of GA_3, in combination with activated charcoal, were utilized. A germination rate of nearly 100% was achieved with the use of charcoal and GA_3 after 24 h of incubation. Sterilization was performed using a solution of sodium hypochlorite (NaOCl) and ethanol (EtOH). Meristem induction occurred on basal media that included different types and concentrations of cytokinins. The addition of coconut water significantly increased the multiplication rate, resulting in up to 1000 shoots generated from a single shoot tip explant within 5 months. Rooting was also successful, with a 100% root induction achieved through the incorporation of auxins (see Figs. 12.1 and 12.2). By carefully regulating factors such as relative humidity, light, temperature, and fertigation, acclimatization was attained after 4 weeks, resulting in a 100% survival rate (see Fig. 12.3).

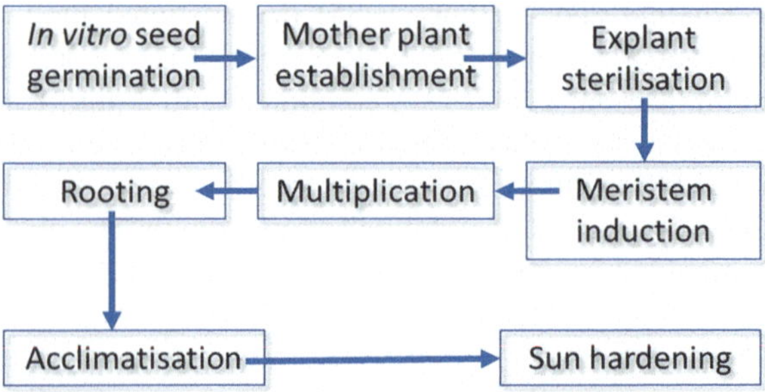

Fig. 12.1 Micropropagation method for *Duboisia*. (Source: Mason et al. 2022)

Fig. 12.2 Meristem culture micropropagation platform for *Duboisia myoporoides* × *Duboisia leichardtii* hybrid. (Adapted from Mason et al. 2022)

12.3.1 Protoplast Isolation and Somatic Hybridization

The global demand for scopolamine and other alkaloids, such as hyoscyamine, is on the rise, creating a pressing need for improved alkaloid content in the industry. Ploidy manipulation through protoplast fusion is a well-established method in in vitro plant breeding used for a variety of agricultural plants. Applying this technology to *Duboisia* could lead to the development of new varieties with larger and thicker leaves, as well as faster-growing plants, ultimately ensuring greater production per cultivated square meter. Scopolamine, a key alkaloid, is stored in the leaves of *Duboisia*, and using polyploid techniques is expected to increase scopolamine

Fig. 12.3 Field ready plants of micropropagated *Duboisia myoporoides* × *Duboisia leichardtii* hybrid. (Adapted from Mason et al. 2022)

yield, thereby enhancing its commercial value. Li et al. (2022) and Xue et al. (2024) outlined methods for protoplast isolation and somatic hybridization to improve *Duboisia* plants. For callus initiation, leaves from in vitro grown *D. myoporoides* and its hybrids (*D. myoporoides* × *D. hopwoodii* and *D. myoporoides* × *D. leichhardtii*) were used. Immature top leaves (1–3) were sampled, and 1 cm^2 leaf pieces were tested with various cytokinins for friable callus induction. For suspension culture, 5 g of callus was placed in media with different concentrations of 2,4-D, and factors like subculture duration and shaking speed were optimized.

Results showed that Kn at 0.01 mg/l significantly enhanced callus induction for *D. myoporoides* and the hybrid *D. myoporoides* × *D. hopwoodii*, while both Kn and Benzyladenine (BA) were effective for *D. myoporoides* × *D. leichhardtii*. Suspension cultures were successfully established in MS medium with 1 mg/l of 2,4-D under shaking conditions, leading to optimal growth. Protoplast isolation achieved successful results using a digestion mixture (0.4% pectolyase, 2% cellulase, and 2% driselase) for 20 h, although a shorter 16-h duration improved yield. This study initiated polyploid development for the three *Duboisia* selections, confirming the importance of hormone concentration for successful callus induction and suspension culture. Further optimization is required for better protoplast purification and fusion (Fig. 12.4).

12.3.2 Micropropagation of Atropa

Zarate et al. (1997) studied the induction and development of adventitious shoots in *Atropa baetica* for propagation. The best results for multiple shoot induction were achieved with MS medium supplemented with 0.75 or 1.25 mg/l of BAP, yielding

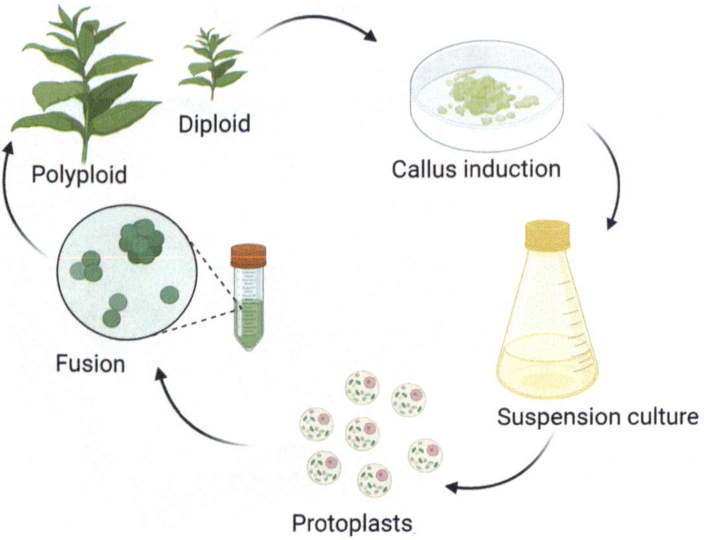

Fig. 12.4 Protoplast isolation and somatic hybridization method in *Duboisia* species. (Source Li et al. 2022)

an average of 5.6 shoots per explant after 31 days. Higher BAP concentrations (1.75–2.0 mg/l) also produced good results but negatively affected subsequent root induction. In contrast, media with only 0.25 mg/l of BAP led to significantly fewer shoots. Root induction occurred spontaneously when the shoots were transferred to MS medium without plant growth regulators and also on media supplemented with 0.125 and 0.25 mg/l of NAA. Acclimatization of plantlets in soil was highly successful, achieving a rate of 95.52%.

In vitro culture, plant regeneration, cryopreservation, and the production of tropane alkaloids in *Atropa belladonna* have been reviewed by Bajaj and Simola (1991). Manjula and Manjula (2024) provided a comprehensive review of the micropropagation of *Atropa acuminata* and *A. belladonna.* Saito et al. (1992) reported the successful formation of *A. belladonna* adventitious shoots from herbicide-resistant hairy roots grown in a B5 medium without the addition of phytohormones. Similarly, Aoki et al. (1997) observed adventitious shoot formation in *A. belladonna* using nodal explants on 12 Macro MS medium, which was supplemented with 1.5% sucrose, 0.5 M NAA, 5 M BA, and 0.2% gelrite. Ahuja et al. (2002) presented a micropropagation protocol for multiplying and conserving *A. acuminata* by inducing axillary shoot proliferation from nodal segment and shoot tip explants on MS medium augmented with 1 mg/l BAP and 1 mg/l IAA. The Revised Tobacco (RT) medium was found to be the most effective for promoting shoot elongation when supplemented with 1 mg/l IAA. Additionally, root differentiation was highly successful on MS medium containing 1 mg/l IBA, and the resulting plantlets were acclimatized under greenhouse conditions. The authors noted that shoot tip explants exhibited a better regeneration response compared to nodal explants. The combination of BAP and IBA produced the best results in shoot tip explants, yielding a maximum of 5–6 shoots per culture after a 40-day incubation period on MS medium with 1 mg/l BAP and 1 mg/l IBA.

Maqbool et al. (2014) successfully produced callus from leaf explants of *A. acuminata.* The leaves were thoroughly cleaned with running tap water and treated with Labolene detergent and Tween-20, followed by rinsing with sterile double-distilled water and a 2% sodium hypochlorite solution. After further washing, the explants were inoculated onto a sterilized nutrient medium. In 2016, Maqbool et al. refined micropropagation protocols using petiole and nodal segments as explants. These were washed with tap water, treated with Labolene and Tween-20, and sterilized with sodium hypochlorite before inoculation. Eighteen-day-old plantlets were transferred to jiffy and earthen pots containing a 1:1 mix of autoclaved soil, sand, and compost. The plants were kept in a greenhouse, achieving an 80% survival rate in compost and 60% in a soil-sand mixture. Maqbool et al. (2018) explored regeneration methods via callogenesis, inoculating leaf explants on MS medium with varying BAP concentrations. The highest callus and shoot induction, with a 100% success rate, occurred at 3 mg/l BAP. In vitro rooting was successful with 0.5 mg/l IBA. Well-rooted shoots were acclimatized in vermicompost, achieving a 100% survival rate under greenhouse conditions. Additionally, methanolic extracts of regenerated plants were analyzed for important secondary metabolites and antioxidant potential.

Khan et al. (2017) successfully regenerated *A. acuminata* in vitro from leaf explants of seed-derived plants. They found that 1 mg/l of TDZ produced the highest callogenic response at 92%. After 35 days, the highest shoot induction rate of 89% occurred with a medium containing TDZ (1 mg/l) and NAA (1 mg/l). Lower concentrations of TDZ, alone or with NAA, promoted shoot organogenesis, while higher concentrations inhibited it. The addition of NAA to Murashige and Skoog (MS) medium with BAP significantly increased shoot induction. TDZ and BA in combination with NAA were the most effective in producing the highest number of shoots per explant. For rooting, MS medium with IBA at 0.5 mg/l achieved a 90% rooting rate, with moderate IBA levels being more effective than higher ones. Notably, NAA did not induce rooting in *A. acuminata.* Rajput and Agrawal (2020) developed a micropropagation protocol for *Atropa acuminata* using root-derived callus via indirect organogenesis. After 45 days of incubation, callus formed on all media, with the highest percentage on 10 μM TDZ. Calli on IBA, NAA, or 2,4-D were slow-growing and non-morphogenic, while those on BA were green but did not grow further.

After 28 days of subculturing on 10 μM TDZ, callus differentiated into an average of 5.54 ± 0.65 shoots measuring 2.38 ± 0.37 cm in 71.5% of explants. The best rooting response occurred with MS + 1 μM IBA, resulting in an average of 8.08 ± 0.19 roots, each 7.28 ± 0.11 cm long. Despite successful shoot development, shoot elongation was inhibited during further subculturing on 10 μM TDZ. After another 28 days, 99% of cultures produced an average of 8.08 ± 0.19 roots per shoot. Dominant root formation was observed without intervening callus, but poor rooting was noted with IAA and NAA. At low concentrations (0.1 and 1.0 μM), roots were thin and minimally branched, while higher concentrations (5.0 and 10 μM) showed short, thin roots with profuse callus. This study emphasizes the effectiveness of TDZ as a plant growth regulator for the in vitro regeneration of *A. acuminata* (Rajput and Agrawal 2020) (Fig. 12.5).

Mera and Takayama (2012) studied the effects of medium components on adventitious bud formation in *Atropa belladonna* hairy roots using hormone-free liquid B5 medium with 3% sucrose. They evaluated B5 and MS media, both containing 3% sucrose and 8% agarose, supplemented with varying concentrations of Kn (0, 0.5, 4.7, and 46.5 μM). Additionally, they assessed how different sucrose concentrations (0, 1, 3, 5, and 7%) affected bud formation when combined with Kn (0–0.5 μM). All experiments were conducted at 25 °C under continuous light.

Al-Ashaal et al. (2013) used leaves of *Atropa belladonna* as explants, which were treated with sodium hypochlorite and ethyl alcohol before culturing in Murashige and Skoog (MS) medium with 3% sucrose and various growth regulators like NAA. Cultures were kept at 26 ± 2 °C with a 16/8 h light/dark cycle, and the callus was subcultured monthly for shoot development. Asha Rani and Prasad (2013) utilized MS media with Iron EDTA, vitamins, 3% sucrose, and phytohormones (NAA, IBA, Kn) for organogenesis from *A. belladonna* leaf explants. They prepared hormone solutions such as [MS + Sucrose + Agar + NAA (0.2 mg/l)], [MS + Sucrose + Agar + IBA (0.3 mg/l)], and [MS + Sucrose + Agar + KI (0.4 mg/l)]. *A. belladonna* thrives at 26 ± 3 °C with a 16-h light/8-h dark photoperiod. The authors also conducted a comparative study on callus induction using fresh and reused explants. Ahmed et al. (2021) outlined methods for the micropropagation of *Atropa belladonna*. They utilized shoot tips and nodes as explants, with Kn and BAP serving as the growth regulators. Among the explants, the shoot tip proved to be the most effective, yielding the highest average number of shoots, the longest shoot length, and the greatest number of leaves. BAP emerged as the most potent cytokinin for micropropagation, as it significantly increased the average number of shoots, shoot length, and leaf count compared to Kn. The optimal concentration for achieving the highest average number of shoots was found to be 2 mg/l BAP.

Fig. 12.5 (continued) (**d**) Emergence of roots from nodal explant of *A. acuminata*; (**e**) Root explants cultured on MS medium augmented with different PGRs; (**f**) Green and swallow root explant on MS + 10 μM TDZ; (**g**, **h**) Proliferation of callus and induction of shoot buds on MS + 10 μM TDZ; (**i**) Elongated shoot buds on MS basal medium; (**j**) Excised shoots on MS medium augmented with different auxins for root induction; (**k**) Induction of roots from cut end of shoots on MS + 1 μM IBA; (**l**) Exposed view of rooted plantlet along with scale; (**m**) Tissue culture raised plants in cups having Soilrite mix; (**n**) Hardened plant of *A. acuminata* grown under field conditions in clay pot containing garden soil (scale bar = 1 cm). (Source: Rajput and Agrawal 2020)

Fig. 12.5 Micropropagation of *A. acuminata*. (**a**) Seeds of *A. acuminata*; (**b**) *A. acuminata* plant grown under in vivo conditions; (**c**) Nodal explant of in vivo grown plant on MS basal medium;

12.4 Micropropagation of *Hyoscyamus*

Saad et al. (2024) developed an in vitro propagation protocol for preserving *Hyoscyamus boveanus*. Seeds were surface sterilized and germinated on a liquid MS medium without growth regulators. Explants from in vitro-grown seedlings—leaves, stems, and roots—were used for direct and indirect organogenesis. They tested various hormone concentrations and found that the combination of IAA (1.0 mg/l) and Kn (0.5 mg/l) was most effective for inducing callus. Leaf explants achieved direct organogenesis, while other explants formed callus that developed into shoots, leaves, roots, and flowers. The plantlets were then transferred to asbestos perlite soil. Plantlets from indirect organogenesis acclimatized successfully, but those from direct organogenesis did not adapt to the ex vitro environment. Successful plant regeneration of *Hyoscyamus muticus* was demonstrated by Sevon et al. (1995) through hairy root-derived protoplasts. High yields of protoplasts were isolated using an enzyme mixture (1% macerozyme and 2% cellulase) with an osmoticum of 0.2 M $CaCl_2$ and 0.6 M mannitol. Initial culture in liquid NT/PRO I medium was followed by semi-solid NT/PRO II agar medium, resulting in 1–9% plating efficiency. Small colonies regenerated shoots at 18% and roots at 70%, with an 8–11-week development time. Ibrahim et al. (2009) explored callus induction using hypocotyl and leaf explants from *Hyoscyamus muticus* seedlings on MS medium. Optimal callus formation occurred with 2 mg/L NAA and 0.5 mg/L Kn, with hypocotyls and roots producing the best roots, while leaves yielded the highest shoots. Quadri et al. (2012) found that 12.5 μM BA enhanced shoot proliferation from *Hyoscyamus niger* shoot tip explants, while 10 μM TDZ only produced disorganized callus. The best rooting was with 4.0 μM IBA and 2.5 μM NAA, and plantlets showed a 50% survival rate in field conditions. Cheng and Raghavan (1985) described somatic embryogenesis in *Hyoscyamus niger*, culturing leaf segments and ovules in a 2,4-D medium, resulting in extensive callus growth. Transfer to an auxin-free medium encouraged root or bud regeneration, and friable calluses formed somatic embryos similar to zygotic embryos.

A high-frequency adventitious shoot regeneration protocol for *Hyoscyamus niger* was developed by Uranbey (2005) using TDZ. Hypocotyl, cotyledon, and stem explants were cultured on MS medium supplemented with various concentrations of BAP and TDZ. The most effective medium was 16 μM TDZ, achieving a 100% regeneration frequency and an average of 19.53 shoots per hypocotyl explant. High rooting and survival were obtained using a half-strength MS medium with 8 μM IBA. Keng et al. (2009) described plantlet regeneration from embryogenic calli derived from leaf, petiole, and root pieces. Small globular somatic embryos formed from leaf explants on MS medium with 6 mg/l picloram and matured on MS medium with 1.0 mg/l BAP. Though these embryos did not germinate into complete plantlets, regenerated roots could be mass propagated for hyoscyamine extraction. Shah et al. (2018) reported micropropagation via shoot tip culture, achieving effective seed sterilization with 0.02% HgCl2 for 20 minutes. Shoot tips grown on MS medium supplemented with varying concentrations of TDZ and BAP showed the best results in terms of shoot number, length, and shooting percentage after 8 weeks. Lai (2003) carried out studies on micropropagation, callus induction and root culture of *Hyoscyamus niger.*

12.5 Micropropagation of *Brugmansia*

Brugmansia species are known for their alkaloids, particularly scopolamine and hyoscyamine, which have clinical significance. Consequently, these plants are cultivated in South America for alkaloid production, and various techniques for their micropropagation have been explored. Bhatt et al. (2004) studied the clonal propagation of *B. versicolor* using shoot tip culture. They utilized an LS basal medium with 3% sucrose and 0.8% agar, adjusting the pH to 5.6 before autoclaving. Shoot tips served as explants, and amongst the tested plant growth regulators, 0.1, 1, or 10 μM BA or Kn, alone or with 0.1 μM IAA, produced an average of 2.4 shoots per culture. A combination of 1 μM BA and 1–10 μM gibberellic acid improved shoot multiplication to 45 shoots per culture. For rooting, a medium with 1 μM IBA was found to be the most effective. The study also assessed the effects of cold storage on shoot viability. Shoots stored at 5 °C for 6 months died, while those stored at 15 °C under light for 12 months achieved a 100% survival rate. Notably, plants from these stored shoots retained their ability to produce scopolamine, indicating that this storage method can help conserve germplasm without losing metabolic capacity.

Additionally, an effective tissue culture protocol for *B. suaveolens* was developed using fresh sprouts as explants (Shekhawat 2012). After surface sterilization, these explants were cultured on MS medium supplemented with 2 mg/l BAP, resulting in 5–6 shoots per culture within 15 days. The highest shoot proliferation, 8–9 shoots per culture, was achieved with a combination of 0.5 mg/l BAP and Kn. Both IAA and IBA effectively induced roots in plantlets, which were then acclimatized in a greenhouse. Research on *Brugmansia* micropropagation emphasizes the significance of specific hormonal combinations for optimal shoot and root development, offering valuable insights for effective plant tissue culture techniques.

12.6 Micropropagation of *Anisodus*

A highly effective protocol for shoot multiplication using shoot tip explants of *Anisodus tanguticus* has been developed by He et al. (2011). This protocol successfully induces multiple shoots when cultured on Murashige and Skoog (MS) medium supplemented with various concentrations of BAP, Kn, or a combination with NAA. The results clearly indicate that BAP is more effective than Kn in promoting shoot multiplication, achieving a remarkable maximum of 3.2 shoots per explant on MS medium containing 2 mg/L BAP. However, adding 0.5 or 1 mg/L NAA to the BAP-containing medium significantly enhances callus formation, while reducing the efficiency of shoot multiplication. Shoots treated with 0.5 mg/l IAA show the highest average root number of 4.08 and an outstanding rooting percentage of 75.2%. The micropropagated plantlets are successfully hardened in a greenhouse and established in soil without issue. To evaluate the genetic stability of the regenerated plants, 25 inter-simple sequence repeat (ISSR) markers were meticulously applied. Impressively, 18 out of 25 markers produced clear and reproducible bands, resulting in an average of 6.7 bands per marker. These results unequivocally confirm that the regenerated plants maintain high genetic fidelity.

12.7 Micropropagation of *Scopolia*

Hiraoka et al. (2004) described a protocol for the cold storage of shoot cultures of *Scopolia japonica*. They found that 100% of the cultures could withstand storage at 2 °C for 1 year, while 70% remained viable after 2 years. After 2 years of cold storage, the shoots were multiplied on freshly prepared Linsmaier and Skoog medium supplemented with 1 μM IAA, resulting in a successful acclimatization rate of 97% for the plants. Kang et al. (2004) established a rapid micropropagation system for *Scopolia parviflora* using rhizome cultures. They conducted experiments in which the rhizomes were cultured on a half-strength B5 liquid medium supplemented with various growth regulators. The optimum condition for shoot multiplication was found in a half-strength B5 medium containing 3% (w/v) sucrose and 5.77 μM GA_3, with each rhizome producing an average of 12 shoots. Shoot elongation and root induction from multiple shoots occurred on growth regulator-free half-strength B5 solid medium. Healthy plantlets were then transferred to a mixture of peat moss and vermiculite for acclimatization, which proved successful. When the rhizome was cut into segments and placed under optimal conditions for multiple shoot propagation, a mature plant could be produced in just 12 weeks. Analysis of the concentrations of tropane alkaloids in different tissues of native plants, in vitro-propagated plants, and acclimatized plants revealed that the levels of hyoscyamine and scopolamine were higher in the in vitro-propagated plants compared to the native plants.

12.8 Micropropagtion of *Schizanthus*

Jordan et al. (2006) developed a rapid in vitro propagation system for *Schizanthus hookeri* that involves two phases: callus initiation and shoot formation, followed by rhizogenesis and plantlet regeneration. A single callus treated with 2.69 μM NAA and 2.22 μM BA produced over 10 shoots, resulting in about 20 plantlets after 90 to 110 days in a growth regulator-free medium. Similar conditions induced callus and root formation in *Schizanthus grahamii,* but while root organogenesis occurred with 4.44 μM BA and 0.54 μM NAA, no shoots were formed. In *Schizanthus tricolor,* only callus formation was achieved with various growth regulators.

12.9 Conclusion and Future Direction

In conclusion, the research on micropropagation of different species, comprising *Datura, Duboisia, Brugmansia, Hyoscyamus,* and *Anisodus*, has provided an appreciated understanding of the optimization of in vitro culture procedures for effective plant multiplication as well as clonal propagation. These investigations advocate that micropropagation is a favorable and efficient technique for clonal propagation of tropane alkaloids yielding plants, addressing the challenges linked with commercial crop growing and warranting the true-to-type plants production. The above studies demonstrated the significance of choosing suitable plant growth regulators, their concentrations and combinations, and explants, such as seeds, cotyledons, nodal segments, and shoot tips, to maximize shoot regeneration and root formation. The usage of media like MS and RT, combined with precise levels of auxins (mainly, NAA, IAA) and cytokinins (mainly, BAP, Kn), has been revealed to increase proliferation rates and guarantee clonal conformity, with fruitful acclimatization to glasshouse environments. Furthermore, the genetic stability of micropropagated plants, as measured by molecular markers (ISSR), is critical to ensure that in vitro grown plantlets retain the desired characteristics.

Future exploration should emphasize further optimizing micropropagation techniques to improve efficiency and decrease expenditures. This comprises exploring species-precise requests for nutrient media and plant growth regulator treatments, as well as developing vigorous acclimatization procedures to improve plantlet survival rates post-transplantation. Also, expanding the genetic diversity of propagated plants through micropropagation could support conservation efforts and the development of new cultivars with enhanced traits. Investigating the potential for integrating micropropagation with other biotechnological approaches, such as genetic engineering, could also open new avenues for improving the quality and yield of these economically and pharmaceutically important species. Additionally, investigating the impact of environmental factors such as light intensity, temperature, and humidity on tissue culture success could lead to more sustainable and scalable propagation systems. Another area of potential growth is the use of advanced molecular techniques, such as CRISPR-Cas9, to improve genetic traits of micropropagated plants, including disease resistance or increased yield. Finally, further research into the conservation of rare or endangered species using micropropagation, along with the study of somatic hybridization and protoplast fusion, may offer solutions to preserve genetic diversity and maintain the ecological balance of critical plant species. Through continued innovation and refinement of micropropagation protocols, the field holds great promise for advancing plant breeding, conservation, and sustainable agriculture.

References

Ahmed ZS, Salim AM, Alsoufi ASM (2021) In vitro micropropagation of *Atropa belladonna* l. using plant growth regulators. Plant Cell Biotech Mol Biol 22(3&4):169–174

Ahuja A, Sambyal M, Koul S (2002) In vitro propagation and conservation of *Atropa acuminata* Royle ex Lindl – an indigenous threatened medicinal plant. J Plant Biochem Biotech 11:121–124

Al-Ashaal HA, Aboutabl ME, Maklad YA, El-Beih AA (2013) Tropane alkaloids of *Atropa belladonna* L.: *in vitro* production and pharmacological profile. Egypt Pharm J 12(2):130–135

Amiri S, Kazemitabar SK (2011) Enhancement of callus induction and regeneration efficiency from embryo cultures of *Datura stramonium* by adjusting carbon sources and concentrations. Afr J Biotech 10(50):10101–10107

Amiri S, Kazemitabar SK, Ranjbar GA, Azadbakht M (2011) In vitro propagation and whole plant regeneration from callus in Datura (*Datura stramonium* L). Afr J Biotech 10(3):442–448

Aoki T, Matsumoto H, Asako Y, Shimomura K (1997) Variation of alkaloid productivity among several clones of hairy roots and regenerated plants of *Atropa belladonna* transformed with *Agrobacterium rhizogenes* 15834. Plant Cell Rep 16:282–286

Asha Rani NS, Prasad MP (2013) Studies on the organogenesis of *Atropa belladonna* in *in-vitro* conditions. Int J Biotechnol Bioeng Res 4(5):457–464

Bajaj YPS, Simola LK (1991) *Atropa belladonna* L.: in vitro culture, regeneration of plants, cryopreservation and the production of tropane alkaloids. In: Biotechnology in agriculture and forestry, vol 15. Springer, Berlin/Heiderlberg, pp 1–23

Bhatt ID, Chang JI, Hiraoka N (2004) In vitro propagation and storage of *Brugmansia versicolor* Lagerheim. Plant Biotechnol 21(3):237–241

Cheng J, Raghavan V (1985) Somatic embryogenesis and plant regeneration *Hyoscyamus niger*. Am J Bot 72:580–587

dos Santos MCF, Esquibel MA, dos Santos AVP (1990) In vitro propagation of the alkaloid producing plant *Datura insignis* Barb. Rodr. Plant Cell Tiss Org Cult 21:75–78

Engvild KC (1973) Shoot differentiation in callus cultures of *Datura innoxia*. Physiol Plant 28(1):155–159

Fan BS, Zhang EH, Wu M, Guo JM, Su DF, Liu X et al (2016) Activation of alpha 7 nicotinic acetylcholine receptor decreases on-site mortality in crush syndrome through insulin signaling--Na/K-ATPase pathway. Front Pharmacol 7:79. https://doi.org/10.3389/fphar.2016.00079

Figueiredo SFL, Esquibel MA (1987) Micropropagação de *Datura insignis* Barb. Rodr. IX Encontro Regional de Botânicos, Vitória, ES, p 41

Figueiredo SFL, Esquibel MA (1988) Organogenesis and calli development of *Datura insignis*. Barb. Rodr. Fenabio. In: Congresso e Feira Nacional de Biotecnologia, 1, Rio de Janeiro, RJ, p 42

Figueiredo SFL, Vieira RC, Esquibel MA, Costa CG (1989) Histogênese de calo de explante caulinar de *Datura insignis* Barb. Rodr. Rodriguésia: Revista do Jardim Botânico do Rio de Janeiro 41:71–77

He T, Wand H, Yang L, Zhao Z (2011) Micropropagation of *Anisodus tanguticus* and assessment of genetic stability of regenerated plants using inter simple sequence repeat (ISSR) marker. J Med Plants Res 5(25):5993–5998

Hiraoka N, Shinohara C, Ogata H, Chang JL, Bhatt ID (2004) Medium term conservation of *Scopolia japonica* shoot cultures and evaluation of the regenerated plants. Nat Med 58:98–103

Ibrahim IA, Guirgis AA, Mostfa AK, Ahmed N, Amira AM, Asama AA (2009) Alkaloid production and organogenesis from callus of *Hyoscyamus muticus* L., in vitro. J Appl Sci Res 5:82–92

Jha M, Pandey RK (2012) In vitro micropropagation of *Datura metel* L. through callus induction from leaf and anther culture. Bioscan 7(1):77–80

Jordan M, Humam M, Bieri S, Christen P, Poblete E, Muñoz O (2006) In vitro shoot and root organogenesis, plant regeneration and production of tropane alkaloids in some species of *Schizanthus*. Phytochemistry 67(6):570–578. https://doi.org/10.1016/j.phytochem.2005.12.007

Kang YM, Min JY, Moon HS, Karigar CS, Prasad DT, Lee CH, Choi MS (2004) Rapid in vitro adventitious shoot propagation of *Scopolia parviflora* through rhizome cultures for enhanced production of tropane alkaloids. Plant Cell Rep 23:128–133. https://doi.org/10.1007/s00299-004-0820-0

Keng CL, Saidon NA, Bhatt A (2009) Somatic embryogenesis and root regeneration in *Hyoscyamus niger* L. for the production of hyoscyamine. Afr J Biotech 8(24):6952–6960

Khan FA, Abbasi BH, Shinwari ZK, Shah SH (2017) Antioxidant potential in regenerated tissues of medicinally important *Atropa acuminata*. Pak J Bot 49(4):1423–1427

Khanam N, Khoo C, Khan AG (2000) Effects of cytokinin/auxin combinations on organogenesis, shoot regeneration and tropane alkaloid production in *Duboisia myoporoides*. Plant Cell Tissue Organ Cult 62:125–133

Kitamura Y (1988) *Duboisia* spp.: in vitro regeneration, and the production of tropane and pyridine alkaloids. In: Bajaj YPS (ed) Biotechnology in agriculture and forestry volume 4, Medicinal and aromatic plants I. Springer, Berlin/Heidelberg, pp 419–436

Kukreja AK, Mathur AK (1985) Tissue culture studies in *Duboisia myoporoides*; 1. Plant regeneration and clonal propagation by stem node cultures. Planta Med 51:93–96

Lai EM (2003) Micropropagation, callus induction and root culture of *Hyoscyamus niger* L., a temperate medicinal plant (MSc. Thesis). Universiti Sains Malaysia, Penang, Malaysia

Lemma DT, Banjaw DT, Megersa HG (2020) Micropropagation of medicinal plants. Int J Plant Breed Crop Sci 7:796–802

Li Y, Hiti-Bandaralage J, Xue Y, Mitter N (2022) New developments in protoplast technology for Australia's top-secret narcotic. Conference: Trop Ag 2022. https://doi.org/10.5281/zenodo.7332390

Lin G-LD (1991) Tissue culture of a *Duboisia* hybrid (*D. leichhardtii* x *D. myoporoides*) and the production of tropane alkaloids. Ph.D. Thesis, School of Pharmacy, The University of Queensland, Brisbane

Liu X, Xia B, Yu HK, Hu LZ, Fan SM, Xiao D et al (2019) Atropine premedication facilitates ultrasound-guided reduction by saline enema in children with intussusception. Front Pharmacol 10:862. https://doi.org/10.3389/fphar.2019.00862

Lockery JE, Broder JC, Ryan J, Stewart AC, Woods RL, Chong TTJ et al (2021) A cohort study of anticholinergic medication burden and incident dementia and stroke in older adults. J Gen Intern Med 36:1629–1637. https://doi.org/10.1007/s11606-020-06550-2

Luanratana O (1997) Micropropagation of *Duboisia* species. In: Bajaj YPS (ed) High-tech and micropropagation VI, Biotechnology in agriculture and forestry, vol 40. Springer, Berlin/Heidelberg, pp 313–331. https://doi.org/10.1007/978-3-662-03354-8_23

Manjula S, Manjula BL (2024) Micropropagation protocol in *Atropa acuminata* Royle ex Lindl. and *Atropa belladonna* L. In: Pullaiah T (ed) Micropropagation of medicinal plants. Singapore, Bentham Science, pp 77–98

Maqbool F, Singh S, Zahoor A, Kaloo A, Jan M, Meraj M (2014) Callus induction and shoot regeneration of *Atropa acuminata* Royle – a critically endangered medicinal plant species growing in Kashmir Himalaya. J Sci Innov Res 3(3):332–336

Maqbool F, Singh S, Zahoor A, Kaloo A, Jan M (2016) A rapid micropropagation protocol of *Atropa acuminata* Royle ex Lindl. – a threatened medicinal plant species of Kashmir Himalaya. Indian J Biotechnol 15:576–580

Maqbool F, Singh S, Wani TA, Ahmad M, Ganai BA, Jan M (2018) Phytochemical screening and antioxidant activity of methanolic extract of *in vitro* raised plants of *Atropa acuminata* Royle. Int J Sci 7:1–11

Mason P, Hiti-Bandaralage J, Xue Y, Mitter N, Henry R (2022) Advancing the *Duboisia* industry for sustainable alkaloids. TropAg, Brisbane

Mera N, Takayama S (2012) Effects of medium components and shear conditions on the formation and growth of adventitious bud derived from hairy roots of *Atropa belladonna* L. Environ Contr Biol 50(4):393–406

Muthukumar B, Arockiasamy DI, Britto J (2000) In vitro propagation of *Datura metel* L. from hypocotyl explants. Plant Tissue Cult 10(1):39–44

Muthukumar B, Arockiasamy DI, Natarajan E (2004) Direct organogenesis in *Datura metel* L. from in vitro & in vivo nodal explants. Indian J Biotechnol 3:449–451

Nithiya P, Arockiasamy DI (2007) In vitro micropropagation of *Datura metel* L. through somatic embryos from root explants. Plant Tissue Cult Biotech 17(2):125–130

Quadri RR, Teixeira da Silva JA, Kamili AN, Shah AM (2012) Effect of 6-benzyladenine, kinetin and thidiazuron on in vitro shoot proliferation of *Hyoscyamus niger* L. Med Aromat Plant Sci Biotechnol 6(1):81–83

Rajput S, Agrawal V (2020) Micropropagation of *Atropa acuminata* Royle ex Lindl. (a critically endangered medicinal herb) through root callus and evaluation of genetic fidelity, enzymatic and non-enzymatic antioxidant activity of regenerants. Acta Physiol Plant 42:160

Saad SF, Ibrahim MT, ElKousy RH, Abo-Elghiet F, El-Dahmy SI (2024) Conservation efforts for the endangered medicinal plant *Hyoscyamus boveanus* (Dun.) Aschers. & Schweinf. (Endemic to Egypt): an in vitro propagation protocol. Azhar Int J Pharm Med Sci 5(1). https://doi.org/10.21608/aijpms.2024.224893.1226

Saito K, Yamaai M, Anai H, Yoneyama K, Murakoshi I (1992) Transgenic herbicide resistant *Atropa belladonna* using an Ri binary vector and inheritance of the transgenic trait. Plant Cell Replication 11:219–224

Sangekar S, Pandhure N (2019) Tissue culture and biochemical studies in *Datura innoxia* Mill. World J Pharma Pharm Sci 8(5):1192–1199

Sangekar S, Devarkar V, Pandhure N (2019) In vitro micropropagation and pharmacognostical studies in *Datura innoxia* Mill. Plant Sci 2(2):32–35

Sevon N, Oksman-Caldentey KM, Hiltunen R (1995) Efficient plant regeneration from hairy root-derived protoplast of *Hyoscyamus muticus*. Plant Cell Rep 14:738–742

Shah D, Kamili AN, Sajjad N, Nazir N, Sumira Tyub S (2018) Micropropagation of *Hyoscyamus niger* L. by shoot tip culture: an important medicinal plant of solanaceae family. Int J Adv Res Sci Eng 7(4):1880–1888

Sharma VK, Jethwani V, Kothari SL (1993) Embryogenesis in suspension cultures of *Datura innoxia* Mill. Plant Cell Rep 12:581–584

Shekhawat M (2012) Direct shoots regeneration from nodal meristems of *Brugmansia suaveolens* (Humb. & Bonpl. ex Willd.) Bercht. & JPresl. Pharma Innov J 1:50–56

Singh A, Singh DV, Rao MR, Shukla YN, Hussain A (1985) Cultivation of *Duboisia myoporoides* R. Brown as source of tropane alkaloids in India. Indian J Pharm Sci 47(3):120–121

Thapa A, Ali Y, Madan S, Verma P, Verma P, Gaurav N (2022) An assessment of in vitro propagation and medicinal properties of *Datura stramonium* (Dhatura). Sci Temper 13(2):293–301

Thomas S (2018) Effect of growth regulators on callus and multiple shoot formation of *Datura metel* L. Int J Appl Res 4(2):216–219

Uranbey S (2005) Thidiazuron induced adventitious shoot regeneration in *Hyoscyamus niger*. Biol Plant 49:427–430

Waoo AA, Khare S, Ganguli S (2013) In-vitro propagation of *Datura innoxia* from nodal and shoot tip explants. World J Environ Eng 1(1):11–14

Xu Q, Zhao X (2017) Progress in clinical application of compound anisodine injection on ophthalmology. Chin J Clin Pharmacol 33:861–864. https://doi.org/10.13699/j.cnki.1001-6821.2017.09.025

Xue Y, Hiti-Bandaralage JCA, Jambuthenne DT, Zhao Z, Mitter N (2023a) Micropropagation of *Duboisia* species via shoot tip meristem. Horticulturae 9:1313. https://doi.org/10.3390/horticulturae9121313

Xue Y, Hiti-Bandaralage JCA, Mitter N (2023b) Micropropagation of *Duboisia* species: a review on current status. Agronomy 13:797. https://doi.org/10.3390/agronomy13030797

Xue Y, Hiti-Bandaralage JCA, Hu Z, Zhao Z, Mitter N (2024) First report on mesophyll protoplast isolation and regeneration system for the *Duboisia* species. Plan Theory 13:40. https://doi.org/10.3390/plants13010040

Yoshimatsu K (2008) Tissue culture of medicinal plants: micropropagation, transformation and production of useful secondary metabolites. Stud Nat Prod Chem 34:647–752. https://doi.org/10.1016/S1572-5995(08)80033-6

Yu YT, Zhu C, Hong YC, Chen L, Huang ZP, Zhou JC et al (2021) Effectiveness of anisodamine for the treatment of critically ill patients with septic shock: a multicentre randomized controlled trial. Crit Care 25:349. https://doi.org/10.1186/s13054-021-03774-4

Zarate R, Cantos M, Troncoso A (1997) Induction and development of adventitious shoots of *Atropa baetica* as a means of propagation. Euphytica 94:361–366

Zayed R, Winka M, Shamy HE (2006) In vitro organogenesis and alkaloid accumulation in *Datura innoxia*. Z Naturforsch 61c:560–564

Zhou W, Wang C, Hao X, Chen F, Huang Q, Liu T et al (2023) A chromosome-level genome assembly of anesthetic drug-producing *Anisodus acutangulus* provides insights into its evolution and the biosynthesis of tropane alkaloids. Plant Commun 5:100680. https://doi.org/10.1016/j.xplc.2023.100680

Chapter 13
Cocaine: A Stimulant Tropane Alkaloid Drug and Its Regulation

Priya Koley, Poojitha Jain, Kiran S. Mawale, and Parvatam Giridhar

Abstract Cocaine, produced from the leaves of the *Erythroxylon* species, is chemically known as erythroxylin, a typical psychoactive tropane alkaloid. Though historical perspectives of coca leaf extracts and cocaine had little medicinal and recreational application relevance, later, it became one of the major addictive drugs. Now, cocaine abuse is an economic burden to many countries. Although many laws and regulations are in place throughout the world, illegal drug trafficking has become a significant factor affecting the people who are addicted to this drug. Various reviews on cocaine highlighted cocaine's physicochemical characteristics, extraction, health issues, and legal and regulation aspects; an update on these facets is covered in this chapter.

Keywords Cocoa · Dopamine · Erythroxyline · Narcotic · Psychoactive · Stimulant

13.1 Introduction

Cocaine, a naturally occurring sympathomimetic alkaloid from the *Erythroxylon coca* plant, has been utilized as a stimulant in South America for over 5000 years, primarily through practices like chewing the leaves or brewing teas. The extraction

P. Koley · P. Jain · K. S. Mawale
Plant Cell Biotechnology Department, CSIR-Central Food Technological Research Institute, Mysuru, Karnataka, India

Academy of Scientific and Innovative Research (AcSIR), Ghaziabad, India

P. Giridhar (✉)
Plant Cell Biotechnology Department, CSIR-Central Food Technological Research Institute, Mysuru, Karnataka, India

Academy of Scientific and Innovative Research (AcSIR), Ghaziabad, India

Food Safety and Analytical Quality Control Laboratory, CSIR-Central Food Technological Research Institute, Mysuru, Karnataka, India

T. Pullaiah et al. (eds.), *Tropane Alkaloids*,
https://doi.org/10.1007/978-981-96-5805-3_13

of cocaine from the leaves began in the mid-1800s when it was viewed as a safe substance and found its way into products such as toothache drops, nausea tablets, energy tonics, and even the original Coca-Cola beverage (Phillips et al. 2009; Drake and Scott 2018). In recent decades, the analysis of cocaine has become increasingly important as illegal use has spread throughout Western Europe and North America. While cultivated coca plants hold significant economic value, they also play an essential cultural role for Indigenous communities in South America (Plowman 1979; Grinspoon and Bakalar 1981; Naranjo 1981; Schultes and Hofmann 1980). Although coca-chewing has been practised for centuries, there is still much to learn about this tradition. It is important to approach discussions around coca and its uses with a balanced perspective, recognizing the historical significance and the contemporary concerns surrounding its regulation. The Erythroxylaceae family encompasses four genera: *Aneulophus, Erythroxylum, Nectaropetalum,* and *Pinacopodium* (Hegnauer 1981). *Erythroxylum* is the most prominent genus, with over 230 species of tropical trees and shrubs distributed throughout South America, Africa, and Madagascar (Plowman and Hensold 2004). Further classifying this genus into 19 divisions was established, providing a valuable framework for comparative phytochemical studies. *Erythroxylum* species, particularly *Erythroxylum coca* and *Erythroxylum novogranatense*, are notable for being the only natural sources of cocaine (Plowman 1984). While there has been significant interest in non-cultivated *Erythroxylum* species regarding their cocaine content, there remains an opportunity for more comprehensive research on the genus. Many *Erythroxylum* species utilized in traditional medicine are not well-documented, suggesting a promising avenue for future exploration and study (Evans 1981). This highlights the potential for further investigation into their medicinal properties and biodiversity conservation within this genus.

Aynilian et al. (1974) made significant strides in understanding *Erythroxylum* species by discovering the presence of cocaine in herbarium specimens from seven varieties. Later Holmstedt et al. (1976) employed advanced techniques like capillary gas chromatography and mass spectrometry to analyze 62 samples from 13 tropical South American species. Their findings highlighted cocaine in the leaves of only two species, *Erythroxylum coca* and *Erythroxylum novogranatense*, indicating that the other 11 species did not contain detectable levels of cocaine. Further advancing this research, Plowman and Rivier (1983) utilized more sensitive assays, successfully identifying trace quantities of cocaine in 13 wild *Erythroxylum* species from five distinct regions. Notably, they found that two species from Venezuela, *Erythroxylum recurrens* and *Erythroxylum steyermarkii*, exhibited cocaine levels comparable to those seen in cultivated varieties. This work underscores the diverse chemical profiles within the *Erythroxylum* genus and opens avenues for further exploring its ecological and pharmacological potential.

13.2 Cocaine and the Initial Scientific Outlook

In the 1800s, medical professionals and others widely used coca (and its cocaine) for its extraordinary ability to treat or heal a variety of diseases, including asthma, colds, eczema, hysteria, melancholy, exhaustion, overwork, the "affliction of shyness in society," impaired neurological systems, despair, digestive illnesses, opium addiction, and drinking. On the bright side, cocaine increased and strengthened pulse, respiration, and urine excretion, encouraged sexual activity, generated pink cheeks, and provided a very delightful sense of being incredibly alive. Chewing coca relieves and avoids excessive tiredness, suppresses hunger and thirst, and does not affect appetite or digestion. After observing coca chewers in South America, the therapeutic use of coca leaves was introduced in Italy. By the end of the nineteenth century, several doctors recommended chewing the leaves for health benefits. Erythroxylin, which is known as cocaine, was first isolated from the coca leaf in 1855 by German scientist Friedrich Gaedcke, and later, in 1859, Albert Niemann refined cocaine (Stolberg 2011). Wilhelm Lossen discovered the molecular formula of cocaine $C_{17}H_{21}NO_4$ in 1865 (Calatayud and González 2003)—the excitement surrounding this remarkable substance led to numerous developments. Various comprehensive historical reviews highlighted interesting medicinal applications of cocaine (Andrew and Solomon 1975; Holmstedt and Fredga 1981; Grinspoon and Bakalar 1981; Mortimer 2000; Gootenberg 2007). Coca's outstanding properties led to proposals for lengthy voyages (Campbell 1889), and more recently, naturally occurring cocaine in the leaf was proven to have insecticidal properties (Nathanson et al. 1993). "Vin Mariani" was the first stable formulation of cocaine blended with Bordeaux wine that was considered the "most effective tonic" (Andrew and Solomon 1975; Mariani 1896). Vin Mariani contained about 0.12 grains of cocaine per fluid ounce, meaning a typical 6-ounce glass contained approximately 0.72 grains, or 46.65 mg of cocaine (Pendergrast 2013). Cocaine's exceptional qualities as medicine, including aphrodisiac qualities, were reported (Freud 1884). Coca-chewing has recently been suggested as a cocaine maintenance therapy (Hurtado-Gumucio 2000).

13.3 Cocaine Consumption

Coca is now available in two primary forms for misuse, each having distinct characteristics and methods of consumption. The first form is cocaine hydrochloride, which is frequently referred to as "coke," "blow," or "snow." This substance is a fine, white crystalline powder that is highly soluble in water. Users typically consume cocaine hydrochloride through intranasal routes, commonly known as "sniffing" or "snorting." It can also be ingested orally or intravenously, making it accessible to different types of users seeking its stimulating effects. The second form of cocaine is freebase cocaine, which is created through a chemical reaction involving cocaine hydrochloride and a base, such as ammonium or baking soda. This process results in a smokable product commonly known as "crack cocaine" or simply "crack." Crack cocaine typically appears as solid, rock-like crystals, which users break into smaller pieces before smoking. The method of inhalation delivers an intense and rapid euphoric effect, contributing to its popularity among users seeking an immediate high. The misuse of cocaine presents significant public health challenges on a global scale, leading to substantial socioeconomic repercussions. As reported by Wang et al. (2021), the addiction and dependence associated with cocaine use have pervasive implications, not just for individual users but also for communities and healthcare systems. According to the World Drug Report published in 2021, approximately 0.4% of the global population aged 15–64, which equates to around 20 million individuals, used cocaine in 2019 (UNODC 2021). In Europe, the European Monitoring Centre for Drugs and Drug Addiction (EMCDDA) highlighted that cocaine remains the second most widely misused substance, trailing only cannabis in prevalence among recreational drug users (EMCDDA 2021). From 2010 to 2019, the United Nations Office on Drugs and Crime estimated that around 20 million people worldwide engaged in cocaine use, underscoring the ongoing and widespread nature of cocaine misuse (UN WDR 2021). These statistics reveal the need for continued public health efforts to address the challenges posed by cocaine addiction and its impacts on society.

13.4 The Natural Occurrence and Chemical Characterization of *Erythroxylum coca*

Indigenous Andean tribes have used coca leaves for centuries, primarily through chewing. These leaves are integral to their religious ceremonies and Inca festivals and serve important therapeutic purposes (Stolberg 2011). Albert Niemann successfully isolated cocaine from coca leaves between 1859 and 1860, marking a significant scientific breakthrough. Subsequent research conducted a century later confirmed that the dry leaves of *E. coca* var. *coca* contain approximately 6.3 mg of cocaine per gram of plant material (Plowman and Rivier 1983). Coca plants are characterized by their broad, thick, dark green leaves, which have an elliptical shape, pointed tips, and small red fruits (Plowman 1979). It is essential to recognize that coca leaves contain around 18 distinct alkaloids, including cinnamoylcocaine, tropacocaine, methylecgonine, benzoylecgonine (BE), and pseudotropine. Notably, these alkaloids are significantly less euphoric and harmful than cocaine itself (Biondich and Joslin 2016; Stolberg 2011) (Table 13.1).

Table 13.1 The content of principal components in several *Erythroxylum* species

Species	Total alkaloids (dry leaves) (%)	Total phenols (dry leaves) (%)	Total tannins (dry leaves) (%)	Total flavonoids (dry leaves) (%)	Total diterpenes (dry stems) (%)	Refs.
E. coca	0.5–1.5					Aynilian et al. (1974)
E. coca var. *coca*	1.05–2.26					Johnson and Foy (1996)
E. novogranatense var. *novogranatense*	1.4–2.4					Johnson and Foy (1996)
E. suberosum		17.97	6.31	3.87		Rodrigues et al. (2015)
E. tortuosum		10	8.4	0.064		Rodrigues et al. (2015)
E. deciduum		12.04	0.87	1.37		Rodrigues et al. (2015)
E. australe					1.8	Ansell et al. (1993)
E. pictum					0.09–1.1%	Ansell et al. (1993)

13.5 Cocaine's Physicochemical Characteristics and Analytical Techniques for Identification

Cocaine is a tropane alkaloid with poor basic characteristics. While cocaine's ionized hydrochloride salt is easily dissolved in water and has high stability at extremely high temperatures, it does not volatilize in smoke. In its free base form, cocaine is unionized and insoluble in aqueous media, with a boiling point of 187 °C (Siegrist and Wiegand 2014) (Table 13.2). Immunoassays are a rapid and widely adopted technique used for the qualitative presumptive assessment of drugs, including cocaine, in various biological matrices such as blood, urine, and saliva. These tests provide initial results that suggest the presence of a drug; however, they are not definitive. Due to their susceptibility to false positive and false negative results, a confirmatory quantitative method must be utilized following a positive immunoassay result (Niedbala et al. 2001). In addition to immunoassays, chromatographic techniques have become essential for detecting cocaine in biological and

Table 13.2 Physical and chemical properties of cocaine

Cocaine form	Cocaine hydrochloride	Cocaine free base
Other names it is known by	'Coke', 'snow', 'blow'	'Crack'
CAS	53–21-4	50–36-2
Molecular formula	$C_{17}H_{22}ClNO_4$	$C_{17}H_{21}NO_4$
Molecular weight	339.8 g Mol^{-1}	303.35 g Mol^{-1}
Boiling point	–	187 °C
Melting point	195 °C	98 °C
Solubility in water	2 g/mL	$1.7 * 10^{-3}$ g/mL
pKa; pKb (at 15 _C)	–	8.61; 5.59
Log P -	–	2.3

non-biological samples. Numerous studies have documented the presence of cocaine and its metabolites in human urine (Cone et al. 2003; Smith et al. 2010). Furthermore, cocaine and its metabolites have been detected in other biological samples such as blood, plasma, sweat, and hair (Menzies et al. 2019), highlighting the compound's prevalence in various bodily fluids. Comparative analyses between high-performance liquid chromatography (HPLC) and gas chromatography combined with mass spectrometry (GC-MS) have demonstrated that both instrumental techniques can effectively detect cocaine and its metabolites in urine samples (Phillips et al. 1996) and human hair (Barroso et al. 2008). The GC-MS technique has advanced to the point of simultaneously identifying multiple compounds, including the "crack"-specific metabolite anhydroecgonine methyl ester (AEME), cocaine, and other significant metabolites like benzoylecgonine (BE) and cocaethylene (CE). Kintz et al. (1997) first described this approach, demonstrating its effectiveness in various biological samples such as plasma, saliva, urine, and hair. This technique is vital for accurately confirming the presence of cocaine and its metabolites in drug testing and forensic investigations. The limits of detection for AEME are clearly defined: 1 ng/mL for the initial three matrices, with a linear range of 5–1500 ng/mL; 0.1 ng/mg for hair, with a linear range of 0.2–25 ng/mg; and 0.5 ng/patch, with a linear range of 2–100 ng/patch. This method has been effectively applied to 1 mL samples of urine, saliva, and plasma, as well as 50 mg of hair and sweat extracted from a sweat patch. Furthermore, Fernandez et al. have validated a robust method employing GC-MS with electron ionization for the detection of cocaine and several of its metabolites in 0.5 mL urine samples. The method employs a straightforward derivatization step following solid phase extraction (SPE), allowing for low limits of quantification between 2.5 and 10 ng/mL (Fernandez et al. 2019). This approach aligns with the successful use of ultra-performance liquid chromatography coupled with tandem mass spectrometry (UPLC-MS/MS) in dried blood samples (de Lima Feltraco Lizot et al. 2019). Additionally, liquid chromatography coupled with tandem mass spectrometry (LC-MS/MS) has proven effective in detecting cocaine and its metabolite, benzoylecgonine (BE), in both urine and oral fluid (Melanson et al. 2020). To optimize sample integrity, blood or plasma samples must be treated with a pseudocholinesterase (PChE) inhibitor, such as sodium fluoride; otherwise, cocaine may hydrolyze spontaneously and enzymatically to BE (Goldstein et al. 2009). For improved stability of cocaine, it is advisable to adjust the pH of samples to 5 using acetic acid and then store them either in a freezer or at a refrigerated temperature of 4 °C. It is important to acknowledge that some degradation may occur over time, even at around 20 °C, emphasizing the need to handle samples carefully. Nowadays, green analytical techniques are in use to detect cocaine in toxicological samples (de Paula Meirelles et al. 2023).

13.6 Extraction Procedure for Cocaine

Cocaine is extracted from coca leaves by harvesting, drying, and grinding them into a fine powder. Grinding increases the surface area for more efficient alkaloid extraction. Different parts of the *Erythroxylum* plants are used for extraction. Crude extracts display a variety of biological activities (Table 13.3).

13.6.1 Solvent Extraction Procedure

Powdered coca leaves are mixed with water and organic solvents like ethanol or chloroform to dissolve alkaloids like cocaine. The mixture is acidified with hydrochloric acid, turning cocaine into its hydrochloride salt form, which is more soluble in water. After the mixture sits for a time, it is filtered to separate solid plant material from the liquid. Sodium bicarbonate is added to make the liquid basic, converting cocaine hydrochloride back to its freebase form for extraction. An organic solvent, such as ether or chloroform, is then added, allowing cocaine to dissolve in the organic layer. After collecting this layer and evaporating the solvent, crude cocaine is left behind, which can be subsequently purified for analysis (de Paula Meirelles et al. 2023) (Table 13.4).

13.6.2 Super-Critical Fluid Extraction (SFE) Procedure

Supercritical fluid extraction (SFE) is an effective method for extracting cocaine from coca leaves, providing a strong alternative to traditional techniques. By using supercritical carbon dioxide as a solvent, SFE enhances extraction efficiency and selectivity. Compared to classical methods, it offers shorter extraction times, lower solvent consumption, and reduced operating temperatures (Brachet et al. 2000).

Table 13.3 Biological activities of crude extracts of *Erythroxylum* plants

Plant source	Extract Source	Crude extracts	Pharmocological activities	Refs.
E. monogynum	Leaves	Chloroform	Antidiabetic	Kumar et al. (2019)
		Ethanolic	Hepatoprotective effects; Nephroprotective effects	Kumar et al. (2019) and Syed and Namdeo (2013)
		Aqueous	Antimicrobial; Antioxidant	Kumar et al. (2019)
		Methanol	Antiplasmodial; Cytotoxicity	Kumar et al. (2019)
E. delagoense	Leaves and bark	Acetonic; Methanol; Aqueous	Antibacterial	Wet (2011)
E. emarginatum	leaves and stems	Acetonic, methanol, aqueous	Antibacterial	Wet (2011)
E. pictum	Leaves and stems	Acetonic, methanol, aqueous	Antibacterial	Wet (2011)
E. pungens	Roots	Ethanolic	Vasorelaxant	Oliveira et al. (2012)
E. caatingae	Leaves	Ethanolic	Myorelaxing effect on smooth muscle tissue	Santos et al. (2016)
	Stems	Methanol;	Antimicrobial activity;	Jaciana et al. (2012) and Oliveira et al. (2011)
		Low-polarity fractions	Cytotoxicity	
E. macrocarpum	Leaves	Acetone/water (70/30, v/v)	Acetylcholinesterase inhibition	Ramhit et al. (2018)
E. areolatum	Leaves	Hydroalcoholic	Antiherpetic activity	Gonzalez-Guevara et al. (2004)
E. confusum	Leaves	Hydroalcoholic	Antiherpetic activity	Gonzalez-Guevara et al. (2004)
E. minutifolium	Leaves	n-Hexane	Hepatoprotective effects	Rodeiro et al. (2008)
E. confusum	Leaves	n-Hexane	Hepatoprotective effects	Rodeiro et al. (2008)
E. ovalifolium	Stems	Ethanolic	Neutralize toxicity of snake venom	Oliveira et al. (2016)
E. subsessile	Stems	Ethanolic	Neutralize toxicity of snake venom	Oliveira et al. (2016)
E. daphnites	Leaves	n-Hexane	Anti-proliferative effects	Elias et al. (2016)

Table 13.4 Chemicals and solvents along with their substances used for solvent extraction procedure

Step	Chemicals	Substitute	References
Coca leaves to coca paste	Kerosene, sulfuric acid, sodium bicarbonate	Gasoline, benzene, hydrochloric acid, sodium hydroxide	U.S. Department of Justice (1991)
Coca paste to cocaine base	Sulfuric acid, potassium permanganate, ammonia hydroxide	Hydrochloric acid, potassium dichromate, sodium hydroxide	U.S. Department of Justice (1991)
Cocaine base to cocaine hydrochloride	Ethyl ether, acetone, hydrochloric acid	Chloroform, methyl ethyl ketone, hydrogen chloride gas	U.S. Department of Justice (1991)

13.6.3 Solid Phase Extraction (SPE) Cartridge Extraction

To prepare the sample, coca leaves are cut into small pieces (50 mg) and mixed with a solvent solution of methanol, acetonitrile, and ammonium formate in a 25:25:50 ratio. This mixture is stirred for 15 minutes to extract the alkaloids. The liquid phase is then separated and loaded into the SPE cartridge, which is washed with a 5% methanol solution to remove non-targeted compounds. Finally, the targeted molecules are eluted using a suitable solvent (Alvarefreire et al. 2023).

13.6.4 Focused Microwave-Assisted Extraction Procedure

The focused microwave-assisted extraction (FMAE) method for extracting cocaine and benzoylecgonine from coca leaves outperforms traditional techniques by being faster and more efficient. In this process, coca leaf powder is dissolved in water or an alkaline solution and then placed in an extraction vessel with the microwave power and radiation time set appropriately (Brachet et al. 2002).

13.7 The Psychoactive Nature of Cocaine

Cocaine is a powerful psychoactive stimulant characterized by its ability to increase levels of dopamine in the brain, which can lead to heightened pleasure and increased energy. However, this also contributes to its high potential for addiction. Cocaine's widespread use is undeniably driven by its highly addictive properties, primarily due to its ability to block dopamine absorption in the nucleus accumbens (Aronson 2016; Isenschmid 2020; Toth et al. 2023). Furthermore, cocaine is a potent sympathomimetic drug that significantly elevates levels of serotonin and norepinephrine (Isenschmid 2020). This catecholamine surge has pronounced adrenergic effects, such as tachycardia and hypertension (Schwartz et al. 2010). Additional side effects of cocaine use include impaired platelet function, endothelial damage, and the release of pro-inflammatory mediators (Bachi et al. 2017). Moreover, in stark contrast to the global COVID-19 pandemic, European authorities have consistently reported an alarming increase in cocaine seizures at seaports in 2020 (UNODC 2021). Numerous case studies emphasize the severe negative effects of cocaine use (Ali et al. 2021). This chapter aims to deliver a definitive overview of the existing data on cocaine's physicochemical properties, detection methods, pharmacodynamics, pharmacokinetics, effects and toxicity, patterns of abuse, and prevalence.

13.8 Use of Cocaine and Few Positive Implications

13.8.1 Cocaine Has Local Anesthetic Properties

Ophthalmologist Carl Koller demonstrated that cocaine could elicit local anesthesia in the eyes by trying it on frogs, guinea pigs, rabbits, and dogs (Goldberg 1984; Koller 1934). Later, cocaine's anesthetic effects on all mucous membranes were discovered by Knapp (1985), and using cocaine in more surgical operations was recommended (Goldberg 1984; Knapp 1985). The pharmacodynamic processes that cause anesthesia include inhibiting voltage-gated sodium channels and the cessation of neural potentiation. However, this activity also provides cardiotoxicity due to cardiac sodium channel blockage.

13.8.2 Cocaine's Ability to Activate the Psychomotor System

Cocaine unequivocally inhibits the reuptake transporters of multiple neurotransmitters in the central nervous system, including norepinephrine (NET) and dopamine (DAT) (Knapp 1985). Prazosin effectively suppresses cocaine's effects in mice, reinforcing the notion that cocaine antagonizes adrenergic receptors. The impact of cocaine and other drugs of abuse on extracellular dopamine levels in rat brains was demonstrated, and the findings decisively established that the limbic system plays a critical role in modulating the effects of stimulants (Di Chiara and Imperato 1988). Building on these foundational discoveries, Chen and Reith (1994) compellingly illustrated cocaine's extensive influence on monoamine levels by documenting the significant increases in other extracellular monoamines, such as serotonin and norepinephrine, following localized cocaine administration in the ventral tegmental area (VTA) of rat brains. It is undeniable that the majority of dopaminergic fibres originate in the VTA, a key site where cocaine exerts its stimulant effects and has a high potential for addiction. Furthermore, research has definitively shown that the preference for a cocaine-associated environment is abolished in mice lacking both the dopamine transporter (DAT) and serotonin transporter (SERT) (Sora et al. 2001).

At low doses, cocaine, much like amphetamine, produces a pronounced alerting response, evidenced by increased exploration, locomotion, grooming, and rearing behaviors (Scheel-Krüger et al. 1977; Snoddy and Tessel 1983). In stark contrast, higher doses in rats result in a clear reduction in locomotor activity and the emergence of stereotyped behavioral patterns (Randrup and Munkvad 1967; Lewander 1977).

13.9 The Economic Burden of Cocaine Abuse

Any use of cocaine is considered abuse because it is an illegal substance. Cocaine creates a strong physical addiction and is becoming recognized as one of the most dangerous illicit drugs abused today. The myth is that cocaine is harmless and non-addictive (Cregler 1989). The use of coca leaves in ancient times has been well established. Multiple archaeological evidence shows that chewing coca was used as a general stimulant, considered a medicine, and a status symbol throughout Central and South America (Stolberg 2011). Over time, cocaine negatively affects every part of the body, with the potential for severe long-term effects. Chronic cocaine use can lead to long-term changes in the brain, an increase in stress hormones, as well as decreased functioning in other parts of the brain. The economic burden of cocaine abuse is immense. Cocaine use affects healthcare systems worldwide, leading to issues such as cardiovascular problems, strokes, and acute intoxications. Long-term cocaine use can also contribute to chronic conditions, including hypertension, mental health disorders, and infectious diseases like HIV and Hepatitis, often due to needle sharing (NSEPSUI 2019). In the United States alone, an estimated $193 billion is spent each year on healthcare, law enforcement, and lost productivity associated with illicit drug use, with cocaine accounting for a significant portion of these costs (USSC 2020). Cocaine abuse is a significant global public health issue that affects millions of individuals each year. According to the United Nations Office on Drugs and Crime (UNODC), approximately 21 million people worldwide used cocaine in 2021, which is about 0.4% of the global population aged 15 to 64. The highest number of cocaine users is reported in North America and Western Europe, where its use continues to challenge public health systems and law enforcement agencies (UNODC 2021). Among different age groups, the highest rates of cocaine abuse are observed in individuals aged 18 to 25, driven by various factors, including socioeconomic influences, peer pressure, and the drug's recreational appeal (Sweta et al. 2024). The Global Report on Cocaine 2023 highlights the shifting dynamics of cocaine markets, including a rise in production in South America and the diversification of trafficking routes. The report emphasizes that as the influx of cocaine increases in a region, the associated public health costs also rise. Additionally, increased cocaine use contributes to the proliferation of organized crime within society, which particularly impacts the youth and further deteriorates community well-being. Emerging markets in Asia, especially India and China, are becoming significant targets for traffickers, underscoring the urgent need for enhanced international cooperation (World Drug Report 2023). Traditionally, India has served as a transit point for cocaine smuggling; however, recent years have seen a rise in domestic consumption. The National Crime Records Bureau (NCRB) 2023 data reveals an increase in cocaine seizures and arrests over the past decade. Although the rate of cocaine abuse in India is lower than in Western countries, this upward trend is concerning, especially as it is linked to affluent urban populations and the growing popularity of nightlife culture (Balhara et al. 2024).

13.9.1 Enhancing Law Enforcement: Addressing Challenges and Identifying Solutions

The foundation of international drug control began with the 1909 Shanghai Conference, which aimed to assess and regulate narcotic production and consumption. This led to the 1912 Hague Opium Convention, the first international treaty on drug control. In 1920, the League of Nations established the Opium Advisory Committee (OAC) to oversee the convention's implementation. This committee later became the Commission on Narcotic Drugs (CND) under the United Nations, which now plays a key role in addressing drug abuse and trafficking globally. Several international laws and treaties address the control of cocaine and other drugs, including the International Convention Relating to Dangerous Drugs. Adopted in 1925, this treaty established uniform controls on addictive drugs like cocaine and opium. The United Nations took over policy-making for drug control in 1946, further strengthening international coordination through pivotal conventions such as the 1961 Single Convention, the 1971 Convention on Psychotropic Substances, and the 1988 UN Convention against Illicit Traffic in Narcotic Drugs and Psychotropic Substances (https://www.unodc.org/pdf/convention_1961_en.pdf;1972). United Nations Convention Against Illicit Traffic in Narcotic Drugs and Psychotropic Substances in 1988 is one of the three major drug control treaties currently in force. These agreements established a unified framework for combating illicit drugs, including cocaine, while permitting flexibility for national adaptations (Sweta et al. 2024). The legal status of cocaine varies by country, with some banning its sale for recreational use while others have decriminalized or legalized it for certain uses. In India, the Prevention of Illicit Trafficking in Narcotic Drugs and Psychotropic Substances Act was passed in 1988 to help enforce the Narcotic Drugs and Psychotropic Substances Act of 1985. The Single Convention on Narcotic Drugs has made recreational use of cocaine a crime since 1961. In the United States, cocaine is a Schedule II drug under the Controlled Substances Act, which means it has a high potential for abuse but has an accepted medical use. Classified by the World Anti-Doping Agency (WADA) as an "S6 Stimulants" class of illicit drug, cocaine is an illegal substance listed on the Prohibited Substances and Methods list under the Substances of Abuse category as a drug which is prohibited for use 'In-Competition' (https://www.sportintegrity.gov.au/sites/default/files/SIA279-0124-Cocaine%20Factsheet-A-8%20%5BDIGITAL%5D.pdf accessed on 19.12.2024). The specified limit for cocaine in the UK is 10 μg/L as a drug. Cocaine is a Schedule 2 drug in the United States. This means that it has legitimate medical uses. In the UK, cocaine deaths were nearly 10 times higher in 2023 than a decade earlier, claiming more than 1100 lives (Cocaine use fuels record high in drug deaths 23 October 2024 https://www.bbc.com/news/articles/cy5l4nknew0o accessed on 19.12.2024). In India, the Narcotics Control Bureau (NCB) under the Ministry of Home Affairs (MHA) and the Department of Revenue is responsible for reducing the availability of illicit drugs in the country. The NCB administers the Narcotic Drugs and Psychotropic Substances (NDPS) Act, 1985, as well as the Prevention of Illicit Traffic in Narcotic Drugs and Psychotropic Substances (NDPS) Act, 1988. The Narcotic Drugs and Psychotropic Substances Act (NDPS), 1985, is the cornerstone of India's legal framework for combating drug abuse, including cocaine. This legislation imposes stringent penalties for the possession, sale, and trafficking of narcotics. For cocaine, the NDPS Act prescribes imprisonment ranging from 10 years to life, depending on the quantity involved, and includes hefty fines (NSEPSUI 2019; NDDTC 2019).

Despite the strong laws against cocaine abuse, enforcement remains challenging due to factors such as corruption, limited resources, and intricate smuggling networks. Law enforcement agencies involved in curtailing cocaine abuse encounter a variety of challenges that can hinder their effectiveness, such as limited resources, community relations, and the rapid pace of technological advancements. Coordinated efforts by agencies like the Narcotics Control Bureau (NCB) have resulted in significant seizures and arrests in recent years. For instance, in 2022, the NCB reported a 25% increase in cocaine seizures compared to the previous year, underscoring intensified efforts to reduce its availability (Balhara et al. 2024). Criminalizing cocaine has led many users to avoid seeking help out of fear of legal repercussions (War et al. 2023; Yadav and Asha 2024). To effectively combat cocaine abuse, a combination of prevention, treatment, and law enforcement is essential. Additionally, investing in cutting-edge technology can significantly enhance the prevention of cocaine abuse. Utilizing artificial intelligence for drug trafficking surveillance and developing community-based rehabilitation programs can make a significant difference. Fostering strong relationships with the community can cultivate trust and collaboration, encouraging proactive engagement. Public awareness campaigns educating youth about the dangers of cocaine and available support options are crucial for reducing usage among vulnerable populations. Collaboration among countries facing similar challenges is vital.

References

Ali SF, Hoglund JR, Gibbs MA, Littman L (2021) Unusual electrocardiographic manifestations of lethal cocaine toxicity. Clin Toxicol 60:408–409

Alvarefreire I, Cabarcos-Fe Ández P, Rub NC, Moreda-Piñeiro A, Tabernero-Duque MJ, Sánchez-Sellero IS, Bermejo-Barrera P, Bermejo-Barrer AM (2023) Detection of coca alkaloids in oral fluid from coca leaf (tea) consumers: using solid phase extraction to improve validation parameters and widen the detection window. Anal Methods 15:6177–6183. https://doi.org/10.1039/d3ay01298k

Andrew G, Solomon D (1975) The Mariani album: 1884–1993. In: Andrews G, Solomon D (eds) The coca leaf and cocaine papers. Harcourt Brace Jovanovich, New York/London

Ansell SM, Pegel KH, Taylor DAH (1993) Diterpenes from the timber of *Erythroxylum pictum*. Phytochemistry 32:945–952

Aronson JK (2016) Cocaine. In: Meyler's side effects of drugs, 16th edn. Elsevier, Amsterdam, pp 492–542

Aynilian G, Duke J, Gentner W, Farnsworth N (1974) Cocaine content of *Erythroxylum* species. J Pharm Sci 63(12):1938–1939

Bachi K, Mani V, Jeyachandran D, Fayad ZA, Goldstein RZ, Alia-Klein N (2017) Vascular disease in cocaine addiction. Atherosclerosis 262:154–162

Balhara YPS, Sarkar S, Ghosh A, Mahadevan J (2024) Perspectives on addiction-related problems in India. Addiction 120:783. https://doi.org/10.1111/add.16711

Barroso M, Dias M, Vieira DN, Queiroz JA, Lopez-Rivadulla M (2008) Development and validation of an analytical method for the simultaneous determination of cocaine and its main metabolite, benzoylecgonine, in human hair by gas chromatography/mass spectrometry. Rapid Commun Mass Spectrom 22:3320–3326

Biondich AS, Joslin JD (2016) Coca: the history and medical significance of an ancient Andean tradition. Emerg Med Int 2016:4048764

Brachet A, Christen P, Gauvrit JY, Longeray R, Lantéri P, Veuthey JL (2000) Experimental design in supercritical fluid extraction of cocaine from coca leaves. J Biochem Biophys Methods 43(1–3):353–366. https://doi.org/10.1016/s0165-022x(00)00062-2

Brachet A, Christen P, Veuthey JL (2002) Focused microwave-assisted extraction of cocaine and benzoylecgonine from coca leaves. Phytochem Anal 13(3):162–169. https://doi.org/10.1002/pca.637

Calatayud J, González Á (2003) History of the development and evolution of local anaesthesia since the coca leaf. Anesthesiology 98:1503–1508

Campbell DL (1889) Hints to lady travellers at home and abroad. Iliffe & Sons, London

Chen NH, Reith MEA (1994) Effects of locally applied cocaine, lidocaine, and various uptake blockers on monoamine transmission in the ventral tegmental area of freely moving rats: a microdialysis study on monoamine interrelationships. J Neurochem 63:1701–1713

Cone EJ, Sampson-Cone AH, Darwin WD, Huestis MA, Oyler JM (2003) Urine testing for cocaine abuse: metabolic and excretion patterns following different routes of administration and methods for detection of false-negative results. J Anal Toxicol 27:386–401

Cregler LL (1989) Adverse health consequences of cocaine abuse. J Natl Med Assoc 81(1):27–38

de Lima Feltraco Lizot L, da Silva ACC, Bastiani MF, Hahn RZ, Bulcao R, Perassolo MS, Antunes MV, Linden R (2019) Simultaneous determination of cocaine, ecgonine methyl ester, benzoylecgonine, cocaethylene and norcocaine in dried blood spots by ultra-performance liquid chromatography coupled to tandem mass spectrometry. Forensic Sci Int 298:408–416

de Paula Meirelles G, Fabris AL, Ferreira Dos Santos K, Costa JL, Yonamine M (2023) Green analytical toxicology for the determination of cocaine metabolites. J Anal Toxicol 46(9):965–978. https://doi.org/10.1093/jat/bkac005

Di Chiara G, Imperato A (1988) Drugs abused by humans preferentially increase synaptic dopamine concentrations in the mesolimbic system of freely moving rats. Proc Natl Acad Sci USA 85:5274–5278

Drake LR, Scott PJH (2018) DARK classics in chemical neuroscience: cocaine. ACS Chem Neurosci 9:2358–2372

Elias ST, Macedo CCS, Simeoni LA, Silveira D, Magalhães PO, Lofrano-Porto A, Coletta R, Neves FAR, Guerra ENS (2016) Cytotoxic effect of *Erythroxylum daphnites* extract is associated with G_1 cell cycle arrest and apoptosis in oral squamous cell carcinoma. Cell Cycle 15(7):948–956. https://doi.org/10.1080/15384101.2016.1151583

EMCDDA (2021) European drug report 2021: trends and development; 1977–9860; European Monitoring Centre for Drugs and Drug Addiction: Luxembourg

Evans WC (1981) The comparative phytochemistry of the genus *Erythroxylon*. J Ethnopharmacol 3(2–3):265–277. https://doi.org/10.1016/0378-8741(81)90057x

Fernandez N, Cabanillas LM, Olivera NM, Quiroga PN (2019) Optimization and validation of simultaneous analyses of ecgonine, cocaine, and seven metabolites in human urine by gas chromatography-mass spectrometry using a one-step solid-phase extraction. Drug Test Anal 11:361–373

Freud S (1884) Über Coca. Centralblatt für die gesamte Therapie 27:289–314

Goldberg MF (1984) Cocaine: the first local anesthetic and the third scourge of humanity: a centennial melodrama. Arch Ophthalmol 102:1443–1447

Goldstein RA, DesLauriers C, Burda A, Johnson-Arbor K (2009) Cocaine: history, social implications, and toxicity: a review. Semin Diagn Pathol 26:10–17

Gonzalez-Guevara JL, Gonzalez-Lavaut JA, Pino-Rodriguez S, Gar Torres M, Carballo-Gonzalez MT, Echemendia-Arana OA, Molina-Torr J, Prieto-Gonzalez S (2004) Phytochemical screening and in vitro antiherpetic activity of four *Erythroxylum* species. Acta Farm Bonaer 23:506–509

Gootenberg P (2007) A forgotten case of 'scientific excellence on the periphery': the nationalist cocaine science of Alfredo Bignon, 1884–1887. Comp Stud Soc Hist 49:202–232. https://doi.org/10.1017/S001041750700045X

Grinspoon L, Bakalar J (1981) Coca and cocaine as medicines: an historical review. J Ethnopharmacol 3:149–159

Hegnauer RJ (1981) Chemotaxonomy of Erythroxylaceae (including some ethnobotanical notes on Old World species). Ethnopharmacol 3(2–3):279–292. https://doi.org/10.1016/0378-8741(81)90058-1

Holmstedt B, Fredga A (1981) Sundry episodes in the history of coca and cocaine. J Ethnopharmacol 3:113–147

Holmstedt B, Jäätmaa E, Leander K, Plowman T (1976) Determination of cocaine in some south American species of *Erythroxylum* using mass fragmentography. Phytochemistry 16(11):1753–1755. https://doi.org/10.1016/0031-9422(71)85082-3

Hurtado-Gumucio J (2000) Coca leaf chewing as therapy for cocaine maintenance. Ann Med Intern Fenn 151(Supp B):B44–B48

Isenschmid DS (2020) Cocaine. In: Levine BS, Kerrigan S (eds) Principles of forensic toxicology. Springer, Cham, pp 371–387

Jaciana A, Rosilma A, Maria R, Kêsia S, André B, Maria G, Steno O, Josean T, Marcelo S, Silene N, Teresinha G-S (2012) Antimicrobial, antiproliferative and proapoptotic activities of extract, fractions and isolated compounds from the stem of *Erythroxylum caatingae* Plowman. Int J Mol Sci 13:4124–4140

Johnson EL, Foy CD (1996) Biomass accumulation and alkaloid content in leaves of *Erythroxylum coca* and *Erythroxylum novogranatense* var. *novogranatense* grown in soil with varying pH. J Plant Physiol 149:444–450

Kintz P, Sengler C, Cirimele V, Mangin P (1997) Evidence of crack use by anhydroecgonine methylester identification. Hum Exp Toxicol 16:123–127

Knapp H (1985) Cocaine and its use in ophthalmic and general surgery. G P Putnam's Sons, New York/London, p 1885

Koller C (1934) Preliminary report on local anesthesia of the eye. Arch Ophthalmol 12:473–474

Kumar CD, Anitha S, Varalakshmi P, Rao DM (2019) Evaluation of potential phytochemicals and phyto pharmacological activities of *Erythroxylum monogynum* Roxb. Biosci Biotechnol Res Asia 16:441–449

Lewander T (1977) Effects of amphetamine in animals. In: Martin WR (ed) Handbook of experimental pharmacology, Drug addition 11, vol 45. Springer, Berlin

Mariani A (1896) Coca and its therapeutic application, 3rd edn. J N Jaros, New York

Melanson SEF, Khaliq PAK, Griggs T, Flood DA, JG. (2020) Comparison of Oral fluid and urine for detection of cocaine abuse using liquid chromatography with tandem mass spectrometry. J Appl Lab Med 5:935–942

Menzies EL, Archer JRH, Dargan PI, Parkin MC, Yamamoto T, Wood DM, Braithwaite RA, Elliott SP, Kicman AT (2019) Detection of cocaine and its metabolites in whole blood and plasma following a single dose, controlled administration of intranasal cocaine. Drug Test Anal 11:1419–1430

Mortimer W (2000) History of coca. "The divine plant" of the Incas. University Press of the Pacific, Honolulu. (reprinted from the 1901 edition)

Naranjo P (1981) Social function of coca in pre-Columbian America. J Ethnopharmacol 3(2–3):161–172

Nathanson J, Hunnicutt E, Kantham L, Scavone C (1993) Cocaine as a naturally occurring insecticide. Proc Natl Acad Sci 90:9645–9648

NDDTC (2019) Magnitude of substance use in India. In: National drug dependence treatment centre report 2019. AIIMS, New Delhi

Niedbala RS, Kardos K, Fries T, Cannon A, Davis A (2001) Immunoassay for detection of cocaine/metabolites in oral fluids. J Anal Toxicol 25:62–68

NSEPSUI (2019) National survey on extent and pattern of substance use in India' 2019. https://www.aiims.edu/index.php/en/national-drug-use-survey-2019, Accessed on 20 December 2024

Oliveira S, Tavares J, Sobral M, Lucena H, Barbosa FJ, Agra Maria De F, Nascimento S, Aguiar J, Gonçalves-Silva T, de Simone C, Araújo-Júnior J, Silva M (2011) Tropane alkaloids from *Erythroxylum caatingae* plowman. Chem Biodivers 8:155–165. https://doi.org/10.1002/cbdv.200900400

Oliveira CA, Sena-Filho JG, Mendes-Júnior LG, Anjos RM, Ribeiro TP, Barbosa-Filho JM, Braga VA, Medeiros IA (2012) *Erythroxylum pungens* elicits vasorelaxation by reducing intracellular calcium concentration in vascular smooth muscle cells of rats. Revista Brasileira de Farmacognosia Braz J Pharmacogn 22(2):436–442

Oliveira CE, Rodrigo A, Rodrigo C, Nayanna D, Mello A, Marcelo S, Luiz C, Simas P, Oswaldo S, Caio F, Rafael G, Leandro R, Andre F (2016) Protective effect of the plant extracts of *Erythroxylum* sp. against toxic effects induced by the venom of *Lachesis muta* snake. Molecules 1350. https://doi.org/10.3390/molecules21101350

Pendergrast M (2013) For god, country & Coca-Cola. The definite history of the great America soft drink and the company that makes it. Perseus Book Group, New York

Phillips DL, Tebbett IR, Bertholf RL (1996) Comparison of HPLC and GC-MS for measurement cocaine and metabolites in human urine. J Anal Toxicol 20:305–308

Phillips K, Luk A, Soor GS, Abraham JR, Leong S, Butany J (2009) Cocaine cardiotoxicity: a review of the pathophysiology, pathology, and treatment options. Am J Cardiovasc Drugs 9:177–196

Plowman T (1979) Botanical perspectives on coca. J Psychedelic Drugs 11:103–117

Plowman T (1984) The ethnobotany of coca (*Erythroxylum* spp., Erythroxylaceae). Adv Eco Bot 1:62–111

Plowman TC, Hensold N (2004) Names, types, and distribution of neotropical species of *Erythroxylum* (Erythroxylaceae). Brittonia 56:1–53

Plowman T, Rivier L (1983) Cocaine and cinnamoylcocaine content of *Erythroxylum* species. Ann Bot 51:641–659

Ramhit P, Ragoo L, Bahorun T, Neergheen-Bhujun VS (2018) Multi-targeted effects of untapped resources from the Mauritian endemic fora. S Afr J Bot 115:208–216

Randrup A, Munkvad I (1967) Stereotyped activities produced by amphetamine in several animal species and man. Psychopharmacologia 11:300–310

Rodeiro I, Donato MT, Martinez I, Hernandez I, Garrido G, Gonzalez Lavaut JA, Menendez R, Laguna A, Castell JV, Gomez-Lechon MJ (2008) Potential hepatoprotective effects of new Cuban natural products in rat hepatocytes culture. Toxicol In Vitro 22:1242–1249

Rodrigues GA, Souza WC, Godinho MGC, Ferreira HD, Vila Verde GM (2015) Determinação de parâmetros farmacognósticos para as folhas de *Erythroxylum suberosum* A. St.-Hilaire (Erythroxylaceae) coletadas no município de Goiânia. GO Rev Bras de Plantas Med 17:1169–1176

Santos KC, Monte APO, Lima JT, Ribeiro LAA, Palheta Junior RC (2016) Role of NO-cGMP pathway in ovine cervical relaxation induced by *Erythroxylum caatingae* plowman. Anim Reprod Sci 164:23–30
Scheel-Krüger J, Braestrup C, Nielson M, Golembiowska K, Mogilnicka E (1977) Cocaine and other stimulants. Plenum Press, New York
Schultes RE, Hofmann A (1980) The botany and chemistry of hallucinogens, 2nd edn. Charles C. Thomas, Springfield
Schwartz BG, Rezkalla S, Kloner RA (2010) Cardiovascular effects of cocaine. Circulation 122:2558–2569
Siegrist M, Wiegand TJ (2014) Cocaine. In: Wexler P (ed) Encyclopedia of toxicology, 3rd edn. Academic, Oxford, pp 999–1002
Smith ML, Shimomura E, Paul BD, Cone EJ, Darwin WD, Huestis MA (2010) Urinary excretion of ecgonine and five other cocaine metabolites following controlled oral, intravenous, intranasal, and smoked administration of cocaine. J Anal Toxicol 34:57–63
Snoddy AM, Tessel RE (1983) Nisoxetine and amphetamine share discriminative stimulus properties in mice. Pharmacol Biochem Behav 19:205–210
Sora I, Hall FS, Andrews AM, Itokawa M, Li XF, Wei HB, Wichems C, Lesch KP, Murphy DL, Uhl GR (2001) Molecular mechanisms of cocaine reward: combined dopamine and serotonin transporter knockouts eliminate cocaine place preference. Proc Natl Acad Sci USA 98:5300–5305
Stolberg V (2011) The use of coca: prehistory, history, and ethnography. J Ethn Subst Abus 10:126–146. https://doi.org/10.1080/15332640.2011.573310
Sweta K, Sunita NK, Palwinder K, Rajvir K, Rajnish B (2024) Examining drug use and misuse in India, and evaluating drug control laws. Afr J Biomed Res 27:156–167
Syed SH, Namdeo AG (2013). Hepatoprotective effect of leaves of *Erythroxylum monogynum* Roxb. on paracetamol induced toxicity. Asian Pac J Trop Biomed 3(11):877–881
Toth BA, Chang KS, Fechtali S, Burgess CR (2023) Dopamine release in the nucleus accumbens promotes REM sleep and cataplexy. iScience 26(9):107613. https://doi.org/10.1016/j.isci.2023.107613
U.S. Department of Justice, Drug Enforcement Administration, Office of Intelligence (1991) Coca cultivation and cocaine processing: an overview. National Institute of Justice
United Nations (2021) World Drug Report 2021 [Internet]. 2021. Available online: https://www.unodc.org/res/wdr2021/field/WDR21_Booklet_4.pdf
UNODC (2021) World drug report 2021. United Nations, Vienna, p 2021
USSC (2020) United States Sentencing Commission Annual Report 2020. https://www.ussc.gov/about/annual-report/archive/annual-report-2020 Accessed 20 Dec 2024
Wang J, Deng X, Wu Y, Huang Y, Hou S, Zhang Y, Qiu T, Tong J, Chen X (2021) Sub-lethal toxicity and elimination of the cocaine metabolite, benzoylecgonine: a narrative review. Ann Palliat Med 10(6936):6947
War AH, Kushwaha AKS, Verma PK, Isha Joshi I (2023) A study on drug abused youth of North Kashmir union territory of Jammu and Kashmir India. Int J Ind Psychol 11(3). https://doi.org/10.25215/1103.223
Wet HD (2011) Antibacterial activity of the five South African Erythroxylaceae species. Afr J Biotechnol 10:11511–11514
World Drug Report (2023). https://www.unodc.org/unodc/en/data-and-analysis/world-drug-report-2023.html. Accessed 20 Dec 2024
Yadav A, Asha A (2024) Rights to health and access to treatment of drug addicted patients in India. J Drug Alcohol Res 13:1–16

The manufacturer's authorised representative in the EU is Springer Nature Customer Service Centre GmbH, Europaplatz 3, 69115 Heidelberg, Germany. If you have any concerns regarding our products, please contact ProductSafety@springernature.com

Printed and bound by CPI Group (UK) Ltd, Croydon, CR0 4YY
01/10/2026
02225930-0006